Openness, Secrecy, Authorship

Openness, Secrecy, Authorship

TECHNICAL ARTS AND THE CULTURE OF KNOWLEDGE FROM ANTIQUITY TO THE RENAISSANCE

Pamela O. Long

The Johns Hopkins University Press

BALTIMORE AND LONDON

Printed in the United States of America on acid-free paper

Johns Hopkins Paperbacks edition, 2004
2 4 6 8 9 7 5 3 1

The Johns Hopkins University Press
2715 North Charles Street
Baltimore, Maryland 21218-4363
www.press.jhu.edu

The Library of Congress has cataloged the hardcover edition of this book as follows:

Long, Pamela O.
Openness, secrecy, authorship : technical arts and the culture of knowledge from antiquity to the Renaissance / Pamela O. Long
p. cm.
Includes bibliographical references (p.) and index.
ISBN 0-8018-6606-5 (hardcover : alk. paper)
1. Technology and civilization—History—To 1500. 2. Communication of technical information—Europe—History—To 1500. 3. Technical writing—Europe—History—To 1500. 4. Intellectual property—Europe—History—To 1500. 5. Learning and scholarship—Europe—History—To 1500. 6. Renaissance. I. Title.
CB478 .L65 2001
001.1'094—dc21
00-010152

ISBN 0-8018-8061-0 (pbk.)

A catalog record for this book is available from the British Library.

For

Bob Korn and

Allison Rachel Korn

Contents

Illustrations

Acknowledgments

WRITING THIS BOOK has been a long-term project during which I acquired numerous debts, used many libraries and archives, both in the United States and in Europe, and developed many friendships. It has been supported by a number of grants and fellowships, many of which arrived at precisely propitious moments. I am indebted to friends and colleagues who have discussed and critiqued my work and to my students, from whom I have learned much, especially those in my seminars and courses at the Johns Hopkins University and in a seminar at the Folger Institute for Renaissance and Eighteenth Century Studies.

The research for this book was supported by three grants from the National Science Foundation (grant nos. SES-8607112, DIR-9112729, and SES-9729871). I also am grateful for a three-month fellowship from the Dibner Library, of the Smithsonian Institution Libraries; and an NEH fellowship for an academic year at the Folger Shakespeare Library. The Folger Library has been an ideal location for both research and intellectual exchange. I thank everyone there, especially Werner Gundersheimer, Lena Cowen Orlin, Kathleen Lynch, Carol Brobeck, and Betsy Walsh. In addition, the Library of Congress has been a crucial research base; Bruce Martin in his capacity as supervisor of research facilities has always been helpful and solicitous of scholars and their work. Carol Armbruster, a European area specialist, has provided both technical assistance and moral support.

A different version of chapter 4 appeared as "Power, Patronage, and the Authorship of *Ars:* From Mechanical Know-How to Mechanical Knowledge in the Last Scribal Age," *Isis* 88 (March 1997): 1-41, © by the History of Science Society. All rights reserved. I thank the University of Chicago Press for permission to reprint.

Friends and colleagues who have helped well beyond the call of duty in various ways include Bert S. Hall, Pamela Smith, Alice Stroup, Mary Henninger-Voss, Robert Mark, Frank A. C. Mantello, and Alex Roland. I discussed many parts of the research with Gaby Spiegel and Mary Fissell, both of whom also read major portions of successive revisions of the manuscript and helped to bring about major improvements. My sister, Priscilla Long, an author of scholarly works, fiction, and poetry, critiqued chapters and in addition discussed with me this work in particular and writing in general during numerous usually late-night telephone conversations. My dear friends Glenn Bugh and Jackie and Ken Berkman have been there for the long haul, and I thank Ken as well for his work on the index. I thank Robert J. Brugger, of the Johns Hopkins University Press, for his encouragement; an anonymous reader, whose extensive comments helped to improve the manuscript greatly; and Joanne Allen, whose astute and rigorous copyediting has saved me from many errors.

Finally, and most of all, Bob Korn has helped with innumerable problems involving computers, modems, and logistics, has read drafts, and has discussed this work with me over the years. Our daughter, Allison, has grown from a kid whose favorite game was hide-and-go-seek to a hardworking and enthusiastic college student. This book is dedicated to them.

Note on Editions and Translations

I HAVE USED standard critical editions of sources wherever possible, and I have utilized English translations when adequate ones were available. When I cite a translation that is given in the bibliography, I use that translation unless I indicate in the note that the translation is mine. When only a non-English source is cited, the translation is my own. I provided direct quotations in the original language in notes only in cases where the original text may not be widely available.

Openness, Secrecy, Authorship

INTRODUCTION

Categories and Key Words

Local Meaning in Long-Term History

THIS BOOK investigates openness, secrecy, authorship, and ownership—what we now call intellectual property. Ranging from antiquity to the early seventeenth century, it draws substantially on writings from technical, craft, and practical traditions. Sources include books on catapults, agriculture, and generalship; evidence derived from artisanal contexts, such as ancient and medieval craft-guild records; treatises on painting, architecture, machines, mining, and pottery making; alchemical treatises, magical texts, and Neoplatonic writings that elucidate esoteric beliefs and practices. Rather than contributing to the substantive content of a particular discipline—say, the history of agriculture or the history of architecture—it explores attitudes and practices involving the possession of knowledge and its transmission. It investigates the cultures of knowledge—whether knowledge is open or secret; which topics become the focus of written authorship and why; when and in what contexts people consider knowledge to be property; and what they take knowledge to be.

The ways in which practical and craft cultures have interacted with learned traditions are a significant focus. During most of antiquity the separation of craft know-how from more theoretical knowledge was characteristic. In late antiquity, however, certain alchemical sources integrated material operations and the quest for higher knowledge, while craft magic harnessed the higher world of the spirit for material ends. In contrast, most late antique Neoplatonic sources indicate a quest for understanding, or gnosis, that was specifically opposed to material corporeality.

The world of material artifacts was created largely by artisans, who before the fifteenth century handed down their craft knowledge and skill apart from the world of books and learning, for the most part orally, through apprenticeship systems either formal or informal. In the context of medieval urbanism from the thirteenth century, for the first time significant evidence emerges for proprietary attitudes toward craft knowledge, indicated both by craft secrecy and by patents for invention.

This book argues that from the early fifteenth century there was closer interaction between the technical arts, political power, and knowledge. Indications for such a development include the great expansion of open traditions of authorship on the mechanical arts and a simultaneous renaissance of Neoplatonic writings that promulgated both secrecy and the use of utilitarian magic aimed at changing the material world. This study contributes, finally, to the longstanding issue in the history of science concerning the nature of the contribution of the mechanical arts to the rise of empirical and experimental methodologies within the new sciences of the seventeenth century.

Openness, Secrecy, and Authorship belongs to intellectual and cultural history as well as to the history of science and technology. Yet it recognizes that *science, technology*, and even *history* are present-day terms whose meanings may be inappropriate or misleading for past cultures even when cognate terms exist in the diverse languages of those cultures. For example, *scientia* in Latin means "knowledge" in a broad sense and has none of the methodological or disciplinary meanings that we today associate with *science* and *scientist*. Similarly, the ancient Greek word *technē* refers to material production of all kinds, from making soup to constructing catapults, and the reasoning associated with that production. The term *mechanical arts*, which came into use between the ninth and twelfth centuries, included painting and sculpture as well as the construction and operation of machines. It engages a complex of meanings quite different from those of the modern word *technology*.[1]

Conceptual categories change from one historical period to another. For example, Aristotle delineated three areas of human activity: first, material and technical production *(technē)*; second, action *(praxis)*, such as political or military action, that requires judgment in contingent or uncertain situations *(phronesis)*; and third, theoretical knowledge or knowledge of unchanging things *(epistemē)*. Aristotle's separation of material production from action and from theoretical knowledge presupposed a hierarchy with technē at the bottom and epistemē, or theoretical knowledge, at the top.[2]

Such categories and the relationships between them exert a crucial influence on the construction of knowledge itself. For example, seventeenth-century experimental philosophers attempted to legitimate claims about the natural world by manipulating machines such as air pumps. Thereby they challenged traditional Aristotelian categories by bringing together technē (manipulation of machines and instruments) and epistemē (theoretical knowledge). Seventeenth-century Aristotelians countered the experimentalists with the argument that this combination was a category mistake involving the improper fusion of separate conceptual entities.[3]

The categories of material production, action, and theoretical knowledge and their relative status vis-à-vis one another have complex histories that need more study. Yet such conceptual categories are often ignored because of traditional conventions of historiography, which separate science from technology and political and moral spheres from both. Categories involving material manipulation, action, and knowledge, as well as their histories and the systems of knowledge surrounding them, have been obscured by disciplinary histories such as the history of technology, the history of architecture, and the history of science. Such disciplines were created in the nineteenth and twentieth centuries on the basis of modern notions of what properly belonged to them and what did not.[4]

Autonomous, idealized disciplinary histories are often based on the presupposition that they contain a core of substance that has remained stable for centuries. More recently, historians have approached texts and disciplines in terms of the particular historical cultures in which they developed and have also investigated how disciplinary categories themselves have shifted over time. Central to this latter approach is the notion that a text—or any other historical artifact for that matter—should be understood within its own culture in terms of the categories that belong to historical actors. The meaning of a text, as Quentin Skinner and others have argued, will be misunderstood without careful, detailed attention to the fabric of meaning that exists within the culture to which that text initially belonged.[5]

This book originated within a very different investigation involving early architectural writings, including *De architectura*, by the ancient Roman architect Vitruvius. I was intrigued by Vitruvius's advocacy of the open, written transmission of knowledge, and I wondered how it might be understood in the context of the early Roman imperial age. At the same time, I was interested in the Renaissance and early modern development of empirical approaches to the natural world and in the apparent connection between the ideal of openness and methodologies of empiricism and experimentation. If one were to write a history of the ideal of openness, I wondered, could Vitruvian openness and the openness claimed in early documents of experimental philosophy find a place in the same book? Did the two ideals, expressed in very different times and places, share areas of meaning, or were similar sentiments being used to express utterly different concepts?

What initially led me to this question was a confrontation with the traditional assumption that science by its nature was open, whereas technology and artisanal craft work were secret. Yet Vitruvius's *De architectura*, which treats machines, the fabrication of colors for painting, and other technical matters, as well as the design of buildings, praises the open, written transmission of knowledge. The expressed Vitruvian ideal was mirrored in the apparent openness of numerous manuals involving material production and technologies of various kinds. Yet historians concerned with craft traditions frequently referred to craft secrecy and seemed to assume its universal presence within all craft practice, often without benefit of evidence. Similarly, until recently openness was taken to be characteristic of science, and there was very little reflection concerning whether scientific practices were actually open and, if they were, what that openness meant.

In recent years, though, much scholarship has focused on the issue of "openness" in science and in early modern natural philosophy. The result is a far more ambiguous and complex picture than one that equates science with openness. Sophisticated, sometimes controversial studies, including those by David Hull, Owen Hannaway, Steven Shapin, Simon Schaffer, Paula Findlen, Mario Biagioli, Jole Shackelford, and Martha Baldwin, have underscored the complexity of scientific "openness" in diverse historical contexts. For early modern Europe, openness has been studied in relationship to issues involving public and private space, patronage and the court, the social status of disciplines, and the witnessing of experiment. Openness as an explicit value has also been linked to issues of authorship.[6]

A study of authorship itself requires attention to the particular local circumstances that produce and shape various kinds of authors and their writings. Books are written by individuals whose activity involves the creation of texts by some means—for example, by the compilation of previously existing works or by new textual expression. Authorial attitudes toward authorship, that is, the views of the individual writer concerning his or her creation of a discrete text, are dependent on particular historical cultures and social circumstances. Who writes a treatise, and why? What is the evidence for attitudes toward authorship within the treatise itself and in surrounding or related documents? What constitutes the prospective readership, and how is the book actually read? The history of writing and the history of reading are intricately related.

Yet this study is not exclusively concerned with authors and readers of texts. It focuses as well on evidence from oral traditions of craft practice. From such craft traditions, I suggest, came not only a large body of tacit knowledge concerning material processes, implements, and machines and their use but also more abstract concepts that ultimately influenced learned culture. The development of the concept of intellectual property, that is, intangible property involving inventions and craft processes (but not the term itself or its modern legal development) occurred in the context of the late medieval craft guild. Further, within traditions of painting, sculpture, and other constructive arts of the fifteenth and sixteenth centuries developed a view of original authorship as involving the creation of unique products resulting from the ingenuity of the creator or inventor.

This study posits material and craft cultures as significant contributors to the history of human thought. It investigates certain kinds of authorship involving texts, as well as authorship involving material objects and inventions. While accepting that meaning is mediated by language, it views material production and use, not as unchanging and inert conditions of human existence or as the invisible and unanalyzable material base of that existence, but as an intrinsic aspect of human culture and meaning, one that confers conceptual as well as material goods.

For both orally transmitted craft cultures and text-based cultures, the broad chronological range of this study allows investigation of actors' categories and assumptions within particular local contexts, as well as an inquiry into the appropriation and creative misunderstanding of such categories by later authors. The combination of contextual methodology and long chronological range allows a study of changing conceptual categories and of appropriations that often elude local studies.

Some of the key words in this study—for example, *openness, secrecy, authorship, priority*, and *plagiarism*—are the focus of complex scholarly debates that have shaped the terms themselves as well as issues of historiography surrounding them. Words have histories. They derive from other words, develop and change in meaning, come into being, and disappear. As Raymond Williams suggests, a study of the emergence of particular terms and their variations of meaning elucidates matters of history and culture. Critiquing Williams, Skinner emphasizes the distinction between word and concept and insists that concepts often develop before the words used to describe them. Using Milton's concept of "originality" as one example, Skinner suggests that Milton possessed the concept long before the term emerged to describe it. Similarly, in the present study I argue that the concept of "intellectual property" developed in the context of the medieval craft guild, centuries before the emergence of the modern legal terminology that describes it.[7]

Openness

Openness refers to the relative degree of freedom given to the dissemination of information or knowledge and involves assumptions concerning the nature and extent of the audience. It implies accessibility or lack of restrictiveness with regard to communication. Openness can occur within speech or writing and can serve different purposes, such as the transmission of information or the display of knowledge in order to enhance one's social status or political power.

The traditional association of science with openness and technology with secrecy in part came out of assumptions about the permanent structures of such disciplines. Robert K. Merton posited one of the "norms" of science to be "communism," by which he meant openness. He suggested that the "institutional conception of science as part of the public domain is linked with the imperative for communication of findings." The secrecy of technology contradicted this ethos. Derek de Solla Price argued that science was intrinsically open, that its goal was to further knowledge gained through publication. In contrast, technology was intrinsically secret; its goal was the production of material things and the profit achieved by competitive marketing.[8] These views ignore the existence of open writings that explicate technologies from the time of antiquity. They also tend to strip craft and technical production of rational and discursive features despite ancient traditions of technē authorship that exhibited such features and also insisted explicitly upon them.[9]

The image of science as a disinterested, truth-yielding process of open inquiry has been revised by complex critiques involving social constructivism as well as by less theoretically driven empirical studies. Likewise, historians have developed pictures of the workings of early modern science that are far more complex than one that describes an open, disinterested search for truth. Studies by Findlen and Biagioli, among others, have shown that openness was associated not so much with expeditious publication for the furtherance of knowledge as with proper timing consonant with the establishment of priority, the felicitous advancement of patronage relationships, and display for the purpose of advancing the legitimacy and power of the prince. Steven Shapin argued that the experiments of the Royal Society were carried out in private, witnessed by "gentlemen," who, being of independent means, were considered disinterested. In contrast, public experiments involved the display of results that were already considered certain. While the specific claims of some of these studies have been the focus of criticism and debate, it is at least certain that most scholars will analyze issues of openness, secrecy, and transmission as far more complex and more enmeshed in specific social contexts than previously.[10]

On the other hand, the unquestioned association of craft and technical traditions with secrecy must be challenged. Oral transmission of craft knowledge within apprenticeship systems, whether in households or in workshops, can be open as well as secret. Openly disseminated writings on technology and crafts appeared in ancient and medieval cultures and became especially numerous from the early fifteenth century. Authors of some of these books vigorously articulated the ideal of open, written dissemination of knowledge, including knowledge of mechanical, craft, and technical matters.[11]

Secrecy

The philosopher Sissela Bok defines secrecy as "intentional concealment," to be distinguished from "privacy—the condition of being protected from unwanted access by others" and from unknown things such as "secrets of nature." She notes that the Latin *secretum,* meaning something hidden or set apart, derives from *secernere,* "to sift apart, separate as with a sieve." Bok points to other words, such as the Latin *arcanum,* and terms in other languages, many of which denote secrecy and at the same time communicate value judgments. For instance, the Greek word *arretos* first meant the unspoken and later came to mean the prohibited and unspeakable. Bok argues that such value judgments, whether positive or negative—"whether of something sacred, intimate, private, unspoken, silent, prohibited, shameful, stealthy, or deceitful"—should be explored rather than inserted into the initial definition of secrecy itself.[12]

For the historian, the issue of secrecy presents problems of evidence, meaning, and motivation. If there is evidence for intentional concealment, what is the context and how does it function? Is it, to mention just two possibilities, the secrecy of the priest of a mystery cult, protecting sacred knowledge from defilement by the common rabble, or the secrecy of the medieval artisan, protecting craft knowledge in the interest of profits? If there is evidence for secrecy, what is its meaning for those who practice it and for those who are excluded? How do secret groups function in different contexts, and how do systems of belief and the practice of secrecy reciprocally reinforce one another?

Several categories and phrases are sometimes placed under the rubric *secret* but actually mean something other than intentional concealment. The word *secret* itself and its cognates in various languages often means "technique," as in "The secret of making a good piecrust is to use ice cold water." Another usage, in the phrase *secrets of nature,* was widely employed in the sixteenth and seventeenth centuries. It usually refers to things about the natural world that are unknown but can perhaps be discovered by observation, experimentation, or some other methodology.[13]

Authorship

On the most basic level, authorship refers to the act or practice of creating something, such as a treatise, a painting, or a material invention. In the most mundane sense, an author of a written work is a writer, who must always do something more than copy another text verbatim; copying two or more texts and putting them together may suffice. Most cultures distinguish the scribe, who is a copyist, from the author, who creates treatises or other kinds of writing. Yet authorship as the practice of creating things had greater cultural significance in some eras than in others. Similarly, concerning the authorship of material things, someone who makes wooden barrels just like those made for generations before might be considered a producer rather than an author. Alternatively, one who paints an "original" painting or invents a new device can be considered the author of that painting or of that device. Yet originality as a value is not always present, while authorship itself has been viewed variously in different historical eras. "Author" was an important actor's category in some historical cultures but not in others.[14]

One kind of "authorship" refers specifically to a self-conscious authorial presence within a given text and is tied directly to notions of originality and the ways in which authors assert themselves within their texts. In a study of origin and originality David Quint investigates the development of the notion of authorial originality in the Renaissance period and ties it to humanist historicism and relativism. Authorial creation becomes a purely individual activity tied to a particular historical culture rather than to timeless origins or truths. Kevin Dunn analyzes authorship in terms of authorial voice within changing concepts of public and private space as reflected in the Renaissance preface. He locates his key texts, including "scientific" texts, in the late sixteenth and seventeenth centuries. Others, such as Joseph Loewenstein and Alvin Kernan, analyze the emergence of authorship within the context of the marketplace of print as that market developed in the seventeenth and eighteenth centuries.[15]

Authorship was a central issue in late-twentieth-century literary criticism. In diverse ways, Roland Barthes, Michel Foucault, and Jacques Derrida aimed to free the study of texts from the traditional approach, which focused upon authors' biographies and authorial intent, and from a view of texts as representations of "reality." Authorship became an issue in an epistemological revolution in which individual writers came to be viewed as participants within a common discursive practice rather than as authors of uniquely original works. Barthes and Foucault in famous essays announced the death of the author. This deceased author was an "Author-God," imagined as a genius creating utterly original texts from the core of his or her unique personality.[16]

A more narrowly conceived historical project among literary historians posits the eighteenth-century rise of the modern author, one who produces uniquely original texts. They tie this author to the emergence of the author's copyright, considered to be a natural Lockean right belonging to authors of original creations. This scholarship proposes that modern copyright law, which appeared initially as the 1710 English law the *Statute of Anne*, was enacted to protect the author and his or her unique original writings. Yet the account of an intrinsically close relationship between the rise of the "modern" author and the legal development of copyright has been criticized by legal historians. David Saunders argues that the development of copyright law had nothing to do with "modern" authorship but developed in the early eighteenth century in response to problems in the book trade. Jane C. Ginsburg points out that most early copyright case law concerned compilation authorship, which did not involve a high degree of originality.[17]

This scholarship on the eighteenth-century emergence of authorship in association with intellectual property law seems overly driven by theoretical models. For earlier centuries, a pertinent focus of investigation might be what historical actors themselves thought they were doing when they wrote a book, painted a fresco, or devised an invention. Such an investigation would involve an examination of the significant context of authorship in its broadest sense as a historical practice. It would also insist on a pluralistic view of authorship that encompasses the creation of texts, ideas, inventions, and material objects.

A useful working definition of *authorship* for a variety of historical periods permits a gradation of meanings between the poles of authority and originality. Thus Neil Hathaway observes that in antiquity honor was accorded individual authors, so that the uncredited use of their works was regarded as theft, whereas in the medieval period the *auctoritas* of texts themselves predominated, so that *compilatio*, the compilation of authoritative texts, proceeded without concern for accurate credit to authorship. Writings driven by values of authority rather than values of authorship include many alchemical texts attributed to prior authorities.[18]

Another approach to authorship occurred within many early modern patronage relationships in which writers were clients of patrons. In these situations, especially in the context of the courts, the patron became the "author" and received credit for works written by clients. The same view of the patron as author occurred in the realm of constructive arts such as painting and sculpture.[19]

Yet the concept of original authorship, that is, the view that a unique product is created out of the ingenuity or individuality of a particular author, developed significantly in the fifteenth and sixteenth centuries. It was evident in some instances on behalf of the authors of books, but it also developed particularly with reference to ideas, craft processes, and inventions within traditions of the mechanical arts, including engineering and painting. For example, the early-fifteenth-century petition for a patent by the Florentine Filippo Brunelleschi argues that he had invented the cargo ship (for which he was seeking a privilege) out of his own individual ingenuity. Originality and novelty began to acquire cultural value. Yet as Bruce Cole points out in an essay on Titian, sixteenth-century notions of originality could be quite distinct from modern attitudes and did not exclude the widespread practice of copying.[20] In another area, the priority conflicts that proliferated from the early sixteenth century within traditions such as cosmology and mathematics involved disputes about original authorship of ideas and inventions, both conceptual and material.

Plagiarism and Ownership

Plagiarism and ownership, or intellectual property, are distinct entities, both conceptually and in practice. Plagiarism, or literary theft, involves issues of credit rather than of ownership. Literary theft can be a concern (as it was in antiquity, for example) without the presence of a concept of intellectual property. Intellectual property involves the notion that some intangible entity—an invention, a craft process, an idea, textual expression—constitutes property apart from the material object with which it is associated and apart from the value of the labor that produced it. The concept of intellectual property exists in certain historical cultures and not in others. For instance, within ancient law there is no evidence for a concept of intangible, intellectual property, no notion that inventions, knowledge of craft processes, or textual expression could be owned as property.[21]

In contrast, ancient sources provide much evidence for literary theft or plagiarism. Literary theft involves affixing one's own name to someone else's writings or inserting someone else's writing within one's own text without acknowledgment. What was being stolen in the ancient world was credit, honor, and reputation rather than property. Yet often in both the ancient and medieval practices of authorship such a procedure constituted, not theft, but the creation of an authoritative text. Within modern legal systems plagiarism and intellectual property are overlapping but not entirely synonymous categories. Sometimes, for example, individual writers claim that ideas or plots have been stolen from them, but ideas and plots do not constitute property under copyright law, which protects textual expression.[22]

Intellectual property is a modern legal phrase that refers to laws that were enacted in the eighteenth century, such as patent, copyright, and trademark laws. Yet I argue that the concept of intellectual property and attitudes toward such property developed well before that time within the context of late medieval craft production. The genesis of intellectual-property values can be found in the context of the thirteenth- and fourteenth-century urban craft guilds. Evidence for such values is manifest in both craft secrecy and patents for invention.[23]

The concept of intellectual property emerged first in the context of craft production rather than in the context of book production. Some years after the development of the printing press in the mid-fifteenth century, limited monopolies, originally called "letters patent" and later called "privileges," were sometimes awarded to printers and occasionally to authors of books. In her detailed study of early French privileges Elizabeth Armstrong shows that such book privileges provide little evidence of protection for intangible property. A privilege gave an individual the exclusive right to publish a book or group of books within a particular locality for a limited period of time, usually ten years or less. Yet such a privilege could be compensation for labor or expenditures rather than for intangible (intellectual) property.[24]

Armstrong shows that privileges were by no means granted for original authorship or for textual expression per se. Rather, available evidence points to grants for time expended (labor) and money spent even in the few cases when authors themselves rather than printers were awarded privileges. Because the book-privilege system was grounded in such tangibles as labor and costs rather than in intangibles such as ideas, inventions, knowledge of craft processes, or textual expression, for the most part it cannot be seen to represent a development of the concept of intellectual property. Privileges were granted for "novel books," that is, books never before printed. Well-known ancient texts long available in manuscript were considered "novel" in this context, meaning that they were being printed for the first time. Limited monopolies given for first-time printing of such books offered protection on the basis of labor and costs rather than out of any concern for original ideas or textual expression.[25]

Evidence from other parts of Europe reveals that book privileges or copyright functioned in similar ways. Venice bestowed the first author's copyright on the humanist Marcus Antonius Sabellicus in 1486 for his history of Venice. Elsewhere printers and sometimes authors received privileges or copyrights for particular books. Yet there is little evidence that an author's privileges were connected to the originality of the author's expression (as in modern copyright law). Usually it constituted a commercial privilege similar to that obtained by the printer, justified by the author's outlay of labor and expense. Because these monopolies applied to limited geographical jurisdictions, they rarely prevented pirating. Moreover, the practice of granting privileges often involved interests beyond commercial ones, namely, efforts to control what was printed, or censorship. The English crown chartered the Stationer's Company, or printers' guild, in 1557 and thereafter used it not only to license books but also to destroy seditious ones. Many governments used the book privilege to grant permission to publish approved works and to deny permission to publish treasonous or heretical books.

Censorship on the basis of religious opinions was practiced by both Protestant and Catholic jurisdictions.[26]

Priority

The identification of first discoverers and inventors was a topic of interest from ancient times. For example, Pliny the Elder's first-century encyclopedia, *Natural History*, contains an inventory of discoverers in the arts and sciences (some of them mythical) and a list of inventors of weapons. Pliny's list is part of a larger catalog of the marvelous—outstanding men in science, medicine, and art, examples of exceptional memory, unusual endurance, and remarkable prices paid for slaves. Roger French points out that marvelous things constituted the proper subject of the Roman *historia*, which focused on the exceptional rather than the average.[27]

The tradition of identifying first inventors and discoverers continued into the early modern period and is tangentially related to the sixteenth-century development of priority disputes. Such disputes developed particularly within certain disciplinary areas, such as mathematics and astronomy.[28] The preconditions for the priority dispute include the view that individual ingenuity lies at the basis of new discoveries, the positive valuation of novelty, and a reward system that prizes first discoveries. The ancient tradition of first discoverers valued the marvelous but did not focus upon the ongoing production of novel things and thus did not provide all of the necessary preconditions for priority disputes. Only in the sixteenth century, when social status, honor, and patronage came to be tied to first discoveries and inventions, did such conflicts emerge.

Recent investigations provide detailed information about specific disputes, such as the one that erupted over Tycho Brahe's priority in constructing a geoheliocentric planetary system that was a compromise between the new, heliocentric system of Copernicus and the traditional, geocentric system of Ptolemy. The princely courts and their complex systems of patronage became important arenas for advancement for natural philosophers as well as others. Priority of discovery could be used felicitously in such an environment. Biagioli's study of Galileo is especially revealing in its explication of the complexity of some of those patronage systems and in its analysis of the way Galileo used the priority of his astronomical discoveries to advance his professional interests.[29] Within the compass of such a concern, I suggest, priority disputes show evidence of the emerging value of unique, original authorship—of ideas, discoveries, and inventions. It is significant that this positive valuation of original authorship, of the novel products of individual ingenuity, was also evident within artisanal traditions of the previous two centuries.

This study explores the numerous intrinsic ways that the values of openness and secrecy and the practices of authorship are linked to technical production, praxis, and knowledge. Chapter 1 investigates ancient writings related to technical production, such as Hellenistic engineering books, as well as writings tied to political and military praxis, including Xenophon's *Oeconomicus* and Roman agricultural writings. The separation of the categories of technē and praxis in the fourth century B.C.E. cuts across the modern category of technology. Agriculture belongs to praxis because it was believed to inculcate good character traits, such as fortitude and persistence, deemed desirable for political and military leadership by elite men. The contexts of authorship for the two kinds of writings were separate. Engineers and other practitioners produced technē writings and dedicated them to patrons of higher status. In contrast, men of elite classes wrote praxis writings for their (often younger) peers. Both kinds of writing were evidently open; some articulated the ideal of the open, written transmission of knowledge.

The most significant sources for secrecy and esoterism are from the second and third centuries C.E. These late antique sources include magical texts, early alchemical writings, evidence from the cult of Isis and other mystery religions, hermetic traditions, and finally the writings of Neoplatonic philosophers such as Iamblichus, Porphyry, and Plotinus. Viewing these diverse texts together contextually, as I do in chapter 2, underscores the late antique conflict between efficacious magic, which aimed to affect the material and corporeal world, and Neoplatonism, which in general abhorred material physicality in a quest for the spiritual, noncorporeal realms of the divine. The second-century rhetorician Apuleius, in his famous novel *Metamorphoses* and in his defense against an accusation of magic, the *Apologia*, offers striking evidence of the conflict between efficacious magic and noncorporeal spirituality. His writings also provide an intriguing view of secrecy in the craft of magic and the very different kind of secrecy within esoteric mystery cults.

Chapter 3 investigates the sources for craft knowledge, such as craft recipe collections and medieval guild regulations. The context from which craft recipes derive, such as a scriptorium or a craft workshop, turns out to be crucially important in evaluating their significance. For example, evidence for secrecy in ancient Mesopotamian glass recipes refers to secrecy in the scriptorium, not in the workshop. Abundant evidence for craft secrecy appears only from the thirteenth century in the context of the medieval craft guilds. I argue that such secrecy is indicative of growing proprietary attitudes toward craft knowledge involving intellectual property. Early privileges or patents for invention around the same time point to similar proprietary attitudes.

Such activities as building catapults or, later, gunpowder artillery, processing metal ores, or making pots were not ordinarily transmitted in writing. Why such topics became the focus of writings in the first place is an important question, as is the reason for their changing significance in the political and moral life of particular cultures. Although some technē writings appeared in antiquity and in the medieval era, such writings greatly increased in number in the fifteenth century under the rubric *mechanical arts*. This expansion, the focus of chapter 4, came about in the context of patronage. Underlying such a development, I argue, was the growing legitimation of political power by the use of technical arts such as building construction, urban redesign, and ornamental arts such as painting and sculpture. In the fifteenth century, making things and political praxis mutually reinforced each other; they became closely intermeshed in a way that they had not been before. Because such writings displayed the arts of princes and oligarchs, they were in fundamental ways openly purveyed and often advocated openness as well.

Yet the fourteenth and fifteenth centuries also saw a rising interest in alchemy and in Neoplatonic traditions, the focus of chapter 5. Neoplatonic traditions in particular became popular throughout Europe, in part as a result of the influential translations, commentaries, and other writings of Marsilio Ficino. Although late medieval alchemical and Neoplatonic traditions relied on older writings, they also developed them in their own ways. Given the general split between efficacious magic and spiritual Neoplatonism in the late antique world, it is all the more striking that in the later period the renaissance of both alchemy and Neoplatonism combined efficacious magic with spiritual Neoplatonism in a utilitarian synthesis. Renaissance Neoplatonists sought the divine realm but at the same time posited magic as a means for ameliorating the ills of physical and material life. Some authors, especially Ficino, wrote openly, while many advocated and practiced certain forms of secrecy and participated in esoteric groups. Whether they advocated openness or secrecy, Neoplatonic and alchemical authors of the Renaissance embraced utilitarian values, giving their books a certain affinity with writings on the mechanical arts that proliferated in the same time period.

Chapters 6 and 7 treat authorship in the mechanical arts in the sixteenth century. Authors explicated such diverse arts as mining, metallurgy, military arts, architecture, painting, and pottery in both vernacular and Latin treatises. Some arts, such as painting and architecture, began a transformation into higher disciplines during this period. Yet even if they remained mechanical arts, written explication brought the know-how of craft and engineering practice into written form, making them resemble, in some cases making them become, learned subjects. I argue that some arts, such as architecture and the military arts, became a common ground of communication between elite, learned individuals and artisan practitioners. *Trading zones* developed in which learned men and artisan practitioners communicated reciprocally, exchanging knowledge. Each learned from the other in substantive exchanges that went beyond the well-recognized mutual benefits of patron-client relationships. Writings on the mechanical arts were usually open and sometimes advocated openness. I suggest that the fifteenth- and sixteenth-century culture of authorship in the mechanical arts form an essential background for developments in experimental philosophy in the seventeenth century.

This investigation is grounded in an interest in understanding the shifting conceptual categories involved in technical production, praxis, and knowledge and the meaning of those categories within particular historical contexts. Such cultures of knowledge are tied to issues of communication and transmission. Historical actors saw their inherited traditions from their own vantage points and communicated them in their own ways. Attitudes toward the possession of knowledge and its communication are part of the substantive discourse of those traditions.

If the history of culture is a history of understandings, misunderstandings, and appropriations of old materials for new uses, then contemporary historians are also involved in this process. The horizons of vision of historians of specific disciplines tend to be shaped by the values inherent in modern disciplinary constellations. What Hans-Georg Gadamer called the "historically experienced consciousness" recognizes that the boundaries of the modern disciplines are not necessarily relevant to, and indeed can hinder, "the experience of history," namely, the understanding of historical cultures on their own terms.[30]

Yet this study goes beyond the textual boundaries of hermeneutics and attempts to understand issues of openness, secrecy, and authorship within the broadest context possible, one that is social, political, economic, technological, and material, as well as cultural. This is not to claim that "reality" can be perceived directly beyond the intricate web of perception and interpretation based on evidence that is itself cultural production. But it does assume the possibility of careful historical reconstructions based on the mediated form of the past preserved in both language and artifacts.[31] And it finds a significant place for material culture, craft production, and oral transmission of craft knowledge in the cultural and intellectual history that is its focus.

Chapter 1

Open Authorship within Ancient Traditions of *Technē* and *Praxis*

IN MID-FIFTH-CENTURY B.C.E. Athens, architecture, sculpture, and other technical arts came to be closely allied with political practice. Pericles' vast program of building construction on the Acropolis supported and indeed was intrinsic to his rulership of democratic Athens. Pride in Athenian democracy, successful political control, and spectacular new building construction all went hand in hand. Yet the alliance of *technē* and *praxis* did not last. A hundred years later, Xenophon and Aristotle explicitly separated material construction and the technical arts (technē) from the praxis of political and military leadership. In the succeeding centuries, throughout Hellenistic and then Roman rule, the separation of the two entities, although never complete or absolute, continued. In general, the ancients thought that rulers and military leaders achieved success because of character traits such as courage and virtue, not because of technology or technique. They viewed technical matters, including weaponry and material production of all kinds, as separate from praxis and subordinate to it.

In this chapter I explore the changing relationships between praxis and technē, particularly as they influenced the development of authorship. These ancient categories differ significantly from modern ones. Modern cultures, for example, usually associate agriculture with technology. The ancients certainly used technologies and techniques in the actual practice of agriculture, but they considered it to be conducive to the development of good character traits in the landholder that would prepare him for political and military action. They believed that agriculture inculcated virtue, training elite males to be good leaders. It was a discipline appropriate to the praxis of political and military leadership, quite separate from lower-status occupations involving the technical arts.

It often has been noted that ancient crafts, carried out for the most part by slaves and manumitted slaves, suffered from a profoundly low status.[1] Yet craft practice and material production seem to have enjoyed great prestige in fifth-century B.C.E. Athens. Although the technical arts never achieved such status again, their cultural role did not remain static. Certain kinds of material invention and production, such as those involved in military engineering and architecture, enjoyed higher standing, though never the highest. These higher technical disciplines came to be explicated in writings. Further, ancient technē authors seemingly wrote openly, belying the view that technological and craft production invariably involved secrecy. Even ancient writings about complex weapons such as catapults reveal no concern for secrecy. The Roman architect and military engineer Vitruvius explicitly advocated the open, written transmission of knowledge. Authors of praxis writings, including those on military topics such as generalship, wrote openly as well.

Distinctions between praxis and technē, and the separation of both from *epistemē*, or theoretical knowledge, significantly influenced the cultures of authorship in the ancient Mediterranean world. Such conceptual categories are relevant to who the authors were, what topics they treated, and their prospective readership. This chapter concerns the cultures of authorship associated with the categories technē and praxis. It does not focus upon the vast majority of technical arts that did not find explication in treatises; nor does it attempt to reconstruct historical narratives about ancient politics or war using praxis writings. Rather, it focuses on authorship per se as it concerns the technical arts on the one hand and political and military action or praxis on the other.

Democratic Rule and Building Construction in Periclean Athens

An alliance between political praxis and the building arts emerged in Athens in the mid-fifth century B.C.E., one result of the Athenian democratic revolution. The revolution began when a prodemocratic party established reforms after gaining the upper hand over an aristocratic faction. Pericles persuaded the Athenian assembly to pass additional laws in the 450s to complete a process of democratization. A crucial reform provided that jurors in the *dikastēria* (law courts) would be paid by the state for their services, allowing citizen participation regardless of economic status. There were limits to the new democracy: women were excluded, as were foreigners (metics), and slaves; only about 30 percent of the male population participated.[2] Nevertheless, the creation of the world's first participatory democracy had great cultural and historical significance.

The generals, or *strategos*, of whom there were usually ten, effectively ruled the city-state. Because they were deemed to possess special skills, they could be elected as often as the people wanted them. Pericles thus won reelection year after year for at least fifteen years. Although he came from a wealthy aristocratic family, his vision for Athens was a democratic one. The newly democratic city-state existed on the basis of a greatly enlarged public sphere. Huge assemblies and law courts convened on a regular basis. The rhetorical arts of persuasion became crucial.[3]

Itinerant teachers known as sophists flourished in the newly democratic city-state. Working in and around Athens, sophists offered instruction for fees, primarily in the art of persuasion and oral argument. Yet they also played a significant role in the development of prose handbooks on a variety of topics. Protagoras, one of the earliest and most famous of the sophists, argued that *logos* (argument, discourse) was relative to particular situations and specific individuals. He created a dialogic methodology that accentuated contrasting arguments and points of view.[4]

Protagoras was a friend of Pericles', apparently charged with training the latter's sons. Edward Schiappa argues that his teachings "functioned ideologically to advance the precepts of Periclean democracy and to oppose the aristocratic implications of Eleatic monism." He believed that citizens deliberated and came to collective decisions through *logos*. Protagoras embraced relativism, the positive valuation of the individual point of view, and the belief that individual citizens could be educated to become virtuous.[5]

Sophists such as Protagoras taught argumentative techniques for use in the praxis of participation in democratic government and the law courts. Their writings on this subject, called *technai*, probably consisted of examples of speeches for different situations. Yet they also included other practices within their range of concerns. Protagoras is said to have taught household economy. He may have thought of the household as a miniature city-state. The *hoplomachoi*, who have been called military sophists, taught the art of generalship and tactics.[6]

The practical and pedagogic orientation of the sophists seems to have encouraged the development of prose writings on practical topics. Sophists initially offered instruction in spoken form for a fee; they also distributed written versions of their lessons as a way of enhancing their reputations, upon which their livelihood depended.[7] Sophist relativism and activity within the arena of civic praxis and sophist pedagogic emphasis on the efficacy of particular learned skills contributed to the development of practical manuals and handbooks.

Within the same Athenian context, there appears to have been a great appreciation for artisans and the work they produced, including architecture and sculpture. Thomas Cole points to an anthropological, human-centered point of view concerning the crafts that gained prominence in the fifth century B.C.E. Technical and craft skills came to be admired as significant human achievements rather than divine gifts. Early-fifth-century evolutionary accounts emphasized the gradual human discovery and invention of the various crafts. Early evidence for such a view exists in a fragment by the late-sixth-century/early-fifth-century Ionian poet-philosopher Xenophanes. An exile who spent some time in Sicily, Xenophanes wrote: "Not all things, by any means, did the gods show to mortals: rather, as time went on, men found improvement by constant searching."[8]

Views similar to those of Xenophanes were well established by the mid-fifth century B.C.E. Anaxagoras spent most of his active life in Athens and, like Protagoras, was a close associate of Pericles'. He distinguished humans from other living beings because of their capacity (among others) for developing the arts and crafts. The atomist Democritus of Abdera articulated a mechanistic view in which "necessity separated" certain arts indispensable for human existence. Humans discovered arts such as weaving and building houses by imitating the animals; they invented weaving, for example, after observing how spiders create their webs. Of the more than sixty books that Democritus reputedly wrote, several concerned practical areas such as medicine, agriculture, painting, and warfare.[9]

A positive view of crafts and technical arts helped to bring about the Periclean building program of the mid-fifth century, which produced the spectacular buildings of the Acropolis. Athenian imperialism made the building program possible. It began in 449 with Pericles' decree to begin the construction of new temples on the Acropolis with an initial five thousand talents from tribute money collected from nearby city-states. The project was to be supported with another two hundred talents annually for fifteen years until the work was completed.[10]

Before Pericles' decree the Athenians had undertaken little public building since the catastrophic leveling of their city by the Persians in 479, in part because of the decision to commemorate that terrible event by leaving their sanctuaries in ruins. Now both the means and the will to undertake a major construction program were at hand. Pheidias, a sculptor and friend of Pericles', supervised the project as a whole. Of the many buildings constructed the most famous was a Doric temple dedicated to Athena in her aspect of Parthenos, warrior maiden. The principal architect of the Parthenon, as it has been called since the fourth century B.C.E., was Iktinos, who wrote a treatise on the subject, assisted by Kallikrates. Pheidias created the colossal cult statue of Athena and was involved in the design of the architectural sculptures for the Parthenon. Ira S. Mark argues that the sculptural program of the east frieze was the product of collaboration between Pheidias, Pericles, and Protagoras and that it represents Protagorean anthropomorphic theological views. Especially notable is the active collaboration that this suggests between sculptor, ruler, and sophist. Craftsmen came from all over Attica to build the Parthenon and other buildings and statuary. As Plutarch reported centuries later, "The workmen eagerly strove to surpass themselves in the beauty of their handicraft [*Kallitechnia*]."[11]

In this atmosphere of appreciation for the technical arts craftsmen began to produce writings devoted to particular arts. Agatharchus, a painter from Samos who created a scene for an Aeschylan drama and was the first painter to use perspective on a large scale, wrote a book on scene painting. Agatharchus's book apparently inspired both Anaxagoras and Democritus to write manuals on perspective as well.[12] Iktinos, the architect of the Parthenon working under Pheidias, wrote an account of the temple.[13] Polyclitus, a well-known sculptor in the second half of the fifth century who worked in Argos and Olympia among other places and imitated the style of Pheidias, wrote a treatise on rhythm and proportion.[14] Hippodamus, a town planner from Miletus, designed the port town of Piraeus for the Athenians. Associated with the development of rectangular grid plans, he wrote a book on town planning.[15]

Authorship on painting, building, sculptural proportions, and town planning points to an appreciation for the value of such topics as disciplines worthy of written explication. Yet these texts are no longer extant; also lost is detailed knowledge concerning the specific context of authorship. The general context includes the circulation of sophist practical manuals on various topics, the intensive building and other craft production associated with Periclean democracy, and an appreciation of the technical arts.

The fortunes of Athens soon declined, in part because of the protracted Athenian conflict with Sparta known as the Peloponnesian War. Pericles himself died in the devastating plague of 429, and his friends failed to prosper. Ancient reports suggest that Protagoras was exiled, whereupon he died at sea, and his books were burned. Pheidias was accused of embezzling precious materials purchased for the statue Athena, and Anaxagoras was charged with impiety. Scholars debate the veracity and details of such ancient reports.[16] Whatever the personal fates of these three friends of Pericles', the open Athenian environment of discussion and argumentation and the intensive building program itself were relatively short-lived.

The end of this era of Athenian openness was marked especially by the trial in 399 B.C.E. of Socrates, who was accused of impiety and the corruption of youth. What also ended, perhaps several decades earlier, was the close alliance between building construction, the technical arts, and political praxis that had prevailed within the Periclean democracy. The spectacular buildings of the Acropolis remain. Much less well known but significant nonetheless is the legacy of authorship on the technical arts that had developed within the same context.

The Fourth-Century Separation of Technē and Praxis

During the fourth century B.C.E. city-states such as Thebes, Sparta, and Athens partook in a series of shifting alliances and wars in which each aimed to establish control over the others, with or without Persian help. The inability of any particular city-state to establish dominance and their collective incapacity to form stable alliances or federations facilitated the rise of the Macedonian state to the north and led to the eventual establishment of Macedonian dominance under Philip II and his son, Alexander. In the final analysis, untenable imperialist ambitions destroyed both the autonomy of the Greek city-states and the democracies that some of them had established.[17]

Aristocratic values and the aristocratic tasks of military and political leadership shaped fourth-century B.C.E. authorship in significant ways. Praxis writings that aimed to augment the knowledge and leadership abilities of elite men became particularly significant. Two authors, Aeneus Tacticus and Xenophon, wrote books illustrative of praxis authorship.

Aeneus Tacticus, who wrote a military handbook, possibly can be identified with Aeneus of Stymphalas, general of the Arcadian League in 367 B.C.E. His treatise, *How to Survive under Siege* (ca. 355 B.C.E.), which is only partially extant, has been associated with the development of military professionalism in the fourth century. It concerns mainly the defense of the walled city-state, reflecting the contemporary political context of autonomous Greek city-states, which were involved in shifting alliances and armed conflict. Emilio Gabba points to the influence of political conditions on the content of the treatise, stressing Aeneus's attention to the defense of cities and to the possibilities of internal revolt in a world in which mercenaries had largely taken the place of citizen soldiers. David Whitehead suggests that Aeneus's prospective audience included generals and other officials charged with responsibility for military defense. A didactic manual for military leaders, it deals with a variety of possible situations, pointing to a prospective readership of military leaders extending well beyond the perimeters of a single city-state.[18]

Aeneus assumes that traitors, plots, and intrigue are an integral and permanent part of the city-state. Deception, counterdeception, secret measures, and strategic tricks are recurring themes. The treatise contains the most elaborate discussion of cryptography in extant ancient literature. Deception and secrecy are to be used in imminent battle situations. It is notable that weapons or other technologies do not appear to require secrecy; military knowledge in general seems openly purveyed. Rather, secrecy is necessitated by specific, imminent military and political situations. Treachery is everywhere. Aeneus instructs Greek aristocratic leaders to encourage insurrection in enemy states and to be alert to treason in their own.[19]

Aeneus's contemporary Xenophon was an Athenian who lived much of his life in exile outside of his native city. Xenophon's later career as a cavalry leader and his thorough knowledge of horsemanship suggests that he had been a member of the Athenian cavalry. His admiration for Sparta and his Spartan patronage points to oligarchic rather than democratic sympathies. He was a prolific author whose works include treatises on hunting, household management (the *Oeconomicus*), and generalship. The subjects of his writings constitute a virtual catalogue of practices—horsemanship and generalship, hunting, estate management—that should be honed by aristocratic males, a group that included his own two sons. He wrote openly to a very specific readership, the younger male members of his own social class. Knowledge of horsemanship, hunting, and related practices distinguished aristocratic men from nonaristocratic men as well as from women. Xenophon's writings encouraged virtue, courage, and social cohesiveness (resulting from the inculcation of shared values) within his own elite class.[20]

Xenophon's *Oeconomicus*, a dialogue on household management and farming, is also a praxis manual in which estate management is viewed as conducive to the achievement of excellence and manliness, qualities needed for governance and military leadership. Household management, rulership, farming, and the management of war together constitute the praxis of an *ariston*, a noble man. In the *Oeconomicus* Xenophon firmly separates household management, which he views as training for political and military leadership, from the technē arts of material production. Such banausic arts *(banausikai technai)* are antithetical to those that citizens should practice. Craft production causes practitioners to sit still all day and live indoors—some spend all day at the fire—softening their bodies and consequently seriously weakening their minds. Those who follow technical arts do not have time for friends or city-state; they are poor friends and poor defenders of their country. Preferable disciplines are agriculture and the arts of war.[21]

For Xenophon, agriculture is essentially a praxis discipline, not based upon technology or techniques in any fundamental way. It bestows benefits of both character and body, benefits particularly suited to aristocratic landowners. It arouses in the landowner a love of hunting and stimulates armed protection of the country. The earth teaches "righteousness to those who can learn; for the better she is served, the more good things she gives in return." Those who receive "rigorous and manly" teaching of agriculture will be "well-found in mind and in body."[22] The aristocratic landholder manages his estate as he also trains himself for political and military action. This training is both physical and moral, shaping mind and body for effective leadership.

An interlocutor of the *Oeconomicus*, Ischomachus, serves as a model of a virtuous landholder. He is a highly successful estate owner, a man both beautiful and good. As Ischomachus provides detailed instructions on a variety of tasks and duties, it becomes clear that household management is not primarily a matter of technique but a form of praxis dependent upon the good character traits of the landholder. Ischomachus presents himself as a moderate, reasonable, orderly, and pious head of his own household and suggests that these traits are among those necessary for success. He describes his daily habits, emphasizing his judicious oversight of subordinates and his habit of exercising himself and his horse sufficiently to ensure readiness for war.[23]

Ischomachus also emphasizes the openness of agricultural knowledge, contrasting it with the secrecy of the technical arts. "Helpful, pleasant, honourable, dear to gods and men in the highest degree, it [agriculture] is also in the highest degree easy to learn." Unlike the technical arts, which the pupil must study until he "is worn out" before he can earn his keep by his work, agriculture is not difficult to master. It can be learned in part "by watching men at work" and in part "by just being told." Ischomachus contrasts the openness of agriculture with the secrecy of the crafts. Whereas "other artists [*alloi teknitai*] conceal more or less the most important points in their own art," the farmer works openly and is pleased to explain his work. Further, agriculture "more than any other calling, seems to produce a generous disposition in its followers." Farming is the noblest art because it is the easiest to learn. Husbandry itself shows you how it is done. A vine climbing on a tree teaches you that it needs support. It grows leaves when its grape clusters are young, showing you that it needs shade. When the clusters are more mature it sheds its leaves, demonstrating that the grapes need sun to ripen and grow sweet. In contrast, arts such as smelting gold, playing the flute, or painting pictures are secret.[24]

Openness is an attribute of agriculture that inculcates generosity in the landholder. As a parallel, Xenophon writes openly for members of his own social class. The author and his prospective readership formed a cohesive social unit. The practices described in Xenophon's treatises formed an essential part of the education of aristocratic young men. They were believed to instill particular character traits needed for political and military leadership. The openness of Xenophon's praxis writings must be understood within the context of his very specific prospective audience.

Aristotle, who was a younger contemporary of Xenophon's, reflects the latter's distinctions in the *Nichomachean Ethics*. As noted earlier, he delineates three categories of human activity, or ways of wisdom, and places them in hierarchical order: epistemē, involving knowledge of unchanging things, at the top; praxis, or action, in the middle; and technē, or craft production, at the bottom.[25] Aristotle separates both theoretical knowledge and political and military praxis from material production and manipulation. Considering Xenophon's similar distinctions in the *Oeconomicus*, Aristotle seems to have described categories that already prevailed within his own culture.

These categories, which were established in the fourth century B.C.E., would have significant influence on the successive cultures of the ancient Mediterranean world. This is not to say that the divisions remained static or that they were always rigidly maintained. Indeed, both within Aristotle's own writings and in subsequent centuries there were slippages between categories as well as occasional attempts to join them together. Yet the Greek categories praxis, technē, and epistemē, along with some of their analogues in the Latin language, remained significant classifications that influenced the practice of authorship.

Engineering and Authorship in the Hellenistic State

Xenophon's disdain for crafts and the technical arts could not have been shared by all of his contemporaries. A more positive view must have prevailed in some quarters, given fourth-century B.C.E. developments in the technology of siege machines. The tyrant Dionysius of Syracuse initiated improvements in siege technology when he recruited engineers to develop new instruments of war. One result was the gastraphetes, a mechanically operated crossbow that eventually could hurl 70-centimeter shafts as far as 640 meters. Innovations in siege weapons continued under the initiatives of Philip II of Macedon and his son, Alexander. Macedonian expansion was accompanied by significant changes in military technology and tactics, including developments in catapults and siege warfare. Alexander wielded these new techniques, both technological and tactical, in his conquests of Macedonia, Greece, and then Phoenicia, Palestine, Egypt, and most of the remaining Persian empire. Developments in weapons technology led to authorship in the military arts.[26]

The military engineers associated with Philip and Alexander wrote treatises on siege artillery. Polyidus, who invented a variety of designs for the ram tortoise, a kind of battering ram, and accompanied Philip of Macedon on his siege of Byzantium, wrote a tract concerning military machines. Polyidus's students Charias and Diades, who accompanied Alexander on his campaigns, also wrote manuals on weaponry. Authorship on siege engines must have been encouraged by several factors, including the active recruiting of military engineers by Philip, Alexander, and other rulers, the high level of skill required by catapult technology, and the mobility of military engineers throughout the Mediterranean. Enticed by the patronage of competing rulers, engineers may have taken pen in hand to increase their chances for employment; writing on military machines may have boosted their reputations as effective engineers. Engineer-authors presumably wrote while supervising arsenals or even on the march.[27] Although we do not know the exact content of their manuals, they seem to have described siege machines in some detail, probably explaining how to construct them.

When Alexander died suddenly in 323, his generals carved up the territories he had conquered. Among the new territorial dynasties was Egypt, ruled by the Ptolemies from the newly established city of Alexandria. The Ptolemaic rulers sought cultural as well as military and political superiority in the eastern Mediterranean. In the early third century B.C.E. Ptolemy I Soter achieved preeminence in part by promoting a culture of books and learning. He founded the Alexandrian Museum and Library, which he modeled on the Aristotelian Lyceum, and he dedicated them, as the Lyceum was dedicated, to the nine patrons of the arts, the Muses. Recruiting scholars from far afield, the Ptolemies provided generous lifetime subsidies. The chosen individuals lived in the Museum, took meals in common, and studied and taught in what became the most famous and most cosmopolitan center of learning in the ancient Mediterranean world. The Ptolemies also collected books, by means of, in Diana Delia's words, "confiscation, copying, and the production of new works and translations."[28]

In their aggressive competition for cultural supremacy the Ptolemies imitated and attempted to surpass Athenian academies founded by Plato (the Academy) and Aristotle (the Lyceum). Astrid Schürmann shows that one way they did this was to expand the traditional canon of the fourth-century academies to include mechanical knowledge, including military technology. Technē, the productive arts for the improvement of society, came to be part of the canon of learning in the Museum.[29]

Hellenistic writings on technical subjects such as siege weapons and pneumatic devices provide evidence for this expanded canon. The earliest Alexandrian author on these subjects was the famous inventor Ctesibius, who probably taught mechanics at the Museum. Reputedly the son of a barber, Ctesibius invented numerous mechanical devices, including a war catapult. In addition to his work on artillery, he made important contributions to pneumatics. His inventions include force pumps for air and water, a hydraulic organ, and a water clock, all of which he described in his *Commentaries*. Although his writings are lost, they were famous in antiquity, known to both the Roman architect Vitruvius and the first-century C.E. technē author Hero of Alexandria.[30]

Ctesibius's successor, Philo of Byzantium, undoubtedly also taught at the Museum. Philo was familiar with engineering on the island of Rhodes and also spent a long period of time in Alexandria, where he recorded conversations with people who had known Ctesibius. He wrote on mechanical matters in nine books, covering, among other things, catapults (the *Belopoeika*), pneumatics, fortresses, besieging and defending towns, and stratagems. He refers to a no longer extant manual, perhaps influenced by Aeneus Tacticus, that he wrote on cryptography. Philo's mechanical books are dedicated to one Aristo, possibly, as A. W. Lawrence suggests, the individual of that name mentioned by Diodorus Siculus (3.42) as being sent by Ptolemy II to explore the southern coast of Arabia.[31]

In his treatise on pneumatics, known primarily through Latin and Arabic translations, Philo discusses his reasons for writing. Addressing Aristo, he writes: "Your interest in ingenious devices has been known to me. You say and urge that you want a book about them. I wrote it and send it gladly." Philo hopes his treatise will aid Aristo's studies of devices and notes that such matters are worthy of the attention of learned men. His reference to learning is justified by his discussions on the nature of air. Alluding to the views of Democritus, Strato, and perhaps Ctesibius, he argues that air is a body not a void. He also reports the view that it is made of particles with empty space between them. He demonstrates the attributes of air by describing various pneumatic experiments with vessels and water. Philo disregards the Aristotelian separation of technē and epistemē in his use of mechanical devices to demonstrate characteristics of air. Yet the manual mostly concerns how to make various kinds of pneumatic vessels that dispense liquids in measured amounts, make sounds, cause artificial birds to flap their wings, and the like. Providing illustrations to accompany his written instructions, he seems to write not so much for learned men as for artisans who wish to make various kinds of pneumatic vessels and other devices for the use and enjoyment of the Ptolemaic court.[32]

In his treatise on artillery machines, the *Belopoeika*, Philo again specifies his reasons for writing. His predecessors have used diverse methods, and they differ concerning the proportions of the various parts of the machine as well as the most important guiding factor, the hole that receives the spring. He therefore will ignore the old authors and use later sources only if their suggestions are effective in practice. Philo asks why some artillery engines are more effective than others that are similar in materials and construction. Citing Polyclitus, a fifth-century B.C.E. sculptor who stated in a treatise on proportions that "perfection was achieved gradually in the course of many calculations," Philo explains that small discrepancies result in large total errors and that special caution is required especially when changing scale. He illustrates his point with an account of the gradual discovery of artillery calibration. In the old days, he says, engineers discovered that the diameter of the hole that holds the spring was the crucial measure for the construction of siege engines. They obtained this diameter by experimentally increasing and decreasing the size of the hole. Later engineers learned from mistakes and looked for a standard factor, using experiments as a guide until they discovered the correct diameter. Philo emphasizes that everything cannot be accomplished by theoretical methods; much can be learned by experimentation and by experimental conclusions handed down from one generation of engineers to the next. Philo's notion of gradual improvement based on experimentation and cooperation points to his acquaintance with anthropological accounts detailing the human acquisition of the arts. His citation of Polyclitus demonstrates his knowledge of fifth-century B.C.E. manuals as well.[33]

Despite the highly technical and military nature of Philo's extant treatises, there is no evidence of attempts to keep the information secret. He makes no admonition or suggestion concerning concealment, nor is there evidence for concealment in other sources. Alex Roland and I have argued that there is little evidence for a concept of weapons secrecy in the ancient world. Weapons were certainly important in warfare, but the ancients considered military victory to be based primarily on the praxis of generalship (involving courage, leadership ability, and good *phronesis*, or judgment, in contingent situations), not on weapons superiority. It is further reasonable to assume, as Astrid Schurmann suggests, that whatever the role of his treatises in the training of engineers, Philo also wrote them as part of the canon of knowledge to be "displayed" in the Ptolemaic court.[34] To the extent that such writings functioned as display, openness was a necessary condition. Philo's writings became part of the important tradition of technē authorship that extended from his own time, in the third century B.C.E., to the era of Hero of Alexandria, in the first century C.E., three hundred years later. Yet before examining later technē writings, it is necessary to look at the broader context of authorship in the Alexandrian Museum and Library.

Attribution and Literary Theft in Ptolemaic Alexandria

Philo's remark that he will ignore old authors and use later ones only if what they say is effective in practice suggests a highly utilitarian attitude toward authorship that subordinates it to the requirements of making military machines. His view stands in contrast to the collection policies of the Alexandrian Library as a whole. The Ptolemies did not "ignore old authors" but collected old books aggressively. The second-century C.E. physician Galen remarked that they acquired books primarily in two ways: their agents purchased books at book markets, those at Athens and Rhodes being the most important, and seized all books on the ships that came into the Alexandrian harbor, had them copied, kept the originals, and returned the copies to the owners. That they wanted *originals* is suggested by Galen's story that they tricked the Athenians into "lending" the originals of the Greek plays for a deposit of fifteen talents and returned the copies instead. Yet *original* here refers only to the copies that were in the possession of the Athenians and says little about how the papyrus rolls themselves were created, for example, whether they were created by the playwrights themselves or, more likely, by copyists. In any case, the competition for books was intensified when the rulers of Pergamon, in Asia Minor, set up a rival intellectual center with its own library. Competition for authentic writings led to the craft of forgery, which thrived.[35]

The Alexandrian librarians were the first to care about the correct attribution of authorship. Our knowledge of the work of these scholars comes from the scholiast tradition of commentators of the late Roman imperial and Byzantine eras. They frequently cite Ptolemaic critics, known to them through the authors of compendia of writings and opinions. Although thirdhand, the evidence of the scholiasts is precious for revealing what the Alexandrian librarians actually did. Apparently, they established editions of writings and wrote commentaries on them. An edition was an individual copy of a work, on which the editor made critical signs to mark lines he believed to be spurious. The Alexandrian librarians pioneered the critical study of literature by their work on problems of textual attribution.[36]

Zenodotus, the first head of the Library (ca. 284 B.C.E.), apparently initiated the use of critical signs, a system greatly expanded by a later librarian, Aristophanes of Byzantion. Aristophanes produced editions of many texts and studied the history of words, which helped him to establish the authenticity of particular lines and texts. Of interest for the history of plagiarism, Aristophanes' work included a book about literary theft that is known to us through a single remark by the third-century Neoplatonic philosopher Porphyry. Aristophanes' book concerned "plagiarism" by the playwright Menander. Porphyry reports that in this book Aristophanes collected "the parallel lines of Menander and the selected passages from which he stole [*eklepsen*] them." Porphyry says that Aristophanes "rebuked Menander but gently on account of his great fondness for him."[37]

Two centuries later the Roman architect Vitruvius related a far more detailed story concerning Aristophanes and literary theft. Vitruvius describes a poetry contest sponsored by the king, Ptolemy Philadelphus, to celebrate the opening of the Library. Aristophanes, who was one of the judges, chose the least popular poet as the winner. He explained to the indignant king and assembly that this contestant alone had recited his own poetry rather than the compositions of others. Proving his point by fetching the appropriate volumes from the Library, he argued that the judges should approve original compositions rather than thefts. Aristophanes was duly awarded the librarianship, and the poetry thieves were punished.[38]

As P. M. Fraser noted, some of the details of Vitruvius's story are incorrect. For instance, Ptolemy Philadelphus and Aristophanes were not contemporaries, and the Library was established well before the time of Aristophanes. Nevertheless, the poetry contest and a concern with original authorship are consonant with what we know about the Alexandrian Library and about the work of Aristophanes himself.[39] Yet, Vitruvius's condemnation of literary theft was uncompromisingly harsh compared with Aristophanes' mild rebuke of Menander. Vitruvius's severity, I suggest, reflects the Roman context more than the Ptolemaic. Ptolemaic librarians were primarily interested in correct attribution and the authenticity of texts. The Romans came to be concerned with something very different: honor to past authors.

Engineering, Architecture, and Authorship in the Roman Empire

When Octavian conquered Egypt in the battle of Actium in 31 B.C.E., he maintained the Library and Museum and continued their patronage. Octavian assumed the name Augustus in 27 B.C.E. and exercised increasingly autonomous power as emperor. He consolidated his position by military, political, and cultural means. His policy included the promotion of traditional Roman religion, a massive rebuilding program for the city of Rome, and the active patronage of literature. He thoroughly understood the role that cultural hegemony and pride could play in the maintenance of political power and the cohesiveness of a state.[40]

The emperor and his family commanded an extensive patronage system that supported numerous authors. Beneficiaries included Athenaeus Mechanicus, author of a treatise on siege machines, and the architect-engineer Vitruvius. Athenaeus Mechanicus, a Greek living in Italy, dedicated his Greek-language treatise to C. Claudius Marcellus, nephew and son-in-law of the emperor. Parts of his treatise are strikingly similar to the section of Vitruvius's *De architectura* that concerns military machines. Eric Marsden suggests that both authors probably used an earlier treatise, no longer extant, on siege machines by one Agesistratus. Marsden also points to the evidence that Athenaeus and Vitruvius benefited from the same patron, Augustus's sister, Octavia; they probably also used the same imperial library.[41]

Addressing Marcellus, Athenaeus asserts that he writes his treatise especially in opposition to those who rebel against "the beautiful laws of the hegemony [of Rome]." Having signaled thereby his own loyalty to the Roman empire, he lays out his views about writing on military machines. Two principles are paramount: brevity and practical utility. Endorsing the Delphic oracle's decree to save time, he criticizes authors who, whether describing something or giving instructions, "become scribblers and squander time with useless words" for the sake of profits.

They publish "repeatedly their smattering of many things" and bequeath "books which they have packed full of digressions." Among these "scribblers" he includes Aristotle and Strato.[42]

Having thereby dismissed Greek theoretical and philosophical writings, Athenaeus points to the great differences between previous books, now lost, on siege machines by Deimachos, Charias, Diades, and Pyrrhus the Macedonian. Emphasizing that writings on machines should be characterized by clarity and brevity, he illustrates his point with the negative example of Isocrates. While the great Athenian rhetorician was polishing a letter of military advice to King Philip of Macedon, the war that was its subject ended, rendering the finely honed epistle useless.[43]

Yet in general Athenaeus's attitude toward his predecessors is one of critical respect. He notes that although Pyrrhus had also written about these matters, he did not believe he was raising his voice against an authoritative predecessor as he had seen many others do in questions of craft. "In effect, all that there is of value in the writings of my predecessors I have examined attentively, bringing there scrupulous care, and I myself have been anxious to add to useful things in the fabrication of machines." Athenaeus emphasizes that knowing the inventions of others is not sufficient: "it is also necessary to exercise the activity which is characteristic of the soul itself in making new inventions."[44]

Despite his emphasis on the importance of new inventions, Athenaeus primarily discusses older devices and machines, including siege towers and other weapons that Diades described some three hundred years earlier. Rather than writing a manual for immediate practical use, he provides a short summary of past knowledge and a commemoration of past inventors and authors of treatises on machines. The essential context of Athenaeus's work is the library, not the battlefield.

Athenaeus's contemporary Vitruvius in *De architectura* addresses the emperor Augustus, praising the emperor's divine mind and power and emphasizing his own service to the imperial family. Vitruvius notes that under Octavian's uncle and adoptive father, Julius Caesar, he had been charged with construction and repair of engines of war. His employment had continued thanks to the recommendation of Augustus's sister, Octavia, and he now enjoyed a lifetime stipend. He was a practitioner close to the most powerful family of Rome. Perhaps as a corollary, he insisted that architecture hold a place well above the banausic crafts. Vitruvius undoubtedly was influenced by the Roman author Varro's no longer extant *Nine Disciplines*, which placed architecture among the liberal arts suitable for free men.[45] Yet Vitruvius's efforts were necessary because in general the discipline occupied an ambiguous position, being not quite a servile craft but not quite a liberal art.

Vitruvius insists that architecture involves both reason and fabrication, *ratiocinatio* and *fabrica*. The architect who "without letters" aims at manual skill does not reach a position of authority corresponding to his labor, whereas those who rely on reasoning and letters only seem to follow a shadow rather than substance.[46] Here and elsewhere in his treatise Vitruvius insists that architecture is not merely a manual art but includes rational and learned aspects; yet he also acknowledges its association with handwork and construction, or *fabrica*.

Clearly influenced by the Greek anthropomorphic account of the development of the crafts, Vitruvius claims that humans became civilized by means of architecture. The rise of civilization began with the discovery of fire, from which resulted human speech and conversation. Humans then discovered the art of building in part by imitating the nests of the swallows. "Since men were of an imitative and teachable nature, they boasted of their inventions as they daily showed their various achievements in building, and thus, exercising their talents in rivalry, were rendered of better judgement daily."[47] The Roman architect articulates the same progressive, experimental point of view that is evident in the writings of Philo of Byzantium.

Vitruvius advocates openness in part as a critique of contemporary architectural practice. Urging the openness of craft knowledge, he complains that some gain commissions because of the public ignorance of craftsmanship; indeed, "the ignorant excel in influence rather than the learned." Openness is a remedy for such injustices. He makes his point by means of a story about the Delphic Apollo and Socrates, offering an idiosyncratic explanation of why the Delphic Apollo declared Socrates to be the wisest of all men. Whether or not Vitruvius was familiar with either Plato's or Cicero's discussion of the oracle's pronouncement, he provides a very different interpretation of the story. He suggests that the Delphic Apollo's opinion was based on Socrates' statement that the human breast should be furnished with open windows so that men "might not keep their notions hidden, but open for inspection."[48] His comment on this novel anatomical restructuring constitutes a plea for the openness of craft knowledge. He wishes that nature had constructed humans thus open to view, that is, in the way described by Socrates. If human bodies contained such windows, the "merits or defects of human minds" could be examined. In addition, "the knowledge of disciplines also, lying under view of the eyes, would be tested by no uncertain judgments; and a distinguished and lasting authority would be added both to learned and to accomplished men."[49]

Unfortunately, since the ingenuity of men is concealed, it is not possible to judge their deeply hidden knowledge of the arts. Artists themselves must vouch for their skill. They are justly rewarded only if they come from a particularly venerable workshop or possess public favor or eloquence. Excellence sometimes goes unrecognized, while good judges are swayed by flattery. If men had windows in their chests, popularity would have no further influence. Rather, those who reached the height of knowledge "by true and certain exertions of learning" would obtain commissions.[50]

Vitruvius makes clear that his own reputation will rest on his knowledge as revealed through authorship rather than on the construction of buildings. He also venerates past authors. In a remarkable passage, Vitruvius praises the ancients for not only writing down their own ideas but also transmitting those of their predecessors. He notes that the ancients wisely and usefully transmitted their thoughts to posterity in commentaries. Thus, their accomplishments were not lost; rather, "increasing from generation to generation, having been published in books, step by step they arrive in a very long time at the highest subtlety of knowledge." Infinite thanks must be given to past authors, Vitruvius insists, because they did not neglect the dead but transmitted their ideas of all kinds through writings.[51]

Emphasizing the importance of credit to past authors, Vitruvius defends his own practice of authorship. He did not write by inserting his own name "after changing the titles of other men's books," nor was his intention "to win approbation by finding fault with the ideas of another." He neither stole from his predecessors nor criticized them. Rather, he expresses "unlimited thanks" to the authors of the past. Their books aid contemporary writers, he says, allowing them to produce "new systems of instruction." In contrast, individuals who "steal the writings of such men and publish them as their own" must be condemned. These men who have robbed other men's goods with violence should receive censure and punishment "for their impious manner of life." Vitruvius here relates the story of Aristophanes and the poetry contest in Ptolemaic Alexandria, which rewarded the poet who read his own poetry and punished the poetry thieves.[52]

Vitruvius condemns literary thieves, but he goes further to suggest that those who criticize the writings of dead authors deserve capital punishment. As an example, he tells the story of Zoilus, a critic of Homer, whom one of the Ptolemies condemned for parricide. (The Ptolemies considered Homer to be the father of authors.) The king ordered Zoilus executed, Vitruvius reports, either by crucifixion, stoning, or being thrown upon a burning pyre. Whichever method was used, he deserved his fate: "The penalty fitted the culprit." Such is the just due of one who accuses men who can no longer defend themselves or show the meaning of their writings.[53]

I suggest that the Augustan revival of traditional Roman religion in the 20s B.C.E. influenced Vitruvius's condemnation of the theft of writings and his castigation of Zoilus as a parricide. A religious meaning is suggested by his use of the word *impious* to describe those who engaged in the theft of writings, as well as by the harshness of his condemnation of Zoilus. Traditional Roman religion involved forms of ancestor worship. Household gods included the Lares, the deified spirits of dead ancestors, who remained deeply involved with the family. Vitruvius, I suggest, may have believed that the ancestors of authors of his day included the writers of the past. Accordingly, he condemns impious acts against them.[54]

The relevance of a religious context for Vitruvius's attitudes toward theft and criticism is underscored by the Augustan religious reform of the 20s B.C.E., the decade in which the Roman architect probably wrote the prefaces to his *De architectura*. Augustus initiated his new religious policy in 28 B.C.E. He ordered the reconstruction and repair of the eighty-two temples of Rome and filled the priesthoods, many of which had been vacant for decades. He chose Apollo and Mars as the gods that should be especially worshiped, emphasizing that Mars was the father of Romulus, the founder of Rome. Augustus's reform involved an effort to reinstate traditional religious values and also to transform a localized and particularized religion (in which the gods of Padua, for example, could not easily become the gods of Rome) into a more universal, Roman version.[55] The emperor's revival of Roman traditional religion constituted the significant context for Vitruvius's pious attitude toward past authors and their writings, for his emphasis on the importance of giving proper credit, and for his harsh condemnation of both literary theft and the criticism of dead authors.

Some fifty years later, about 62 C.E., Hero of Alexandria authored the last significant corpus of writings on the technical arts in antiquity. Under the Romans, Hero taught at the Alexandrian Library and Museum, carrying on an apparently centuries-long tradition of engineering pedagogy. He wrote treatises on pneumatics and the dioptra (a surveying instrument); the *Mechanics*; the *Belopoeika*, a tract on a bow and two catapults; the *Automata*, on automatic theaters; and the *Caloptrica*, on plane and curved mirrors.[56]

In the *Pneumatica*, a treatise concerning pneumatic instruments and devices, Hero reflects that pneumatics was studied zealously by ancient philosophers and mechanicians—philosophers explaining their theory, mechanicians explaining "through demonstration of experiment." Hero continues this tradition by setting forth in an orderly arrangement "those things that have been transmitted to us from the ancients" and by adding his own ideas and inventions. Marie Boas Hall underscores the striking way in which Hero integrates both traditions—of philosophers and mechanicians—to which he is heir.[57]

In his *Automata* Hero discusses a previous book on the subject, no longer extant, by Philo of Byzantium. He wants to write something new concerning the stationary automaton (the work treats a standing theater with moving figures representing the myth of Nauplius the Navigator). Nothing better had been found on the subject than the writings of Philo, but there are problems in his account. One difficulty concerns a suspension mechanism containing the figure of Athene; the second involves Philo's silence on the creation of the lightning and the noises representing thunder. "Although we examined many examples of his writings we have not found any record concerning [these noises]." Hero cautions that he has not slandered Philo with his reproach. He notes that by explaining the thunder and lightning mechanism he has not scorned Philo's writings. Rather, it is to the reader's advantage to be shown "the correct statements of the ancients" and then to have explained anything that they overlooked or that since has been improved.[58] Hero criticizes as he also respectfully augments the work of the prior treatise.

Elsewhere, Hero presents the writings of predecessors in a way that will achieve greater clarity. His *Belopoeika*, a book on catapults, includes the name Ctesibius in the full title: "Heron's edition of Ctesibius' *Construction of Artillery*." Marsden suggests that the treatise is based closely on a section of Ctesibius's lost writings. Hero points to numerous prior writers on artillery who dealt with measurement and designs. Not one of them, however, described "the construction of the engines in due order or their uses; in fact, they apparently wrote exclusively for experts." He therefore supplements their work and describes artillery engines, "even perhaps those out of date," in a way that can be "easily followed by everyone."[59]

Hero writes within the tradition of technē writings represented by Ctesibius, Philo, and Vitruvius. Yet in the *Pneumatica* he uses mechanical knowledge to a much greater extent than his predecessors to explore theoretical questions concerning the nature of air. He attempts to integrate both technē and epistemē, or theoretical knowledge about the world. Hero was a teacher of engineering who also wrote for the nonexpert reader. He represents an exception to the general rule that ancient writers on the technical arts avoided broad questions about the nature of the world. Yet he did not have an immediate successor in the centuries that followed.

Ancient technē authors expressed positive attitudes toward the technical arts and toward the books of previous authors. Vitruvius highly praised the authors of the past, while condemning both plagiarism and criticism of dead authors. I have suggested that his views on authorship were shaped by the context of Augustan religious reform. Other authors such as Philo of Byzantium, Athenaeus Mechanicus, and Hero of Alexander critiqued past authors or used them selectively, but all accorded them honor and respect. Most honored the crafts themselves and viewed the progressive accumulation of knowledge as dependent upon experimentation and the handing down of experimental results. Technē writers invariably came from a group of relatively high-status practitioners, engineers and architects. They dedicated their writings to elite men of higher status and benefited from the patronage of those men.

Praxis Authorship and Virtue in Republican Rome

In contrast to technē writers, Roman authors of praxis treatises, writing in the tradition of Xenophon, occupied the elite strata of Roman society. They were governors and military leaders who wrote for their own peers. The Romans considered general knowledge concerning agriculture, tactics and generalship, history, and natural history to be both useful and necessary to ruling elites.[60] Although the Romans are well known for their practical and engineering achievements, elite rulers were not expected to be experts but rather to possess general knowledge of certain disciplines. Most praxis writings belong to an encyclopedic tradition; in general they consist of convenient summaries of current knowledge. Praxis writers usually did not attempt to further specialized knowledge, nor did they write for skilled experts. They often included materials relevant to the inculcation of virtue and other desirable character traits necessary for the tasks of leadership. Praxis authors wrote for owners of large landed estates, governors, officials, and military leaders, men who might consult their books for edification during leisure from their essential activities of military leadership and governance.

An important topic of authorship for these Roman elites was agriculture. The earliest extant example of continuous Latin prose is an agricultural treatise, the *De agri cultura* by Cato the Elder. Cato, a renowned political and military leader, was also a prolific author who expounded traditional Roman virtues such as frugality, simplicity, austerity, and patriotism. Cato praises farming over commerce. Roman ancestors praised a good man by calling him a good farmer and a good husbandman. The farming class produced "the bravest men and the sturdiest soldiers." Cato addresses landholders, not the slaves and other low-status individuals who actually work the land. He views agriculture as a pursuit that shapes in desirable ways the future military leaders of the Roman state.[61]

Kenneth D. White points to the social and economic revolution that occurred after the Second Punic War (218–201 B.C.E.) as the essential context for Cato's treatise. Changes involved the decline of small proprietary farms as a result of the military demands of the state; the great extension of public land resulting from the confiscation of the estates of disloyal communities; the ready availability of that acreage for sale to investors; the influx of capital from the profits of conquests in the East; and finally, the swelling of the slave population by prisoners of war. Although Cato endorsed the small proprietor working his own land as morally superior to the merchant or moneymaker, in fact he wrote his manual for the average Roman patrician who was a large-scale absentee landowner whose farm was run by a steward and worked by a large number of slaves for the greatest possible profit.[62]

More than a century later, in *De officiis,* Cicero also extolled agriculture as the most appropriate activity for the Roman ruling classes. Cicero specifies categories of work that are demeaning and servile: any kind of paid labor, retail sales, and all handicrafts because "there can be nothing well bred about the workshop." Arts such as medicine and architecture are honorable "for those who belong to the class that they befit" (not the highest classes, it should be noted). Conceding that large-scale trade, unlike retail, might be appropriate, Cicero concludes that there is "no kind of gainful employment that is better, more fruitful, more pleasant, and more worthy of a free man than agriculture." He refers not to the work of the farm laborer but to the role of large landowners such as himself.[63]

Cicero's contemporary Marcus Terentius Varro, like most members of his class, participated actively in Roman civic life, including involvement in the Roman civil war in the mid-first century B.C.E. Unlike Cicero, Varro survived the Roman civil war and lived on to become one of the most prolific authors of his era. The titles of fifty-five of his works are known, although only two are extant: *De lingua latina* exists in part, whereas his treatise on agriculture, *De rerum rusticarum,* is complete.[64]

In the *De rerum rusticarum,* written in dialogue form, Varro treats the cultivation of plants and trees, animal husbandry, and small livestock such as fowl, bees, and fish. He explains that his prospective readership includes his wife, Fundania, who has recently purchased an estate and wishes to make money from it. (Fundania is the only female in antiquity to have a book dedicated to her; Varro probably viewed her as the guardian of his patrimony.) Varro writes for her and other farm owners who want productive estates. An octogenarian, he compares himself to the Sibyl whose prophesies provide benefits even after her death because men continued to consult her books. Since the Sibyl aided strangers through her writings, he must also write to help his friends and relatives both while he lives and after his death.[65] By specifying his readership as his own family and friends, Varro connects his work to the prophetic literature of traditional Roman religion.

Varro sees agriculture as the means not only to gain profits but also to inculcate good character traits and religious values into wellborn males who will become officials of the state. He alludes to traditional Roman religious customs by invoking the twelve councilor gods, patrons of agriculture. The differences between the old and the new, Varro makes clear, are the result of the dichotomy between the country and the city. He says that modern times are centered in the city and are characterized by the civil wars fracturing the Roman republic. The first book of his treatise is a dialogue that takes place in the temple of the earth goddess while the interlocutors wait for the sacristan of the temple, who has gone to Rome. It ends with the dismaying news that the sacristan has been murdered, having been mistaken for someone else. The friends leave the temple "rather lamenting the mischances of life than being surprised that such a thing had occurred in Rome." Later Varro suggests that living in the country is a far more ancient custom than living in the city. The country was created by divine nature, whereas the city was made by human art. Country life is far more noble.[66]

Through the interlocutor Scrofa, Varro defines agriculture as both an art *(ars)*—"not only an art but an important and noble art"—and a science *(scientia)*, one that "teaches what crops are to be planted in each kind of soil, and what operations are to be carried on, in order that the land may regularly produce the largest crops." Agriculture is knowledge; its elements *(principia)* are the same as those that make up the universe—water, earth, air, and fire. Equipped with this knowledge, the farmer aims at two goals, utility and pleasure.[67]

Varro contrasts scientia with the knowledge of the Greeks. Referring to Theophrastus's *History of Plants* and *Causes of Vegetation*, Stola notes that they "are not so well adapted to those who wish to tend land as to those who wish to attend the schools of the philosophers," although he concedes that they are still "profitable and of general interest."[68] Roman agriculture, in contrast, is a body of knowledge, or scientia, that is tied to utility and to the direct benefits of life.

The writings of Cato and Varro and the remarks of Cicero on agriculture underscore the essential connection between agriculture and the virtues that the Romans considered desirable for rulership. These authors believed that traditional virtues such as hard work and physical strength, frugality, courage, loyalty, and the rejection of luxury were greatly lacking. Books on agriculture written for Roman elites might serve to reestablish those values as well as provide an encyclopedia of agricultural knowledge.

Praxis and Authorship in the Early Imperial Age

Between the death of Augustus in the year 14 and the death of the emperor Hadrian in 138 the expansion and consolidation of the Roman Empire continued, while republican forms were increasingly disregarded. Yet the emperors ruled with the assistance of elite Greeks and Romans who continued to hold office and to assume the duties and prerogatives of those offices. These wellborn men wrote and read books on subjects they deemed appropriate and useful for the praxis of the governing class, including manuals on agriculture, military strategy, generalship, and natural history. Authors and readers were peers, members of the same elite social and political classes. The Roman encyclopedia encompassed much that was practical and encouraged cultural and political hegemony among members of the ruling classes. Encyclopedic authors in general promulgated not the specialized knowledge of particular disciplines required by experts or detailed theoretical concerns but general overviews in convenient form for rulers, generals, and governors. They organized and consolidated knowledge in a way that paralleled the expansion and consolidation of the empire itself. Authorship in the encyclopedia was intrinsically related to the civic orientation of elites within the empire.[69]

Authors of encyclopedic treatises almost invariably served as highly placed governors, military leaders, and other officials of the empire. Often they wrote on more than one part of the encyclopedia. For example, Sextus Julius Frontinus served in a variety of Roman offices before he became the Roman governor of Britain (probably 74–78). On his return to Rome he assumed further offices, including that of commissioner in charge of the Roman water supply *(curator aquarium)*, a post traditionally reserved for the wellborn. He wrote on military strategy, Greek and Roman military science, and surveying, as well as on aqueducts.[70] Pliny the Elder wrote extensively while pursuing a demanding equestrian career in both military and civilian positions. He wrote treatises on cavalry maneuvers and on oratory and grammar, a history of the Roman campaigns against the Germans, and a history of Rome, as well as his only fully extant work, the *Natural History*.[71]

Columella provides an exception to this practice of eclectic authorship in that he seems to have written only one book, the *Rei rusticae*, on agriculture. Columella came from a landowning family in southern Spain and eventually possessed his own extensive landed estates in central Italy. As a young man (about the year 36) he served as a tribune with a legion of the Roman army, probably in Syria. His agricultural treatise in twelve books treats soil, trees, vines, land measurement, animal husbandry, gardening, and the duties of both the overseer and the overseer's wife, who was charged with household production, including making wool cloth and food preservation. Columella wrote the tenth book, on gardening, in hexameter verse in imitation of Vergil's fourth *Georgic*.[72]

Columella dedicates each of the twelve books to one Publius Silvinus, presumably a landholder, from his native Spain. His procedure was to write a book and then send it to Publius, who read it aloud with a small group of friends or students and then returned the book with comments and suggestions. These readers' critiques clearly influence Columella's discussions in subsequent books. For instance, he begins his second book by noting Silvinus's question concerning an issue discussed in the first: Why does he reject the widely accepted notion that the soil is declining? Similarly, he opens the fourth book with a highly informative statement concerning the practice of reading, which begins: "You say, Publius Silvinus, that when you had read over to several students of agriculture the book which I have written on the planting of vineyards . . ." He continues that while they praised most of it, some criticized his treatment of trenches and props. In a later book Columella treats the measurement of land, prompted by Silvinus's complaint that he had omitted the topic despite his belief that surveying ordinarily belongs to the surveyor or architect, not the landholder. Finally, Columella claims that it is at the insistence of Silvinus that he includes a book on horticulture in verse, imitative of Vergil.[73]

For Columella, attachment to the land and interest in its good management is intrinsically tied to the moral fiber of the governing classes and to the strength of the empire itself. He complains that he often hears "leading men of our state" condemning the barrenness of the soil, sometimes adding the theory (taken from the Epicureans) that it is exhausted from the overproduction of earlier days. This reasoning, he argues, is false and impious. The earth is both divine and everlasting and has always brought forth things in abundance. Supporting thereby the Stoic doctrine of the divinity of nature, Columella suggests that lack of productivity is "our own fault." He explains that the best Roman ancestors treated agriculture with the greatest of care, while in the present day it has been "delivered over to all the worst of our slaves, as if to a hangman for punishment."[74]

Columella laments that today "we ply our hands in the circuses and theatres rather than in the grainfields and vineyards." Providing a catalog of the licentious habits of his contemporaries, he contrasts those habits with "that true stock of Romulus," which hunted constantly and also toiled in the fields. As a result, they had real physical strength. Hardened by "the labours of peace," they easily endured the hardships of war. In addition, Columella insists, they "always esteemed the common people of the country more highly than those of the city." Five hundred years after Xenophon, agriculture, virtue, and military praxis remained closely tied. Yet Columella's respect for the ordinary people of the countryside signals a change brought about in part by the influence of Stoicism, which promulgated a positive view of labor and a view that all people could achieve virtue.[75]

For Columella's contemporary Pliny the Elder the natural order and the human moral order were intrinsically related. Pliny's only fully extant work, the

Natural History, contains thirty-seven books, beginning with a preface in the form of a dedicatory letter to Titus, the son of the emperor Vespasian. Pliny assumes that Roman political power and the encyclopedia of knowledge go hand in hand. The encyclopedia includes sections on the universe, geography, man, other animals, botany, botany in medicine, zoology in medicine, metals and stones (including their medicinal use), art, and architecture. Influenced by Stoicism, he views nature as a divine benefactor. Humans are also part of nature, but their greed and desire for *luxuria* cause them to abuse it. Andrew Wallace-Hadrill suggests that the notion of luxuria is central to the purpose of the *Natural History*. Nature supplies everything humans need. Yet they are blinded by greed and abuse nature, turning it into a tool of self-destruction. The *Natural History* reveals the proper use of nature and thus saves the human race.[76] If life is lived according to nature (in conformity with the old Roman values) rather than against it (in the pursuit of luxury and waste), nature and humans will coexist in harmony and be intrinsically joined.

In an intriguing letter, Pliny's nephew, known as Pliny the Younger, describes his uncle's activity as an author. Replying to a correspondent who wishes to obtain his uncle's books, he describes the elder Pliny's continuous authorship carried out in the course of an active life of service to the state. He would arise in the middle of the night (in winter often at midnight or an hour later) to study, often dozing during his work. Before daybreak he would visit the emperor Vespasian, who also worked at night, perform his official duties, and then return home to continue his studies. After his meals, when he was not too busy, he would lie in the sun while a book was read to him and make notes and extracts. His thoroughness was notable: he "made extracts of everything he read." During dinner, a book was read aloud, while "he took rapid notes." He continued his routine of study "in the midst of his public duties and the bustle of the city." In the country, the only time he took from his work was for his bath, and while he was being rubbed down and dried, he "had a book read to him or dictated notes." When traveling, he gave every minute to work; "he kept a secretary at his side with a book and notebook." In the winter he wore long sleeves to protect his hands so that even the bitter cold would not cause him to lose time. When in Rome he had himself carried in a litter, enabling him to continue his studies while going from place to place.[77]

The fundamentally oral and social character of Pliny's authorship is notable. It involved listening to readers, at least some of the time with others, and dictating. His authorship depended upon the service of slaves, not only to read and take notes but also to carry, serve, and towel him dry, among other numerous tasks. Both intellectually and as a practical activity Pliny's authorship was firmly grounded in the privileges and duties of his social class.

Pliny claims that to create his encyclopedia he perused "about 2000 volumes" in order to collect "20,000 noteworthy facts" obtained from a hundred authors. In addition, he added numerous facts that had been ignored by previous writers or had been discovered only by subsequent experience. Pliny does not doubt that other things have escaped him, but he is human, "beset with duties," and has worked on this in spare moments. For these reasons, he makes no promises concerning the work but suggests that "many objects are deemed extremely precious just because of the fact that they are votive offerings."[78] His books are offerings, he seems to suggest, to the Roman gods.

Having implied that his own authorship involves acts of piety, Pliny emphasizes the importance of credit to past authors: "You will deem it a proof of this pride of mine that I have prefaced these volumes with the names of my authorities." The first book of his encyclopedia consists of an extensive list of topics, and under each topic, a list of the authorities upon whose works his own account is based. He explains that it is "a pleasant thing and one that shows an honourable modesty, to own up to those who were the means of one's achievements." He reveals that most authors have stolen from others. "For you must know that when collating authorities I have found that the most professedly reliable and modern writers have copied the old authors word for word, without acknowledgement." Pliny condemns such acts: "Surely it marks a mean spirit and an unfortunate disposition to prefer being detected in a theft to repaying a loan—especially as interest creates capital."[79] He seems to suggest that if one repays the interest on the loan of openly transmitted writings by acknowledging the debt to past authors, an even greater supply of knowledge will result.

Pliny's method involved a complex system of compilation. He gathered relevant materials from hundreds of past authorities, arranging them by subject. Similarly, his contemporary Frontinus gathered excerpts from numerous past writings to create his treatise on strategy. Frontinus wrote the *Strategemata* to summarize the strategies of generals in succinct sketches in order to provide commanders with examples of advice and foresight that might enhance their abilities to conceive and carry out similar deeds. Generals will not fear their own stratagems if they compare them with those that have already been successful. They will make improvements after reading additional examples from past writings. Successful generalship depends crucially on the ingenuity and leadership abilities of the commander. Although military technology is a factor, Frontinus believes it to be a constant one, not capable of change or improvement. He informs readers that he is "laying aside also all considerations of works and engines of war, the invention of which has long since reached its limit, and for the improvement of which I see no further hope in the applied arts." He deals instead only with various strategies to be used within siege operations, such as surprise attacks and inducing treachery.[80] Frontinus thereby follows the ancient view that generalship, involving the ingenuity and leadership of the military leader himself, is what requires extensive thought. Weapons are a factor, but a static one requiring no special attention.

Yet Frontinus also wrote a technical manual—his treatise on aqueducts. He wrote the manual following his appointment by the emperor Nerva to the position of water commissioner, *curator aquarium*. Citing his own diligence, devotion, and sense of responsibility, he explains that he is writing on this topic in order to familiarize himself with it. The position of commissioner has always been held by eminent men, says Frontinus, and concerns not only the convenience but also the health and safety of the city. Frontinus believes that "there is nothing so disgraceful for a decent man as to conduct an office delegated to him, according to the instructions of assistants." Yet if the person has little experience related to the matter at hand, he must inevitably depend on the practical knowledge of subordinates. Frontinus notes that although he wrote his earlier books for his successors, he writes this one for himself as well.[81]

Frontinus served as water commissioner under the reform emperors Nerva and Trajan; perhaps following their instructions, he attempted to correct abuses involving aqueduct regulation. His treatise is itself an instrument of reform in that it includes a record of each aqueduct's yield according to imperial records and a notation of its much greater actual yield according to new measurements. Frontinus lists numerous examples of the theft of public water for private use. His own record of actual yields was intended to inhibit further abuse.[82] He wrote both for himself and for future supervisors of the aqueduct.

In sum, wellborn men, most of whom held positions of high honor and responsibility in the imperial government, wrote encyclopedic works for readers in their own social and political class. Often, as evidence from Varro and Columella attests, their readers came from a close circle of family and friends. Written tracts provided guidance for political and military leadership and summaries of knowledge considered appropriate for men in positions of responsibility in government. Agriculture, natural history, generalship, and tactics were considered to be moral and political subjects. Authors wrote treatises on such topics not only to transmit knowledge but to promote the character traits believed necessary to effective leadership and governance. Some of the treatises contained technical information. Yet, with the exception of Frontinus's work on aqueducts, they were not primarily technical treatises; none were written for engineers or for overseers or laborers on landed estates. They pertained to praxis, leadership and action by heads of landed estates and by governors and generals, high-born males to whom such responsibilities devolved primarily by virtue of birth in the early Roman empire.

Authorship in the ancient world was very different from its homologous activity in modern or even early modern times. Jens Erik Skydsgaard describes one ancient method of writing a book. It involved reading earlier books, or more likely, being read to, and excerpting relevant passages from them. Authors created their own writings by a process of selection and elaboration. Their procedures were facilitated by scribes, who copied passages onto clay tablets and then transferred them to papyrus rolls. The goal was to collect the best and most up-to-date information on a subject and expand on it when necessary.[83] To this I would add that often the author did not write but rather dictated, as the example of Pliny the Elder demonstrates. Authorship tended to be a social activity involving various individuals in which the author was just as likely to be speaking to a scribe as actually writing. In the writings investigated in this chapter the values of encyclopedism predominated over those of originality. Moreover, writing as well as reading was embedded in oral culture to a greater extent than has often been assumed. Just as reading aloud was the usual form of reading, so dictating was a common form of writing.

E. J. Kenney notes that for all of antiquity, once the first copy of a book had been made and distributed its fate was beyond the author's control. There was no legal or practical way to safeguard the integrity of a text or limit the number of copies made of it. Moreover, it was impossible to ensure that an author's corrected or amended edition of a book would supersede the first distributed version. After a book's first distribution it was not unusual for parts of it to be excerpted into anthologies. It might also suffer "adulteration" in various other ways, including distribution under the name of a new "author." Neither Greek nor Roman laws contain any notion of intellectual property. Yet accusations of theft or plagiarism were commonplace in both the Hellenistic and the Roman world.[84]

Care must be taken not to project modern ideas about intellectual property onto ancient cultures. Concern about plagiarism in the Hellenistic context involved primarily an interest in correct attribution for the books collected in libraries, a concern for "authentic" copies. Roman condemnation of literary theft centered on the pious honoring of past authors and was tied to the revival of traditional Roman religious values.

Recent scholarship questions the traditional assumption of widespread literacy in the ancient world, just as it modifies the meaning of the term *publish* in antiquity. Rosalind Thomas convincingly suggests the fundamentally oral character of Athenian culture, bringing into question longstanding assumptions about the great extent of Athenian literacy. William V. Harris argues that traditional scholarship greatly exaggerates the extent of literacy in both the Greek and Roman worlds. The conditions for mass or large-scale literacy, including the general availability of elementary schooling, simply did not exist. This fact is especially relevant to understanding the authorship and readership of lengthy prose books of the kind considered in this chapter. Harris argues that authorship and readership of such books were largely limited to upper-class male readers. He further suggests that the most common method of distributing books was not through the market "but through gifts and loans among friends."[85] Ancient authorship and readership were embedded in oral culture and involved a very limited circulation of books.

The openness of Greek and Roman praxis writings must be qualified by the narrow social range of prospective readers. Praxis authors wrote openly for members of their own class. Their writings, including those on agriculture and generalship, served hegemonic functions and promoted character traits thought to be appropriate for the duties of governance and military leadership. Although both the Greeks and the Romans could be astute users of military technology, they believed that the crucial requirements for military victory were the general's ingenuity and leadership abilities rather than military technology per se.

Open writings on the technical arts, such as the books of Ctesibius, Philo of Byzantium, Vitruvius, and Hero, emerged from a context of authorship different from that of praxis writings. Periclean democracy produced a brief alliance between the technical arts and political praxis. In contrast, technical production in the fourth century B.C.E. and thereafter occupied a sphere separate from praxis and was subordinated to it. Most technē treatises were written by skilled practitioners, often military engineers. Invariably they wrote within the context of patronage by institutions, including state libraries (e.g., in Ptolemaic and Roman Alexandria), or by personal patronage relationships within ruling families. Writings on the technical arts were also open. They may have been written in part for other engineers and practitioners, but they also helped to display disciplines such as military engineering and architecture for the enhancement of kings and emperors.

Attitudes toward authorship and toward past authors within both technē and praxis writings encompass a range of values. Technē authors from Philo of Byzantium to Hero of Alexandria express respect for past authors; yet they used past writings selectively and regarded their own authorship as a means of improving upon them. In contrast, the librarians at the Alexandrian Museum and Library treasured authentic writings regardless of any utilitarian value. These librarians carefully distinguished forgeries and false attributions from authentic lines by known authors. While the Hellenistic librarians called attention to literary theft, the Romans, especially Vitruvius, harshly condemned it. Both Vitruvius and Pliny viewed such theft as an affront to Roman piety, a failure to accord honor to past authors. I suggest that the essential context of their attitudes toward authorship involved the Augustan revival of traditional Roman religion.

Despite their differences, both technē and praxis writings in Greek and Roman antiquity seem to have been written openly; with regard to these texts no evidence for concealment exists. Both kinds of authorship developed under the purview of ruling elites. Practitioners dedicated their technē books to elite patrons or wrote them within the context of institutional patronage of the Alexandrian library; elite authors wrote praxis books for their own peers. Openness would seem to have been appropriate for both.

The traditional ruling classes of the Roman empire experienced declining effectiveness during the crises of the third century C.E. A new bureaucracy arose derived from groups traditionally lower on the social scale, and a new religion, Christianity, spread rapidly throughout the Mediterranean world. One result, as Paul Veyne argues, was that wellborn families began to substitute friendship and loyalty within the family circle for traditional values involving civic duty to the Roman state.[86] In this turn away from the public realm toward privacy the values of secrecy and the practices of esoterism found fertile ground.

Chapter 2

Secrecy and Esoteric Knowledge in Late Antiquity

LATE ANTIQUITY, between about 200 and 500 C.E., was an era of changing boundaries, including psychological boundaries. For many individuals the personal significance of the extended horizontal expanse of the Roman empire contracted in favor of small, intimate groups that functioned outside the formal structure of the state. Yet closer horizontal boundaries were compensated by greatly expanded vertical ones. That is to say, individuals greatly augmented their small (albeit deeply involving) earthly spheres by intensifying their relationship to the immense, immaterial realms of demons, gods, and spirits in the heavens. Elite groups increasingly constructed knowledge in terms of gnosis, understanding of the divine.

Mystery religions such as the Isis cult flourished, while alchemical practitioners were much in evidence. A variety of esoteric groups emerged, including Neoplatonic circles devoted to deified figures such as Hermes Trismegistus and Pythagoras and groups of devotees who followed philosophical and spiritual leaders such as Plotinus. Some groups dedicated themselves to the god or goddess of a temple; others centered on charismatic individuals. Magicians and alchemists often combined particular kinds of craft practice with nonmaterial goals. Traditional folklore, medical remedies, Hermetic practices, alchemy, and magical formulae were often combined into a rich syncretism in which the distinct strands of particular traditions are difficult to distinguish.[1]

The traditions and practices treated in this chapter can be traced to diverse ancient origins. Yet magic, mystery cults, alchemy, and other esoteric practices flourished partly in response to the cultural and political shifts of late antiquity. In the era of the Roman republic and early empire the scaffolding of civic life profoundly supported and shaped the identity of wellborn men, who in turn received ample compensation in their steady climb up the *cursus honorum* of public office. The upheavals of the third century led to the replacement of traditional privilege by a meritocracy. Many of the wellborn, deprived of their public roles, turned to an enlarged private sphere. Paul Veyne argues that a transformation occurred in the character of the Roman family, from a public, civic one in republican Rome to a more private and personal one. Marriage, Veyne suggests, moved away from its civic moorings toward private intimacy and friendship between partners. Peter Brown notes that the power of the Roman empire to absorb the energies of the aristocratic males whom it had traditionally rewarded diminished significantly. Sorcery became an interest and occupation of marginalized elite groups.[2]

The apparent expansion of magic and esoteric practices supports the notion of an enlarged private sphere, a turn from civic life toward more intimate personal and spiritual concerns. In the republic and early empire, for example, state religion was a public matter marked by careful attention to ritual formalities carried out by publicly appointed priests. In contrast, mystery religions, which enjoyed a surge of popularity in the imperial age, were based on the close bonds of initiates who shared in the devotion of secret cults. A rich variety of new spiritual movements emerged—Manichean, Mithraist, gnostic, Christian, Hermetic, and alchemical, among others.[3] Each in its own way offered deeply shared intimate, human associations and purification of body and spirit to both aristocrats and ordinary people, including women. They often offered as well an enthralling pathway through and beyond physical suffering and death.

Although late antique spirituality was influenced by Platonism, and although various groups shared certain assumptions, the era is better characterized by spiritual diversity than by uniformity of belief. It was a creative, if also traumatic, transitional era during which an exceptionally rich cornucopia of spiritual and philosophical choices became available. Within the documentary remains of this cornucopia lie the most important early sources for the investigation of secrecy and esoteric groups. Magical texts, alchemical treatises, Hermetic writings, and accounts of the lives of Pythagoras and Plotinus offer fascinating evidence about how esoteric groups functioned. Magical and alchemical texts reveal not only recipes and formulas but also clues concerning the individuals and groups involved, their practices of concealment and dissemination, and their attitudes toward authorship.

The Craft of Magic

The magical crafts involved often complex recipes and processes, elaborate incantations, and detailed rituals. Their goal was to harness spiritual powers for human ends. In a study of the "location" of magic in ancient culture Richard Gordon points to the power of magic in its secret knowledge, which promised "to renegotiate the boundaries of human existence," and to its inherent subversiveness vis-à-vis dominant social and political groups. He posits a range of elite responses to magical claims and practices, from outright rejection to the ambiguous representation of magic, the depiction of magic as spatially distant or foreign and therefore suspect, and finally an acceptance of the reality of the magician's power. Gordon rightly emphasizes that to reject the basic premises of magical practice would involve the unacceptable necessity of repudiating the presuppositions of ancient religion as well. I would add that a sharp distinction between religion and magic is not appropriate for this era.[4]

Important evidence concerning the actual practice of magic comes from a collection of magical texts of Egyptian provenance known as the *Greek Magical Papyri*. Dating from the second to the fifth centuries, the collection most likely came from a tomb or temple library in Thebes; many items may be from a single collection, perhaps one belonging to an ancient magician. Hans Dieter Betz, editor of the English edition of the corpus, describes it as diverse in nature and origin. It includes spells for acquiring spiritual assistants or demons, recipes for obtaining revelations, attracting or binding lovers, curing illnesses, and inflicting harm, as well as fragments of hymns and invocations. Most are written in Greek, but some are in Demotic (a form of Egyptian) and some are in Coptic (Egyptian written with the Greek alphabet supplemented by Demotic signs for non-Greek sounds).[5]

The magical papyri appear to be the working papers of practicing magicians, of which, Betz suggests, there may have been two very different kinds. The first sort were those associated with the temples of Greek and Egyptian deities. In Egyptian practice the magician was "a resident member of the temple priesthood." A second type, known from a Greek context, was "the wandering craftsman," who used and adopted material from numerous religious traditions, creating a syncretic belief system that was "more than a hodge-podge of heterogeneous items," effectively "a new religion altogether, displaying unified religious attitudes and beliefs." Both kinds of magician emphasized the deities of the underworld and the forces of the universe and employed elaborate technologies to control and influence them.[6]

Magical practice was prescriptive and utilized recipes that included physical, scribal, and verbal ingredients. It often required that the magician embody certain personal characteristics, such as physical purity. The goal of a magical procedure usually involved the creation of some effect in the world. An example (*PGM* 1.1–42) is a prescription for creating a helpful demon. The demon will act "as an assistant who will reveal everything to you clearly and will be your [companion] and will eat and sleep with you." You begin the procedure by taking ingredients from your person, "two of your own fingernails" and "all the hairs [from] your head." Then you take a falcon, and "deify it [i.e., mummify it] in the [milk] of a black [cow] after you have mixed Attic honey with the milk." Then you wrap the falcon in an undyed piece of cloth and place it beside your fingernails and hair. Take a piece of papyrus and inscribe it in myrrh with certain figures. Set the papyrus beside the fingernails and hair, then plaster it with frankincense and old wine. Drink the milk and honey before the rising sun, "and there will be something divine in your heart." Set the falcon up as a statue in a shrine of juniper wood, make an offering of nonanimal foods, and have old wine on hand.[7]

The instructions provide the magician with the precise sounds of the spell to be chanted and the words of a prayer to Orion. The recipe further instructs that when the magician is dismissed, he should go without shoes, walk backwards, and then enjoy dinner and the prescribed food offering. He will then come "face to face as companion [to the god]." The rite requires "complete purity." The final instruction admonishes, "Conceal, conceal the [procedure and] for [7] days [refrain] from having intercourse with a woman."[8]

As the above instructions reveal, the process by which the magician creates the helpful demon involves a complex mix of actions and abstentions and an elaborate combination of substances, procedures, chants, and inscriptions. The corporeal aspects of the procedure are notable—items from the body of the magician, a falcon that is mummified, and the magician's drinking of the liquid used for that mummification. The prayer alludes to sexual creation myths associated with the Egyptian god Osiris, and the process as a whole requires sexual abstinence. Purity is essential. A successful outcome is the creation of a personal demon who will work for the magician's benefit and at his behest. The person reading the instructions is enjoined to secrecy. Clearly such secrecy is essential to the ritualized procedure; it allows the magician access to powers not available to others. Secrecy ensures the potency and efficacy of the entire procedure.

Evidence in the *Greek Magical Papyri* concerning prospective readers invariably points to apprentice magicians. For example, Pnouthios, a sacred scribe, provides Keryx with a recipe for acquiring a helpful demon: "Pnouthios to Keryx, a god[-fearing man], greetings." Pnouthios remarks that "as one who knows," he has prescribed a spell to prevent Keryx from failing. He has done this after detaching "all the prescriptions [bequeathed to us in] countless books." He is showing Keryx one of them. Now, he continues, "I have dispatched this book so that you may learn thoroughly." Detailed instructions follow, including the promise that "he [i.e., the demon] will serve you suitably for [whatever] you have in mind, O [blessed] initiate of the sacred magic." The teacher ends with the enjoinder to "share this great mystery with no one [else], but conceal it, by Helios, since you have been deemed worthy by the lord [god]."[9] Pnouthios addresses an apprentice and shows him one magical recipe from a collection of many. He deems the apprentice worthy of the knowledge and commands him to conceal the recipe from the unworthy.

The practice of secrecy is evident in other recipes as well. A formula for discovering the meaning of dreams ends with the injunction, "Keep it secret, son," again, presumably an admonition to an apprentice.[10] Further evidence of secrecy appears in a magical text that provides a list of names of ingredients and name substitutions, for example, "A snake's head: a leech," for the purpose of concealment. The list is accompanied by a discussion of the precautions taken by the temple scribes to prevent ordinary people from learning magic: "Because of the curiosity of the masses they [i.e., the scribes] inscribed the names of the herbs and other things which they employed on the statues of the gods, so that they [i.e., the masses], since they do not take precaution, / might not practice magic. . . . But we have collected the explanations [of these names] from many copies [of the sacred writings], all of them secret."[11] This somewhat obscure passage indicates that the temple scribes wrote the ingredients of magical recipes on statues of the gods. In order to prevent common people from practicing magic successfully, they wrote name substitutions; for example, they wrote "snake's head" when they actually meant "leech." The list, in Greek translation, includes both the code name and the name of the actual item needed. The author of the Greek text suggests that he and others gathered the actual meanings from sacred (Egyptian) writings.

The complex process of transmission seems to presuppose groups of texts collected, handed down, and amended. Some spells contain evidence that magicians or scribes made changes in an original text. For example, a spell that renders one invisible at will requires "a plant of peony." A later scribe adds, "(he means the rose)." Another spell for summoning a demon includes an alternative method—"In another [text] I have found the following"—that suggests an author compiling magical prescriptions from various sources. In yet another example a spell for revelation is interrupted at the point of itemizing the ingredients, wheat meal and the herb calf's-snout, with the interjection, "(but I have heard from a certain man of Herakleopolis that he takes 28 new sprouts from an olive tree / which is cultivated, the famous one)."[12] Such interjections suggest a tradition of working magicians using written recipes and formulas handed down to them who occasionally augment or amend their material.

The *Greek Magical Papyri* reveal craft practices transmitted through apprenticeship systems that clearly involved oral as well as written instructions. In almost every case a set of carefully followed procedures yields a material or social result or allows accurate predictions or interpretations. Bodily purity is sometimes required. Magic involved the manipulation of divine forces to achieve very concrete material or social results: the acquisition of certain kinds of knowledge, namely, foreknowledge or revelation; the production of spells and charms, such as love spells; and the curing of ailments, such as headaches, fevers, coughs, hardened breasts, or swollen testicles. Temple magicians practiced careful, ritualized procedures that harnessed spiritual powers to effect certain results in the physical world. Evidence of secrecy suggests a kind of craft secrecy that kept knowledge of magical practices and recipes carefully concealed from the vulgar crowd.

The *Greek Magical Papyri* provide evidence of magical beliefs and practices from the point of view of working magicians. They demonstrate that the craft was handed down from master to apprentice and that it required secrecy. The papyri reveal the magician's viewpoint, in which the efficacy of magic was taken for granted and the central concern was accurate technique. The goal was to effect some change, material or psychological, in the world of the here and now.

Yet, from a broader perspective, magic held an ambiguous place in the Roman imperial age. Magic, sorcery, and alchemy were illegal, attesting both to a general belief in their efficacy and to their association with treasonous acts. In addition, the materialist, corporeal goals of magic sometimes were at odds with the quest for the noncorporeal divinity, which was central to some mystery religions, to philosophical Hermeticism, and to Neoplatonic philosophers and their followers.[13]

Apuleius between Magic and Mystery

The writings of Apuleius of Madaura provide rich evidence of the conflict between efficacious magic and the spiritual aims of the mysteries. Apuleius was an orator whose writings display both fascination and deep ambivalence toward magical practices. In 158 C.E. he was accused of being a sorcerer. His defense against this charge, the *Apologia*, is a revealing document in which he denies any dealings with magic and the "black arts" while proudly disclosing his membership in mystery cults.[14]

Because the practice of magic was illegal in the Roman empire and punishable by death, the charge of sorcery was an extremely serious one. After Apuleius married a widow, Prudentilla, his wife's relatives brought the charges against him. Probably with their eyes on her estate, they charged that he had used magic to make her fall in love with him. Apuleius vigorously contests the arguments of his accusers, describing himself as a philosopher who has been initiated into several of the mysteries but has not practiced magic. Accused of keeping magical talismans in a room of his house, he explains that the carefully wrapped objects (which are never identified) are not magical instruments but valued, secret tokens of his initiation into the divine mysteries. Apuleius acknowledges with pride that he has been initiated into such mysteries (he does not specify which ones) but vociferously denies that he is a magician and that he has practiced any of the "black arts."[15]

Neither Apuleius nor his accusers questioned the efficacy of magic, and indeed the illegality of the magical arts in the Roman empire suggests a widespread belief in their power. Given that this power often involved demons, it is relevant that Apuleius himself wrote a tract on demonology, *De deo Socratis* (On Socrates' god), which discusses the nature of demons in detail. Developing ideas articulated in the earlier writings of Plutarch, Apuleius argues that the divine gods in the celestial ether are too exalted to have anything to do with humans but that the air between those regions and the earth is filled with various kinds of demons that function as intermediaries between humans and the gods and play crucial roles in oracles, prognostication, and other kinds of divination. Apuleius was a Platonist, and his writings manifest the central concern of the middle Platonists to reconcile their philosophy with Egyptian religious ideas.[16]

In his famous novel *Metamorphoses,* Apuleius portrays the intense attractions of magic as well as its ultimate rejection in favor of the cult of Isis. *Metamorphoses* is the story of how Lucius the man, using the wrong ingredient in a secret magical recipe, accidentally changes himself into an ass. Lucius relates his trials and hardships as a four-footed creature and divulges as well the many tales that he as a dumb animal is privileged to hear. Stories within stories, including the famous tale of Cupid and Psyche, imminent death, and sudden escape carry the narrative forward until in the final book Lucius the ass becomes a man again thanks to the goddess Isis. Subsequently Lucius the man, who may represent Apuleius himself, is initiated into the Isiac mystery cult.[17]

In *Metamorphoses,* Apuleius describes both the powerful attraction of magic and its ambiguous status. At the outset, Lucius the man is on a journey and finds lodging in Hypata, a town in Thessaly famed for its witches and magical powers. He lodges with Milo, Milo's wife Pamphile, who turns out to be a witch, and their lovely servant, Photis. In the morning he tours the city in a state of impassioned anxiety, looking for signs of magic. He searches the city "dumbfounded by my torturous longing," that is, his longing to learn the secrets of the magical arts.[18]

Burning to discover the secrets of his hostess, Lucius tries to gain access through Photis, and they become lovers. Lucius eventually confesses to Photis his "most passionate desire to know magic at first hand" and persuades her to hide him in

Pamphile's "laboratory" so that he can discover her secrets. With Photis hidden beside him, he watches in fascination as the witch rubs ointment over herself, changes into an owl, and flies out of the window to find the man of her desires. Lucius enlists Photis's help to effect a similar transformation in himself. Reluctantly complying, Photis gives him an ointment that he rubs all over his body. Much to their dismay, he changes into an ass, a mistake resulting from the use of the wrong unguent.[19]

The subsequent tales in *Metamorphoses* recount the misadventures and sufferings of the ass that Lucius has become and record his failure to find and eat the roses that are required to change him back to a man again. In one notable incident Lucius engages in a long night of lovemaking with an aristocratic woman attracted by his huge donkey penis. Intrigued by the moneymaking potential of the situation, the donkey's owner plans a public show featuring a similar event. Just as Lucius is about to be subjected to public humiliation at a carnival by forced copulation with a condemned woman and to probable death by the wild beasts who are then to be set upon her, he escapes. He gallops to the shore, purifies himself by bathing seven times in the sea, and prays to the "Queen of Heaven." He falls asleep, and Isis appears in a dream vision with promises to help him. The next day, after eating a wreath of roses carried by a priest in an Isiac procession, Lucius changes back to human form. Not technical magic per se but Isis herself releases him from donkeyhood and from the vicissitudes of fortune that have governed his life as an ass. In turn, he accepts celibacy and dedicates himself to the service of the goddess. He ends his story with an account of his successive initiations into the cults of Isis and Osiris. The initiations are preceded by visionary dreams and involve purifications, entrance into the secret recesses and rituals of the temples, examination of a sacred, hieroglyphic book, and other matters too holy for Lucius to reveal to his uninitiated readers.[20]

The transformation of Lucius and his initiation into the Isis cult involve both secrecy and public display. His metamorphosis from ass to man occurs in a crowd during a ceremony in which the priests of Isis consecrate a new ship before its first day of sailing. The goddess has told Lucius that he will live in happiness and glory and that he will constantly obey and worship her, serving with "determined celibacy." Subsequently, he finds himself in a crowd of happy worshipers watching a great procession, including "the crowds of those initiated into the divine mysteries . . . men and women of every rank and age." While in the procession, he is transformed into a man, having eaten the required roses handed to him by a priest. Joining the throngs as a man, he enjoys special recognition: "The whole city knew about me and I was the centre of attention as people pointed their fingers and nodded at me. Everyone was talking about me." He arrives at the temple, where a priest reads from a book, offers prayers, and opens the navigation season. Lucius dedicates himself to the service of Isis and repeatedly entreats the high priest to let him be initiated into "the mysteries of the holy night." The priest "gently" puts him off until the goddess determines that he is ready.[21]

Apuleius describes Lucius's initiation in vivid detail, at the same time deliberately concealing its most essential and holy features. A priest leads Lucius by the hand to the entrance of the temple. After an opening ceremony and sacrifices, the priest brings out books written in hieroglyphic script and reads the preparations that have to be made for the initiation. The priest then places Lucius at the feet of the goddess and "secretly" gives him "certain instructions too holy for utterance." Yet, concerning certain matters pertaining to restrictions of food and drink the priest teaches him openly.[22]

On the day of the initiation, "suddenly crowds flowed in from every direction, in accordance with the ancient practice of the mysteries, to honour me with their various gifts." Afterward all the uninitiated are dismissed. Lucius is wrapped in a linen robe and taken into the innermost part of the sanctuary. "Perhaps my zealous reader," Lucius says, "you are eager to learn what was said and done next. I would tell if it were permitted to tell; you would learn if it were permitted to hear." But, he continues, the tongue would incur guilt for "unholy talkativeness," the ears for "their unbridled curiosity." Yet, he does reveal to his readers something about initiation since they may want to know because of "religious longing." He promises that he tells to the uninitiated only what is permitted.[23]

The morning after the completion of the ceremonies, Lucius comes forth "wearing twelve robes as a sign of consecration." He notes that "this is very holy attire, but no obligation prevents me from talking about it, since at that time a great many people were present and saw it." He then stands on a wooden platform, the focus of attention of a large crowd. He is decorated further, and then "the curtains were suddenly opened and the people wandered around to view me." He celebrates his "birth into the mysteries" with a "delicious banquet and cheerful party."[24]

Apuleius suggests through Lucius that the search for magic is asinine because it involves unbridled, illegitimate curiosity, unseemly lust, violence, and subjection to the winds of fortune. The alternative is celibacy, service, and the higher love of the divine. The setting for Lucius's devotion is the established temple. No longer does he wander alone, relying upon his own very limited donkeylike resources. Rather, he is led by the goddess herself and is surrounded by her many priests and other initiates. He is guided into the center of the temple and its rites by those who went before him.

The secrets of the mystery cults are shared by the initiated, who have undergone the requisite spiritual and ritual purification. Yet the temple is surrounded by throngs of admiring outsiders, who are shown just enough of the secret rites to whet their appetites so that their own path might lead them toward the secret, holy initiation that obscures the boundary, it is hinted, between life and death. Secrecy in the cult serves several purposes. It binds the initiated together, fostering their closeness within the temple. At the same time, it creates a radical distinction between those who have been initiated and outsiders. The latter become ardent admirers who long to become part of the sacrosanct group.[25]

Pythagoras, Plotinus, and the Esoteric Circle

Studies of late antique esoteric practices suggest a chronological development from a predominance of groups associated with temples to an increasing number of esoteric groups centered upon free-roving magicians and divine or holy men.[26] Two striking portrayals of individual charismatic figures and their followers exist from the third century. The Neoplatonic philosopher Iamblichus of Calchis provides a vivid portrait of Pythagoras, who lived hundreds of years earlier; and the philosopher Porphyry describes his own teacher, Plotinus. Whether or not the descriptions are historically accurate, the two accounts provide a window into the esoteric philosophies of the late antique world.

In his portrayal of Pythagoras and his followers, Iamblichus provides a detailed description of an esoteric group in which initiates were sharply separated from outsiders. Iamblichus was an important early Neoplatonist who did much to join Platonic doctrines to ancient Egyptian and Chaldaean wisdom, creating a synthesis of religious and philosophical doctrines. The head of his own school in Syria, he wrote the *Pythagorean Way of Life* as the introduction to a ten-volume study, no longer extant, of Pythagorean thought. Leaving aside the complex issue of its prior sources and its relationship to earlier Pythagorean traditions, early Christianity, and various schools of philosophy, the work in itself constitutes an important document for understanding late antique esoterism.[27]

Iamblichus presents Pythagoras as the founder of the ethical tradition of Greek philosophy, in which "philosophy was not simply a set of doctrines, but a whole way of life." Pythagoras envisioned a unified universe that would allow him to combine knowledge of the cosmos, practical judgment concerning family and civic life, personal piety, and character. The mathematical harmonies of the spheres, the harmony of a properly functioning family and city, and the peace of a pious soul were part of a single sympathetic system. Pythagoras preached the exoteric aspect of this system to various groups and thereby brought about improvements in family and civic life. Another part of his system, concerning philosophy and the divine, was esoteric, reserved for a select group of followers who had undergone strict, long-term, rigorous tests of character and spirit.[28]

Pythagoras was a lawgiver who in his exoteric teachings "wholly abolished sedition, discord, and in a word, difference of opinion," not only among his students but among the cities of Italy and Sicily as well. He counseled segments of the population according to their specific characteristics and needs. Entering the city of Croton, for example, he spoke to the youths, urging them to honor their parents, cultivate friendship among themselves, and work for temperance and education.[29] He advised the men to govern with the understanding that they were trustees of the majority of citizens and told them that they must act justly or they "may appear simultaneously to wrong the whole cosmos." Moreover, they should honor their wives and treat their children in such a way that they would be loved for their kindness. They should reject laziness and perform every action at the appropriate time.[30] He advised the women to pay attention to goodness so that the gods would be ready to hear their prayers. He urged them to sacrifice in the proper way (with barley cakes and other foods made with their own hands and not with the slaughter and death of animals), and he praised their piety.[31]

Other matters were esoteric, reserved for Pythagoras's disciples and no one else. The master chose his students with extreme care and subjected them to years of observation and discipline before teaching them his philosophical doctrines. When young people asked to study with him, he did not agree immediately but examined them carefully. How did they get along with their parents and relatives? Did they engage in "untimely laughter"? Did they talk too much? What were their desires, how did they deal with others, and what did they do with their free time? What things gave them joy and pain? What was the nature of their physiques and their bodily movement. He studied their features, "by which their nature is made known," and then "took the visible things as signs of the invisible character traits in their souls."[32]

If they passed muster, a long period of training ensued. Pythagoras had them supervised for three years to test for stability, love of learning, and resistance to popularity and fame. Then he ordered five years of silence to discover whether they could manage self-control in the most difficult area, mastery of the tongue. During this time, household possessions were held in common. After the five-year silence, those deemed worthy became "esoterics." They heard Pythagoras "within the curtain" and also saw him, whereas before this they had heard him but never seen him. The long period of training involved many "ritual celebrations and initiations in so many sciences" and "many and great cleansings and purifications of the soul" emanating from "so many complex theories."

Those who failed along the way were expelled, loaded with gold and silver from the common store, and thereafter considered dead.[33]

The Pythagorean code of secrecy was tied to levels of training and to purity of the soul. Violations were condemned. Iamblichus relates an account of a violation in which a Pythagorean, Hipparchus, "shared doctrines with uninitiated persons who had attached themselves to him without training in the sciences and theory." Another Pythagorean, Lysis, reprimanded Hipparchus, saying that philosophizing in public with ordinary people was the very thing Pythagoras believed unworthy: It was pious to remember the precepts of the "famous one" and "not to share the good things of wisdom with those who have their souls in no way purified." Those who divulged Pythagorean secrets were "unjust and impious" because it took so long to prepare the soul for the highest learning. Just as dyers prepared cloths with a mordant so that the dye never faded, so also "the divine man prepared the souls of those in love with philosophy," making sure that he was not mistaken about anyone whom he hoped would be noble and good.[34]

Iamblichus reports that Pythagoras furthered the secrecy of his most esoteric teachings by writing in symbols, an ancient method used especially by the Egyptians. He valued in particular the process by which one gained an understanding of these symbols. He considered it significant when someone could elucidate their "meanings and secret conceptions." That person would then understand how much righteousness and truth these symbols contained when they were freed from their enigmatic forms. Pythagoras's disciples followed the same esoteric principles. In all their conversations and writing, they did not make themselves clear and intelligible "in a common or popular manner." Rather, "they engaged in divine mysteries and methods of instruction forbidden to the uninitiated, and through symbols, they protected their talks with one another and their treatises."[35]

When the symbols were misunderstood, Pythagorean teachings appeared trivial and laughable to ordinary people, "full of nonsense and rambling." On the other hand, when they were explicated according to their meaning, "they reveal marvelous thought and produce divine inspiration in those scholars who have grasped their meaning."[36] Here Iamblichus fully articulates esoteric principles of authorship. Writings, produced in the form of obscure symbols, are meant solely for those who have been initiated. They alone can understand their true meaning.

When they wrote books, the Pythagoreans followed the principle of attribution to the master on the basis of his authority. Many wrote books that they attributed to Pythagoras. Iamblichus insists that nothing whatever concerning human knowledge had been omitted in Pythagorean writings. Some books in circulation were by Pythagoras himself, and some had been written by his followers on the basis of his lectures. In general, Pythagoreans assigned their writings to Pythagoras and only very seldom claimed "personal fame for their discoveries."[37] The Pythagorean custom of attributing writings to the master was perfectly in harmony with an initiation requiring the long-term abrogation of speech and the consequent minimization of personal opinion and expression.

The group around Plotinus as described by Porphyry was in many ways very different from the Pythagoreans; nonetheless, it too struggled with issues of secrecy. Porphyry, a follower and friend of Plotinus's, collected and edited the master's writings, the *Enneads*, appending a short biography, "On the Life of Plotinus," in which he vividly describes Plotinus and his circle in third-century Rome.[38] Plotinus particularly valued talk, lecture, and discussion rather than silence. Eventually he wrote tracts on his philosophy, leaving it to Porphyry to edit and arrange them. His authorship was closely attached to the oral discussions that occurred at his school.

Porphyry describes the early education of Plotinus as closely bound to the influence of a single teacher. He tells us that Plotinus went to study with teachers in Alexandria but that he became so depressed that he confided his sadness to a friend, who sent him to a new teacher named Ammonius. Plotinus stayed with Ammonius for eleven years, acquiring a complete training, and then traveled to Rome, where he opened his own school. Plotinus and two companions vowed "not to disclose any of the doctrines of Ammonius which he had revealed to them in lectures." Unlike his companions, Plotinus kept the agreement, and although he talked with people who came to him, he "maintained silence about the doctrines of Ammonius." Plotinus himself wrote nothing for ten years after his arrival in Rome but "began to base his lectures on his studies with Ammonius."[39]

Aside from his silence concerning his teacher's doctrines, Plotinus seems to have maintained an open, informal atmosphere in his own school without sustaining marked distinctions between students or adhering to a rigid hierarchy. It was "open to anyone who wished to come." Moreover, Plotinus "encouraged his students to ask questions," and as a result, according to the criticism of one student, "the course was lacking in order and there was a great deal of pointless chatter." Porphyry describes Plotinus as a man who "had many hearers." The terms with which he describes these auditors indicate warm personal relationships: "friend," "close friend," "admirer," and "on terms of great intimacy." They seem to have been a large, socially prominent group. His lectures were attended by many members of the Roman senate, one of whom, Rogatianus, gave up his position and all his property and possessions to take up a life of philosophy. These auditors and followers included women, "who were greatly devoted to him" as well as to philosophy. Indeed, Porphyry describes himself as "one of Plotinus's closest friends," to whom he had entrusted the editing of his writings.[40]

Plotinus seems to have been a man of kindness and warmth. Men and women of the highest rank who were approaching death would bring him their children to raise, "considering that he would be a holy and god-like guardian." As a result, his house was full of "young lads and maidens," whose welfare, education, and property he carefully tended for their own benefit. He was "gentle" and "at the disposal of all who had any sort of acquaintance with him." He lived in Rome for twenty-six years, and although he served as the arbiter of many people's disputes, he "never made an enemy of any of the officials." Porphyry himself reveals that once in his own house he was thinking of removing himself "from this life," when Plotinus unexpectedly appeared and told him that his lust for death was not rational but the result of a "bilious indisposition" and urged him to go on a holiday, which he did.[41]

Within his vivid account of Plotinus's life and community Porphyry also discusses Plotinus's authorship. He presents an ambiguous picture of a sometimes reluctant author whose writing is closely connected to his immediate teaching and to his own close-knit philosophical community. For ten years after his arrival in Rome, Plotinus wrote nothing. Then he began to write on the subjects that came up at the school. By the time Porphyry arrived on the scene ten years later, the master had completed twenty-one treatises. Yet Porphyry reports that "few people had received copies of them." Indeed, distribution seems to have been troublesome: "The issuing of copies was still a difficult and anxious business, not at all simple and easy; those who received them were most carefully scrutinized."[42] Clearly, Plotinus carefully controlled the dissemination of his writings, reserving them for suitable persons.

Yet Porphyry does not seem to have shared his teacher's diffidence. He arrived in Rome as a disputant, first presenting himself to Plotinus by sending a treatise in opposition to his teachings. Plotinus assigned his student Amelius to write to correct Porphyry's misunderstandings. Soon Porphyry wrote a recantation and became a follower and a regular participant at the school, and he began to urge the master to write down his doctrines. Plotinus not only continued to write but also eventually asked Porphyry to edit his writings. Porphyry's move to Sicily to cure his melancholy does not seem to have interrupted his editorial role. Plotinus sent him new treatises for editing until his own death in 270.[43]

As Porphyry describes it, Plotinus focused not on the writing but on the thought itself. Once he had completed a written work, he could not bear to read it over because of poor eyesight. He worked out "his train of thought from beginning to end in his own mind" and then wrote it down, writing as if "copying from a book." He could maintain a train of thought even if he were interrupted by a conversation with someone. He would pick up exactly where he left off, not backtracking at all.[44]

Porphyry and Plotinus's student Amelius both seem to have viewed writings as a way of protecting the reputation of Plotinus as well as establishing the originality of his doctrines. Porphyry reports, for example, that people from Greece had begun to claim that Plotinus was appropriating the ideas of one Numenius. Amelius immediately wrote a letter on the differences between the two men's thought and dedicated it to Porphyry, who in turn included it in his *Life of Plotinus*. They say, Amelius reports, that Plotinus is "a big driveller" or that he "is a plagiarist" or that his "fundamental principles are the meanest of realities." Amelius states that he writes to put Plotinus's doctrines in a form easier to remember, and he hopes that Porphyry will correct him if he has misrepresented any of the doctrines of "our spiritual home."[45]

Porphyry himself edited, arranged, and on occasion provided commentary on the writings of Plotinus. He divided the corpus into six sets of nine (which he thus called the *Enneads*) because the perfection of the number six along with the nines gave him pleasure. He also provided testimony, most importantly from one of Plotinus's critics, the philosopher Longinus, on the outstanding nature and originality of the master's thought and the absurdity of the accusation of plagiarism from Numenius, whose works were "nowhere near the accuracy" of those of Plotinus. Porphyry wanted to establish Plotinus's reputation and to provide, through the arrangement of his writings, a clear exposition of his doctrines and a defense of their originality and importance. At the same time, he maintained his own critical point of view; for instance, he noted that Plotinus's early writings were not fully formed, that his middle writings were his strongest, and that his latest showed signs of decline.[46]

Plotinus's attitudes toward openness, secrecy, and authorship were seemingly ambivalent. On the one hand, he was reluctant to reveal his teacher Ammonius's teachings, wrote nothing himself for ten years after he opened his school in Rome, and when he did begin to write, attempted to carefully control the distribution of copies to suitable persons. On the other hand, he appears to have been concerned to have his thought preserved accurately in writing and under his own name. He encouraged his students, especially Porphyry, to organize and arrange his writings. It is notable that both Porphyry and Amelius defended Plotinus against charges of plagiarism and strove to record his teachings with accuracy and to ensure that he received credit for his writings and for the originality and significance of his thought.

In sum, the accounts of Iamblichus and Porphyry reveal a wide range of difference between two groups each led by a charismatic leader. Pythagorean initiates observed years of silence, saw their leader only after successfully passing this lengthy term, and learned to read writings written in obscure hieroglyphic symbols. Pythagorean authors often attributed their writings to their master. Plotinus's group, in contrast, was open, friendly, and based primarily on oral communication. Yet at its heart were secret doctrines, particularly those taught by Ammonius. In addition, Plotinus tried to regulate the dissemination of his written doctrines. Yet with his encouragement his followers also explicated his doctrines, defended their originality, and attempted to ensure their correct attribution.

Esoteric Knowledge in the Philosophical *Hermetica*

In ways different from Iamblichus and Porphyry, the Hermetic corpus also illuminates late antique notions of esoteric knowledge. The Neoplatonic writings comprise a diverse compilation of anonymously authored tracts centered on the divine figure of Hermes Trismegistus. Dating from the second to the fifth centuries, they derive from a Graeco-Egyptian milieu. Traditional scholarship distinguishes texts mentioning Hermes that deal with technical magic and the so-called philosophical Hermetic corpus. Yet scholars recently have emphasized the overlap of the two categories of writings and the complex, nonunitary origins of both. Garth Fowden suggests that they originated in Egypt and exhibit syncretic religious beliefs that involve the merging of the Egyptian god Thoth (the god of writing, the moon, and messages) and the Greek god Hermes.[47]

Many of the writings take the form of a dialogue in which a master instructs one who is being initiated. The master, sometimes described as a father, sometimes as a god, reveals divine and cosmological mysteries in an atmosphere of intense emotion that has transformative power. The writings describe a process whereby the pupil moves away from the material world and ascends to the divine, noncorporeal realm with the help of the master. The *Hermetica* do not describe an esoteric group per se, but their dialogues, involving masters and initiates, make them particularly useful sources for understanding the experience of initiation into an esoteric religious philosophy. The ancient initiate was probably a member of a Hermetic group that followed a spiritual leader. He or she would have listened to the *Hermetica* being read along with a small group of fellows.

In the first discourse, the narrator, Hermes Trismegistus himself, describes his own initiation by Poimandres. Poimandres had come to him, he reports, "when thought came to me of the things that are and my thinking soared high and my bodily senses were restrained." He describes Poimandres as "an enormous being completely unbounded in size" who appeared to him, called his name, and asked, "What do you want to hear and see; what do you want to learn and know from your understanding?" First-person narration intensifies the immediacy of the account: "'Who are you?' I asked. 'I am Poimandres,' he said, 'mind of sovereignty; I know what you want, and I am with you everywhere.'" Poimandres then revealed to Hermes a remarkable vision of cosmological transformation in which all became light, darkness descended, and the cosmos changed into a watery realm and was agitated like a fire. He inquired whether Hermes understood what it meant and then explained its complex meaning, which involved the unity of the divine and the human.[48]

Hermes describes the intense emotions, including terror and love, that he felt while being taught by Poimandres. Poimandres said: "Understand the light, then, and recognize it." Then, Hermes recounts, "he looked me in the face for such a long time that I trembled at his appearance." As the vision was further revealed, "I was terrified, out of my wits." Poimandres explained the vision seemingly to alleviate this terror. Later, he revealed the mystery of the creation that occurs as the result of intense love. Man was created in God's image. God, who was "really in love with his own form," bestowed craftworks on humans. When Nature saw man, she "smiled for love." When man "saw in the water the form like himself as it was in nature, he loved it and wished to inhabit it; wish and action came in the same moment, and he inhabited the unreasoning form. Nature took hold of her beloved, hugged him all about and embraced him, for they were lovers."[49]

Hermes heard the mystery, one "that has been kept hidden until this very day." He urged his teacher on: "O Poimandres, now I have come into a great longing, and I yearn to hear; so do not digress." Poimandres replied, "Be silent; I have not yet unfolded to you the first discourse." Poimandres gradually revealed his hidden knowledge. Hermes questioned Poimandres and then replied to the questions of his teacher. A rhapsodic catechism disclosed profound mysteries.[50]

Hermes himself was a deity who revealed this divine knowledge to other initiates. Many of the dialogues involve conversations between Hermes and his own son, Tat. Most concern the creation and nature of the cosmos and the divine mind. These discussions between father and son involve crucial private communications transmitted within an intensely loving relationship. Some are explicitly secret, for example, "A secret dialogue of Hermes Trismegistus on the mountain to his son Tat: On being born again and on the promise to be silent." As Hermes reveals his teachings, he instructs Tat to "keep silence and say nothing" and then tells him that "the powers of god purify you anew for articulation of the word." Silence brings knowledge of the divine, mental knowledge seen not through the eyes but through the mind. "This," Hermes tells him, "is rebirth; no longer picturing things in three bodily dimensions." Hermes emphasizes the privileged nature of his communication. This discourse on being born again, he explains, "I have noted down for you alone to avoid casting it all before the mob but [to give it] to those whom god himself wishes."[51]

At the end of the dialogue Hermes reveals to his son a hymn that must be kept secret. To divulge it has not been an easy choice, he says, and he only does so "at the end of everything." It must not be taught; "it is a secret kept in silence." After hearing the hymn, Tat says a prayer, telling Hermes, "I say what I see in my mind, father. To you, god, genarch of progeneration, I, Tat, send speech offerings." Hermes ends the dialogue by rejoicing that the truth has yielded good fruit, and he admonishes silence. "Now that you have learned it from me, promise to be silent about this miracle, child, and reveal the tradition of rebirth to no one lest we be accounted its betrayers."[52]

The philosophical *Hermetica* provide an unusual view into the experience of initiation. The reader or auditor of the Hermetic texts is in the privileged position of receiving divine truths and vicariously participating in the intense emotionality of the experiences as they are described. The initiated were probably members of small groups who would have heard the sacred Hermetic texts read to them. It is revealing that certain Hermetic texts, those found in the Nag Hammadi collection of gnostic writings, contain evidence of cultic practices, such as a ritual embrace after the prayer and a cultic meal.[53] Shared understanding within the closeness of a small group of initiates, perhaps accompanied by a deep emotional attachment to the leader, and participation in prayers, hymns, and rituals may describe the immediate context of the Hermetic group. Intense emotional involvement undoubtedly paralleled the experiences and revelations related by the interlocutors of the corpus itself. Secrecy would have served to reinforce the intense closeness of the group, giving them a bond of shared knowledge from which outsiders were excluded.

Alchemy: Material Practice and Spiritual Quest

Like the devotees of Hermes, alchemists also transmitted their doctrines within small esoteric groups as they created and handed down a corpus of texts. Alchemy focused on transformative processes, both physical and spiritual. Its goals might include "the transmutation of baser metals into silver and gold; the creation of an elixir of life to prolong it; the creation of a human being *(homunculus)*"; and finally, the purification and perfection of the alchemist's own soul. Alchemists utilized craft processes, including metal processing, distillation, dyeing, and other procedures involving physical transformations, and they employed equipment such as alembics, furnaces, and distillation apparatus. Although alchemy originated within a particular group of artisanal crafts, it combined a concern for physical transmutation with a quest for spiritual transformation.[54]

The craft roots of alchemy are evident in an important early collection of craft and chemical recipes known as the *Leyden* and *Stockholm Papyri.* Associated with early alchemy, the papyri are separate codices dating from the late third century. These codices are related to the *Greek Magical Papyri,* discussed above, in that the ink of the *Stockholm* and *Leyden Papyri* is identical to that of some of the magical papyri; some were copied in the same hand.[55] A reasonable surmise is that the chemical recipes and magical texts came from the same Egyptian temple.

The *Leyden Papyrus* contains ninety-nine chemical recipes, of which eighty-eight pertain to gold and silver and ten pertain to dyeing. In addition, there are sections describing minerals that have been pulled from the first-century *Materia medica* of Dioscorides. The *Stockholm Papyrus* contains nine recipes concerning silver, seventy-eight pertaining to precious stones, and seventy on dyeing. There is some duplication of recipes between the two codices and an occasional repetition within each. The texts of both codices appear to be compilations from diverse sources whose contents partially overlapped.[56]

Robert Halleux points out the relationship of these writings to the Hellenistic industry devoted to fake "precious" metals and fine dyestuffs for a "petit bourgeoisie" with large pretensions and small means. He suggests that the origin of alchemy lay in the meeting of this industry with doctrines from Greek philosophical and mystical traditions. Although the papyri have been designated as early alchemical texts because of their many recipes concerning the imitation or faking of gold and silver, they contain no trace of philosophical alchemy.[57]

The context in which the papyri were used and the purpose for which they were copied are difficult to pinpoint. Their codex form would have been convenient for workshop reference. Yet the copies contain no trace of spilled chemicals, which might indicate such a use. Halleux calls them library copies and makes the intriguing observation that both documents prescribe the use of artisanal tools in a way that suggests an amateur, rather than artisanal, readership. For example, a direction to use the "crucible of the goldsmith" would hardly be made to a prospective audience of goldsmiths. On the other hand, he suggests that three of the recipes in the *Stockholm Papyrus* may have been abstracted from a dyer's manual.[58]

Secrecy is urged in a *Leyden* recipe for creating an imitation of electrum, an argentiferous gold, the Egyptian word for which is *asem.* The recipe instructs to combine tin, copper, and a small amount of asem. It states that the resulting metal will fool even craftsmen: "The metal will be equal to true asem, so much so as to deceive even the artisans." Citing a modern jewelry maker, Halleux notes that the difference between imitation and faking has to do not with the product itself, which is the same in either case, but with the presentation of the product in the marketplace and the intention of the seller;[59] and I would add, with the understanding of the buyer. The above recipe could pertain to either a fake or an imitation.

Other recipes contain word substitutions for ingredients. Yet such substitutions do not constitute clear evidence of secrecy. In the few cases where symbols or substitutions occur, the intent is ambiguous. In recipe 88, line 494, of the *Leyden Papyrus*, for example, the astrological signs for the sun and the moon are substituted for the words *gold* and *silver*. Although such signs for various substances in later alchemical writings are well known, they were not exclusively the symbols of alchemy in this early period. The association of the sun and the moon with gold and silver, respectively, and that of the planets with other metals were ancient ones that presupposed an astrological influence of the celestial bodies on the growth of the metals named after them.[60] This tradition led to the substitution of the names of the planets for the names of metals but does not justify an assumption of secrecy any more than would the use of the symbol H_2O for water in modern times.

Yet the admonition to secrecy is explicit in a recipe for purple dye, the color used for the clothing of royal personages. The instructions begin: "Keep this as a secret matter because the purple has an extremely beautiful luster." Halleux points out that some of the instructions in the recipe, such as one to "take scum of woad from the dyer" and another to dissolve the "color prepared by the dyer," imply that the author of the recipe was not a dyer. Neither, I would add, would the prospective readership seem to have included dyers.[61] Particularly because the recipe concerns the color purple, we can imagine a priestly or royal context, in which the instructions may have been transmitted, perhaps by scribes in a palace or temple. Clearly, the author believed that the recipe should be kept secret to protect knowledge of how to produce the remarkable color that resulted.

The *Stockholm* and *Leyden Papyri* comprise exclusively craft recipes, whereas other early texts offer evidence of the spiritual side of alchemy. One of these, the *Physica et mystica*, by Pseudo-Democritus, is an intriguing compilation of recipes and other materials. The recipes include instructions for making purple dye, gold, and asem. The narrator also recounts the problems faced by a group of initiates, to which he belonged, when their master died. As the narrator tells it, the group had received ideas from "our master," and "recognizing the diversity of matter," they were obliged to "harmonize their natures." Unfortunately, the master died "before we were initiated, and at a time when we were still occupied with the knowledge of matter." Not knowing how to proceed, they attempted to call the master back from the underworld. The narrator relates that he called the dead master several times, demanding to know how he could "harmonize the natures." Finally, the master answered that it was difficult to speak without permission of the demon and said only that "the books are in the temple."[62]

The narrator and his companions went to the temple to seek these books, which the master had never mentioned before his death. The cause of his death was a matter for speculation. Possibly he had taken poison to part his body from his soul, or, "as his son declares," he had swallowed poison by mistake. The master had intended to show such books to his son only when the latter came of age. The narrator and his companions found nothing. They wanted to know "how substances and their natures unite and are blended." They performed some operations (exactly what operations is unclear) on "the composition of matter." Then the time arrived for a ceremony, which they performed together. Suddenly, in the innermost part of the temple, a column opened. At first neither the master's son nor the rest of the group saw anything inside. The son advanced to the column, and the rest followed. They saw a precious formula: "Nature rejoices in nature, nature triumphs over nature, nature dominates nature." After relating the story, the narrator tells his readers that he has come to bring to Egypt "the doctrine of the things of nature, so that you may be raised above the curiosity of the vulgar and the confusion of matter."[63] He aims to help separate them both from ordinary people and from the confusion of matter.

As Michèle Mertens explicates, the precious formula reveals in cryptic form the notion that the material world is governed by two different "natures." The reference is to a doctrine of sympathy and antipathy that governs the combination and separation of all bodies in the physical realm. "Nature rejoices [*terpetai*] in nature" means that one substance has an affinity for another. "Nature triumphs [*nika*] over nature" suggests that one substance imposes its qualities on another. Finally, "nature dominates [*kratei*] nature" denotes that one substance prevents another from acting and thereby neutralizes it.[64] Aside from its intriguing story of the leaderless initiates, the *Physica et mystica* consists mostly of craft recipes, which are often followed by the formulas indicating the universal principle of such transformations.

In contrast to the context of the temple, the alchemical writings of Zosimos point to small groups of alchemists working in their homes. Zosimos, from Upper Egypt, lived about 302 C.E. and was influenced by both Hermeticism and gnosticism. His most important extant treatise, the *Authentic Memoirs*, provides intriguing clues concerning the social context of early alchemy. He addresses some sections of the work to Theosebeia, who according to later testimony was his "sister"; whether she was his natural as well as his spiritual sister is unknown. Certain alchemical, Hermetic, and gnostic communities habitually addressed fellow members as "sister" and "brother." Perhaps a spiritual rather than a sibling relationship is indicated by a passage in which Zosimos addresses Theosebeia with the formal rather than the personal form of *you:* "Staying once in your house Madame . . . ." This usage corroborates other evidence to suggest that Theosebeia may have been part of a society or confraternity of alchemists. Zosimos's description of his stay at her house offers further clues to the social context. He tells her that he admires the activities of those whom she calls *strouktorion* (servants or slaves who organized banquets). Zosimos describes how much he learned during his visit by observing the cooking of a chicken in a strainer above bouillon, which vaporized and then penetrated the chicken meat. He relates this cooking process to dyeing and alchemical operations. He mentions perusing Theosebeia's books as he thought about the process, looking especially "on the shelf of the Jewish books" next to those containing the technical details concerning the *tribicos* (a distillation apparatus).[65] We can surmise that Theosebeia was an alchemist who owned an alchemical library and invited other alchemists to her home.

Near the beginning of *On the Letter Omega,* addressing Theosebeia directly, Zosimos suggests antagonism between his views and those of outsiders. He laments that success in tinctures and dyeing (as a result of opportune moments) has turned a book titled *On Furnaces* into a mockery. Some think that their success in dyeing is a result of favorable astrological dispositions and the good influence from their personal demons. Only when destiny changes, when an evil demon takes over, when their art and good fortune are scattered, do they recognize that there is something more to be considered, namely, the alchemical technique explicated in *On Furnaces.* The precise circumstances of the controversy to which Zosimos refers are unknown. Yet it is clear that he advocates alchemical techniques as a way to overcome the dictates of fate and astrologically determined events.[66] Whether the controversy involved dyers and alchemists or two groups of alchemists, the transformative process of dyeing is taken by Zosimos to be subject to the workings of alchemical knowledge and skill rather than astrological influences or the intervention of demons.

In a later section of the treatise, Zosimos scorns successful dyers who laugh at "the great book *On the Furnaces.*" He insists that people succeed in diverse ways in an art that is unique and that they practice this art in a variety of ways. Both human dispositions and astral arrangements are diverse. One artisan may rely on conjunction (i.e., astral influences), another may be a simple artisan, another may just be dragged along, and another, even worse, may be incapable of progressing. Practitioners of these arts use different instruments and procedures; they might differ in intelligence and in success.[67]

Taking an analogy from the art of medicine, Zosimos gives the example of a person with a broken bone. If he goes to a priest bonesetter, appealing to his devotion, the bonesetter reknits the bone until he hears the two parts grinding back together. But if this is not successful, he does not just give up but goes to doctors, who have books with figures and drawings with crosshatching of all kinds (presumably illustrating different kinds of fractures). The bone is then tied with an apparatus that causes it to heal. Zosimos here points to the varying philosophies of traditional Egyptian medicine and the quite different medicine of the Greeks. By analogy with medicine, he strengthens his argument in favor of learning about furnaces: people who fail in their dyeing die of hunger because they do not want to understand and realize the drawing of the structure of the furnace, whereas those who do, conquer poverty.[68] Clearly, Zosimos identified with the medical tradition, which provides an analogy to the uses of alchemical apparatus and techniques in dyeing. Yet he argues not specifically for the superiority of one method over another but for a diversity of methods based on an assumption of a diversity of causes and for the value of studying the treatise on furnaces.

Zosimos maintains his interest in apparatus and in actual processes of transformation throughout the *Memoirs.* Unlike the author of the *Physica et mystica,* he does not consider equipment and physical operations to be the first stage of a longer spiritual journey, nor does he believe that material processes are subsumed under the spiritual realm. Rather, the physical and material aspects of alchemy are integrally related to its spiritual component. One chapter of the *Memoirs* describes a *tribicos,* an alembic, or distillation apparatus, with three tubes for discharge coming from the bottom and three vases for reception of substances at the top. Zosimos reports that he has taken this description from the writings of Marie the Jewess, and he provides an illustration. Elsewhere he describes the *kerotakis,* a reflux apparatus closed at the top with a hemispherical cover. A substance usually containing sulfur or mercury was placed at the bottom and heated, giving off vapors that attacked metals placed at the top. Some of the vapors combined with the metal, and some condensed on the cover, flowing back in liquid form to the bottom. Zosimos also provides recipes and instructions for the "fabrication of waters."[69]

Zosimos joined his interest in apparatus and the fabrication of substances to a sustained interest in the world of spirit. He describes the conflict between the demonic and spiritual worlds and, in a syncretic amalgamation of disparate traditions, the conflict between good and evil. The good constitutes the realm that he identifies as that of the luminous, "pneumatic," truly Adamic, and Promethean, the spiritual realm that leads to true paradise. This noncorporeal realm is in conflict with the Epimethean, the terrestrial bound, the corporeal. Conflict between the two domains is exacerbated by the "counterfeiter" demon, who envies "pneumatic" and "luminous" individuals and wishes to lead them astray. This demon comes to ravish whether secretly or openly, secretly advising, killing the true Adam. In this view, humans are part of a duality between body and spirit that constitutes an arena of conflict in which "counterfeit" demons operate often in clandestine fashion. Secrecy and deception belong to the world of demons as well as to the corporeal world of human action.[70]

Zosimos greatly admired alchemical authors of the past. Apparently Theosebeia wrote to request information on the applications of some apparatus. He expresses astonishment that she writes to obtain information from him that she must not ask. He points to a passage in Pseudo-Democritus's *Physica et mystica* in which the ancient author notes that he does not speak of the "mountain of water" (i.e., the alembic apparatus) because he has done so in other writings. Zosimos makes clear that he is shocked by the implication in Theosebeia's request that he can provide the information better than a revered ancient author: "Do not believe that I have written in a way more noble than the ancients; know that I would not be able to do it."[71]

In a later section Zosimos chides Theosebeia concerning her desire to know about the technique mentioned above: "I have laughed at you," he tells her. She has surprised him by not supporting the "envy" (i.e., guarding the secret) of the technique and by blaming "the philosopher" (i.e., Democritus) for saying that he will omit certain things because they are discussed at length in his other writings. She has been humiliated into tireless disparagement and has blamed the philosopher foolishly, Zosimos insists, because she has misunderstood him. The philosopher mentions in his other writings not the "fabrication of waters" but their ascendancy (referring to their ascent in the alembic). Fabrication is one thing, ascendancy is another. The ascending has been described at length and in depth. However, concerning fabrication, no one has exposed it because it was a manifest secret, vigorously hidden.[72]

Certain aspects of Zosimos's remarks (which in part refer to a previous discussion, now lost, between him and Theosebeia) are unclear. Yet it is plain that aspects of the alchemical art have been vigorously guarded as secret, even from certain other alchemists, whereas other elements have been explained. Zosimos proceeds to give Theosebeia directions for the "fabrication," an indication that whatever secrecy existed in the past had neither excluded him from the knowledge nor prevented him from passing it on to her.[73]

Early alchemical writings offer significant clues about the early contexts of alchemy. Craft recipes were probably transmitted within an Egyptian temple. A temple also seems to be the setting for the *Physica et mystica*, which reveals a small group working with a master who hope to comprehend the nature of material processes and then go on to higher understanding. Zosimos's writings point to an ongoing interest in physical procedures and to a long tradition of written transmission of which some is secret. His writings provide evidence that alchemists worked singly or in small groups in their homes.

Taken together, the writings treated in this chapter constitute the most significant evidence of secrecy and esoteric groups from the ancient world. Although the random loss of texts must be taken into account, I suggest that the significance of the late antique provenance of these sources involves a move away from Roman civic concerns to more intimate and private groups and to spiritual preoccupations. (Although the spread of Christianity is outside of the purview of this work, it also points to this same development.) In most of the sources treated here secrecy is linked to the intimate spirituality of esoteric groups. Despite their various philosophical orientations, these groups generally aimed to move away from the corporeal world toward nonmaterial, spiritual realms. The craft of magic was different in that it aimed primarily at material and psychological manipulation in the contemporary physical world. Alchemy could use material manipulation as a step the alchemist took to reach higher spiritual teachings (as in the *Physica et mystica*); or it could remain a constant activity within a process of spiritual endeavors (as seemingly in the writings of Zosimos). In the *Stockholm* and *Leyden Papyri* we find recipes involving the fabrication of fake and imitation gold, as well as purple dye. They come from a Graeco-Egyptian temple context and represent the craft basis from which alchemy developed.

In many of these writings, higher forms of knowing involved a rejection of materiality, a turning away from the bodily realm to the vast realms of spirit. While the craft of magic involved similar assumptions concerning the structure of the cosmos, it in contrast attempted to manipulate spiritual realms for utilitarian goals, whether they involved love, healing, or some other concrete aim. As the writings of Apuleius reveal most strikingly, a split developed between the low aims of magic, with its associated materiality, and the higher pursuits of spiritual knowledge. This split is evident in many of the sources treated in this chapter, including the Hermetic. Although such a split is not universally present, it represents, I suggest, the norm of esoteric groups of late antiquity.

In contrast to the traditions treated here, there is little evidence for craft secrecy within the ordinary crafts of the ancient and late antique worlds—crafts such as carpentry, glassmaking, pottery, masonry, stonecutting, leatherwork, clothmaking, and the thousand other artisanal crafts that created the material basis of daily life. This is not to make the rash claim that there was no craft secrecy whatsoever. It is to say that the lack of evidence for craft secrecy suggests that if it did exist, it was probably not widespread or culturally significant.

Chapter 3

Handing Down Craft Knowledge

MAKING THINGS involves a complex set of activities that developed gradually and intermittently along with human evolution itself. The construction and use of tools, the fabrication of textiles, pottery, and implements of various kinds, and the working of stone, metal, and other materials require intricate operations, each of which has its own history. Complex craft technologies developed millennia before the invention of writing, which occurred in Mesopotamia and Egypt around the third millennium B.C.E. Some crafts, such as the fabrication of stone tools, were developed by hominids before the evolution of Homo sapiens and before the full development of speech. As paleontologists have increasingly made clear, technological activities and human evolution should be seen as intrinsic to each other rather than as separate developments.[1]

Despite the existence of a few craft writings from antiquity, there is no doubt that the preponderance of craft knowledge in premodern times was transmitted orally. Skilled persons would have taught the craft to others, who ordinarily would have been younger. After Neolithic times (8000 to 6000 B.C.E. in the Mediterranean region), with the emergence of villages and agriculture most crafts would have been carried out in households and transmitted by formal or informal apprenticeship. Apprentices, whether learning a household craft or training in a workshop, would have learned both by verbal instructions and by practicing aspects of the craft until they mastered the whole.[2]

No evidence allows us to assume an automatic correlation between the oral communication of craft knowledge and secrecy. Most crafts at one time or another constituted the required skills of household production. Even after the development of urbanism about 3000 B.C.E., household craft production was highly significant; it is well to recall that in all premodern times craft production necessarily would have been much more extensive than it is in the modern era. In most ancient and premodern cultures, acquisition of the knowledge and skill of particular crafts would have depended on gender and to a lesser degree on class but would not necessarily have been the result of a particularly privileged or secret course of instruction. Most ancient peoples would have been far more familiar with craft production than their modern counterparts, even those whose social class might lead them to disdain the value of handwork.[3]

We know much more about the products that ancient artisans made and even about the ways that they worked than we do about artisans themselves, their attitudes toward their work, and how they transmitted their knowledge to others. Our knowledge of ancient crafts has been obtained largely through the study of artifacts. Increasingly sophisticated archaeological methods reveal much about how and when a thing was made, but they tell relatively little about the people who made them or how those people communicated their knowledge and skill to others.

On the basis of some evidence we can assume that craft skills often were handed down through families, from father to son, from mother to daughter, whether within the context of households, craft guilds, or other organizations. Family transmission of craft skill is well attested in neo-Babylonian craft documents from the sixth century B.C.E. For Greece and Rome, Alison Burford confirms that "craft secrets" often were handed down within families from one generation to the next. She points both to literary references and to personal records showing that fathers often trained their sons in the crafts into which they had been born. "By far the most numerous records showing the handing down of skills for several generations are those of the sculptors and painters, but there is no doubt that family interest pervaded every other craft." Evidence of family transmission includes the well-documented custom of craftsmen's adopting young relatives or friends as both apprentices and heirs when they lacked family members of their own to fill such roles. For another context, the Roman port town of Ostia, Russell Meiggs also points to evidence of craft transmission within families.[4]

Evidence of craft secrecy in antiquity is exceedingly sparse, an indication, in my view, that it was not a significant aspect of ancient artisanal culture. The meager evidence of ancient craft secrecy is most often associated with the magical crafts or with the crafts of the Egyptian temples, such as manufacture of purple dye (see chapter 2). For most of the hundreds of other crafts in the ancient world there is no evidence of secrecy. Yet evidence of ancient artisans, a group that was lowborn, is sparse. Inscriptions and other sources mention the occupations of deceased artisans and give hints concerning artisanal organizations, pointing to officers and banquets, patrons, and the gods cultivated, but do not provide evidence of secrecy. Nor can craft secrecy be postulated on the basis of modern economic assumptions concerning competition. Ancient societies and peoples simply did not possess modern economic ideas.[5]

For every craft there is a significant body of knowledge that can be "known" only by actually practicing the craft with one's own hands. Written or verbal instructions and physical demonstrations can introduce the craft but cannot fully transmit it. The ability to describe craft processes precisely was, moreover, far more limited in premodern times than it is now. To give an example from metallurgy, smiths possessed practical knowledge based on experience of working with, say, iron forged in a charcoal fire to make steel. The materials used would have been of variable composition; obviously, smiths were not able to describe either the process or the materials in the precise terms of modern metallurgical analysis. The characteristics of the ore, the temperatures of the furnace, the quality of the bloom, the force of the hammering, the amount of quenching, all represented a kind of hands-on knowledge acquired through experience.[6] Ancient metalworkers could manufacture the finest of steel swords, but they could not have explained the process in precise detail even if they had wanted to.

The issue of craft secrecy in the ancient world is best approached with caution: lack of evidence does not mean that it did not exist. Yet the assumption that widespread craft secrecy prevailed is not justified. The results of present-day anthropological studies of apprenticeship should reinforce a cautious approach. For a variety of crafts practiced in diverse cultures secrecy is sometimes evident, sometimes not. When secrecy does exist, the nature of the information that is concealed varies considerably. For example, Roy M. Dilley found that a group of West African weavers practiced secrecy, but not concerning weaving techniques. Rather, the Tukolor weavers concealed weaving "lore," including the sixteen names of the mythical ancestor of the weavers, the line of descent from weaver to ancestor, myths and legends associated with the craft, weaving origin myths and their relationship to contemporary practice, and verses, incantations, and spells. The nature of these weavers' secrets suggests a concern with issues of social identity and status within society, as Dilley points out.[7]

Ancient artisans sometimes belonged to guilds and associations organized around particular crafts. The sources for such groups, consisting mostly of inscriptions, provide some evidence of their functions and activities. Social and mutual support functions seem to have dominated ancient craft associations at least until the late imperial age, when many of them came to be controlled by the Roman state.

Negative Evidence of Craft Secrecy within Ancient Guilds

Guilds generally were organizations of artisans based on particular crafts and technologies. They developed independently in many areas of the world and in many different periods of premodern history. Although the appearance of guilds presupposes a certain development of the crafts beyond a household industry, as well as some degree of urbanism, these two conditions did not invariably lead to their genesis. In the ancient Mediterranean world the development of guilds varied from place to place. Our knowledge of craft guilds and their activities during most of antiquity is fragmentary at best. What is clear is that ancient guilds were not primarily economic organizations. Nor do they seem to have been concerned with the maintenance of secrecy within particular crafts.[8]

An early student of ancient guilds, Mariano San Nicolò, insists that there is little evidence of guild organization independent of the state in the ancient Near East. Although a few scholars have contested this view, arguing for the existence of independent guilds, they have failed to produce convincing evidence. Carlo Zaccagnini notes that the view that artisanal guilds existed in the Neo-Babylonian period "has been generally disputed" and that "the use of the term 'guild' is historically anachronistic, since it denotes a kind of labor organization that is not to be found in the socio-economic structure of the Near East during the sixth–fifth centuries B.C."[9]

There is little indication of widespread guild activity in classical Greece; there is far greater evidence of craft guilds in the Hellenistic age. Craft associations became commonplace throughout Roman territories only from the time of the emperor Augustus, in the 20s B.C.E. According to Jean-Pierre Waltzing, whose work on the Roman guilds remains fundamental, in the fourth century C.E. the guilds of the empire were transformed from voluntary associations with government approval and encouragement into "compulsory public service corporations entirely controlled by the state," the members coming "to constitute one of the hereditary castes into which the population of the late empire was divided." Although this view of complete state control has been challenged, most scholars agree that with the decline of urbanism in the Latin West, guilds for the most part disappeared. The merchant and craft guilds of the Latin medieval West arose independently with the reemergence of urbanism in the eleventh century.[10]

The origins of ancient artisanal associations, or *collegia*, in the Roman republic are obscure; evidence of their activities and structures, provided mainly by law codes and inscriptions, is sparse. Historians generally agree that Plutarch's description of the founding of the Roman collegia by the king Numa is false, being too early. Yet the early development of these associations of artisans and employers is obscured by a virtually total lack of evidence. Waltzing argues that the Roman guilds were originally private, free religious associations that included the cult of a deity and provided burial for impecunious members. They could also be manipulated for political ends, as they were in the tumultuous days of the Cataline conspiracy (64 B.C.E.). It was at this point that the Roman government attempted their suppression and/or regulation, an effort that first succeeded under Augustus. If the association was not viewed as a threat to the social and political order, authorization was given in exchange for some public service.[11]

Collegia were associations of artisans and small tradesmen that included both slaves and freemen and sometimes women. The membership did not usually include all of the workers of a particular craft and often consisted of the workshop owners and managers rather than the actual laborers, many of whom would have been slaves. The associations did sometimes act to defend or advance the material interests of members vis-à-vis the state whether or not the matter concerned the craft itself. Many collegia had one or more wellborn private patrons to help with provisions and to defend members' interests. Yet the primary functions of the collegia seem to have been social rather than economic. The associations had officials, including the *quinquennalis*, or chief, who served for five years, and a treasurer. Each association venerated and attended to the cult of a deity, often the goddess Minerva. They maintained halls *(scholae)* for feasts and meetings. Graffiti from Pompeii tell us that sometimes the collegia supported particular political candidates. There is no evidence that they regulated apprenticeship (which seems to have been carried out informally) or attempted to exert quality control over products or to monopolize production. Evidence of craft or trade secrecy within the ancient guilds is lacking, as is any evidence of an interest in proprietary rights over craft knowledge or processes. This lack of evidence is to be expected if it is true that ancient guilds for the most part fulfilled social rather than economic functions.[12]

In the early empire many craft associations took on public services that came to be seen as obligations. From the time of the early Principate the government depended upon private associations to provision the city of Rome. The model of development established by Jean-Pierre Waltzing in 1895, and accepted until recently, describes an increasingly autocratic state that established complete control over the guilds beginning in the third century. In this view, associations (now called *corpora*) were charged with increasingly onerous obligations; failure to fulfill these obligations resulted in the confiscation of individual members' property. The associations became branches of the state; first the provision of goods and services and then hereditary membership became compulsory. Private, social, religious, and funerary functions disappeared. Waltzing's model has been challenged by Boudewijn Sirks's study of the legal and bureaucratic structure of the empire as it pertained to levying grain and transporting it to Rome and Constantinople. Sirks shows that in addition to the imperatives of Roman law, a free market functioned throughout the imperial age. His critique of Waltzing's autocratic model brings into question the development of an artisanal caste system and the transformation of the collegia into departments of an autocratic bureaucracy. Yet his study does not demonstrate craft secrecy or proprietary attitudes concerning craft knowledge.[13]

During the early medieval centuries the Roman world divided into the Byzantine East and the Latin West. In the East, which remained urbanized, the guilds continued to function and to be regulated by the state, although with some changes from the time of Diocletian (ruled 284–305). Urbanism declined in the West. There is little evidence that the Roman guilds continued to exist in the Latin West in the early medieval period.[14]

Evidence from the ancient guilds tells us very little about issues of openness and secrecy or about proprietary attitudes toward inventions or toward craft knowledge. Yet the reasons for this lack of information are revealing in themselves. The ancient guilds served primarily as social, religious, and funerary organizations. Evidence is lacking that craft knowledge per se, apart from its application to particular tasks and apart from labor, had economic value in the ancient world. To put it in anachronistic terms, evidence of an ancient concept of intellectual property is lacking.

This is not surprising especially given the dominant ethos that accorded most artisans low status. Many artisans within the Roman republic and empire were slaves or freedmen and -women (slaves who had been manumitted); both groups lacked legal paternity and the social status that derived from that paternity in Roman society. Slaves were valued only as property. Of course, if they possessed a high degree of a needed skill, they would have increased value for their owners; that is, their persons as possessions would be more valuable. Free and freed men and women with highly developed craft skills could be well remunerated for their labor either from the sale of their products or from wages. In a study of occupational inscriptions, Sandra R. Joshel has argued that in the face of low social status and marginality workers themselves, including artisans, identified in a positive way with their labor and skill. Yet, craft skills and artisanal knowledge were not generally viewed as commodities separate from labor and handwork. Inventions and novelties were not valued as such; however, this should not suggest that there were no novelties and inventions. It was perfectly possible to value finely crafted objects, especially if they were made of precious materials such as gold, silver, or ivory, without conferring particular value or social status upon the maker or the maker's knowledge apart from the object.[15]

Plutarch's views exemplify the negative tradition subscribed to by many Roman elites. He remarks that many times, "while we delight in the work, we despise the workman." For instance, he continues, we delight in perfumes and dyes but regard dyers and perfumers "as illiberal and vulgar folk." Plutarch believed that work with one's own hands, because of the labor expended on useless things, showed indifference to higher things. "No generous youth," after seeing the Zeus at Pisa or the Hera at Argos, actually wants to be Pheidias or Polyclitus, he suggests. Just because "the work delights you with its grace" does not mean that "the one who wrought it is worthy of your esteem."[16] Plutarch separates praxis, associated with virtue and good character ("the generous youth"), and technē, technical skill.

Although Plutarch's views should not be taken to represent all of antiquity, they do serve as a reminder that modern notions of original artistic genius and creativity, which had their origins in the early modern period, should not unduly influence our perceptions of ancient views. Early modern and modern attitudes place great value upon individual pieces of artwork and upon artisanal knowledge of craft processes and inventions. Such attitudes were not in evidence in the ancient world. In the very different value system that prevailed then craft secrecy as a way of maintaining a monopoly over craft knowledge and processes would have had no particular meaning, and there is little evidence of it. Similarly, there is no evidence to support the notion that artisanal knowledge may have been viewed as separable from labor or from craft products and considered an intangible property on its own.

When they were free associations guilds could undoubtedly have an ameliorating effect on low-status occupations by providing a space for conviviality and support apart from the strictures of a rigidly hierarchical society.[17] Yet there is no evidence that the associations provided a pathway to higher social status outside of the craft or that they could have done so. The uses of secrecy to enhance social cohesiveness and status (as in the example of the Tukolor weavers of Africa, noted above) could not have operated in the ancient Roman world because they could not have overcome the negative influence of the low status of handwork in the wider society or of the slavery with which handwork was associated.

Craft Recipes as Evidence for Craft Transmission

Some evidence concerning the technical arts is embedded in craft recipe writings. Collections of such recipes constitute a genre of great antiquity; the earliest extant collection is from the second millennium B.C.E. Continuity between collections separated by hundreds of years is suggested by similarities between recipes and even almost identical recipes. Yet such similarities should not obscure the very diverse cultures from which craft recipes were copied or recorded. They range from ancient Mesopotamian glass recipes to, for example, the twelfth-century collection of the monk Theophilus. Similarities between collections may result from actual continuities of practice and of textual transmission or from identical requirements derived from identical tasks, such as that of making red glass. Differences result from variations in craft procedures and from the differing origins of particular recipe collections, such as a scriptorium or a workshop.

Among the earliest collections of recipes are Mesopotamian instructions for making glass that are written on cuneiform tablets from the fourteenth to twelfth centuries B.C.E. Another important group of glass recipes date from the seventh century B.C.E. A. Leo Oppenheim's scholarship provides a basis for the study of these Mesopotamian texts on glass. Oppenheim describes them, "strange as it may seem, as literary creations within a complex literary tradition." Recipe tablets were "like all other cuneiform writings subject to certain stylistic requirements." He places them within a group of writings that he calls "procedural instructions," which include mathematical tables and problems, mathematical astronomy, ritual and medical texts, including the earliest pharmaceutical texts (in Sumerian), ritual instructions that were part of certain prayers, and a group of twelfth-century B.C.E. tablets dealing with the preparation of perfumes.[18]

Oppenheim distinguishes the instructions for making perfume and glass from mathematical and astronomical tablets. It is unlikely that the makers of glass and perfume were literate, in contrast to the mathematical astronomer or authority on religious ritual. He argues that texts on perfume and glass belong to a specific event, whose nature is unknown to us, in which "the technical lore of certain artisans which catered to the need of the court was fixed in writing, presumably upon a royal order." Once such texts had become part of the corpus of traditional writings, "tradition-conscious" scribes continued to copy them and to keep them in private or royal libraries. As they were copied and recopied they became ever more remote from working glassmakers and from actual glassmaking processes and procedures.[19]

The historiography of secrecy as it concerns the Mesopotamian texts on glass is particularly germane to this study. Early scholarship on one of these texts by C. J. Gadd and R. Campbell Thompson presents the text as an example of cryptographic writing meant to disguise craft secrets from nonglassmakers. These scholars explain that the writer, "guarding his secrets with true professional jealousy," has purposely disguised his meaning "by artifices of writing which amount to a form of cryptography." They go on to explain that the intention of the writer was to give directions "to members of his Guild," who had access to the cryptographic writing. Campbell Thompson elsewhere expresses his view of such secrecy, calling it "the outrageous custom" of concealing knowledge "from the lay world in a fog of jargon, a pomposity of mannerisms." He continues that each guild closely guarded knowledge of technical methods, exhibiting "a natural selfishness aided often by the illiteracy of the artificers."[20] Gadd and Campbell Thompson assumed the existence of a craft guild and believed that the cryptographic writing found in cuneiform recipe tablets revealed craft secrecy among glassworkers.

In his 1970 edition and translation of the same recipe text Oppenheim rejects the notion that the tablet was written as a form of cryptography by and for craftsmen. He notes that the unsystematic use of substitution values for certain signs was characteristic of many first-millennium Mesopotamian literary texts, "when the scribe composes the subscription to a tablet, or writes his own name, that of his father, and his professional titles." Oppenheim suggests that the recipe for making red glass would have been "perfectly intelligible to any learned scribe" and adds that only scribes of similar backgrounds would have read and copied such a text. He concludes that the motive for such a system of writing was "a display of erudition rather than a desire for secrecy." The importance of scribal culture to an evaluation of the text is underscored by Oppenheim's suggestion, supported by a detailed linguistic and historical analysis, that the scribe signed it with the name of a famous predecessor who had lived centuries earlier.[21] Far from representing secret transmission within a glassmaker's guild, the tablet reveals practices of writing and authorizing within a scribal tradition.

The later group of glass recipes comes from the famous library of the Assyrian king Assurbanipol at Ninevah, the capital of the Assyrian empire. These tablets contain more than five hundred lines of cuneiform text written in Akkadian. Oppenheim concludes that the extant recipes derive from two different written traditions, "which consolidated from forty to sixty recipes in specific sequences." Because of the way in which the recipes overlap, he views the manuscripts as compilations of prior compilations. He emphasizes the conservativism of the scribes: they would copy a second version of a recipe even if the differences between two renditions were trivial.[22]

The influence of scribal transmission also can be observed by examining ritual instructions. As Oppenheim notes, one recipe instructs that the foundations of a kiln be set up only in "a favorable month for a propitious day" and that images of *kûbu* (divine beings about which little is known) be placed there. Thereafter, strangers or unclean persons must not enter the building or pass in front of the images. Then one must regularly perform libation offerings before them, as well as other rituals, before using the kiln. Another recipe includes ritual instructions, such as making offerings "to the dead masters," a reference to the spirits of dead glassmakers. Yet most recipes do not contain references to rituals, an indication, in Oppenheim's view, that some of the source traditions had undergone a process of "de-ritualization."[23]

A very different kind of scribal intervention involves the assignment of credit. For example, one recipe instructs: "One mina of *zûku*-glass, fifteen shekels of [*tuzkû*-glass], ten shekels of lead (are) the ingredients for Elamite [red glass] according to (a written recipe of) *Wa*[. . .]." Oppenheim notes that the first two signs of the name Wa—— (the last part is broken off) indicate an old-fashioned name, one from the Old Babylonian or early Middle Babylonian period. He suggests that credit was being given "to the genius of the inventor of a new technique" and says that "it matters little whether the name was fictitious or not; it most likely was, and the very selection of an old-fashioned personal name may express the desire to have a recipe dated to a long past era." A second recipe contains traces of a personal name, also Middle Babylonian. Both of these texts are the only ones in the Ninevah group to require lead, utilized only in the fabrication of Elamite red glass or *dušû*-glass. Oppenheim proposes that the names suggest the wish to trace the origins of certain glassmaking traditions.[24]

With one exception, Mesopotamian technical recipes contain no evidence for secrecy. The exceptional text concerns metallurgy. Although it is fragmented, Oppenheim says that "it describes a method of producing a silver-like alloy from base metal ingredients . . . The purpose of the operation is to deceive." The recipe contains explicit instructions for secrecy: "Do not be careless (with respect to these instructions); do not [show] (the procedure) to anyone." Oppenheim points to recipes in the third-century *Leyden* and *Stockholm Papyri* (see chapter 2) that "parallel to an astonishing degree" the instructions of the cuneiform tablet. The context that encouraged metallurgical deception involved the development of coinage in Mesopotamia and the parallel craft of producing fake coins that appeared to be made of silver or other metals. Oppenheim notes that the Mesopotamian recipe provides evidence of the centuries-old craft tradition from which the later Greek recipes emerged. I would add that although direct links cannot be established, the Mesopotamian metallurgical recipes suggest not only a continuity of procedures but perhaps a long tradition of recipe writing as well.[25]

Mesopotamian clay tablets are far more durable than the papyrus used as a writing surface by the Egyptians, the Greeks, the Romans, and other Mediterranean peoples. The extent of the practice of writing down craft recipes in ancient Greece and Rome is unclear because the evidence is very sparse, either because of the perishability of papyrus or because of infrequent production. Robert Halleux suggests that most such recipes seem to have been preserved from medical and veterinary writings, but he also points to traces of recipe handbooks in the treatise on stones by the late-fourth-century B.C.E. author Theophrastus.

Although *On Stones* is theoretical and descriptive, Theophrastus seems to use the terminology of technical manuals. From the first century, Pliny the Elder attests to technical manuals on precious stones and their coloration. André Festugière identified a category of ancient technical writings as *baphika,* that is, related to coloring and dyeing. Recipes are a central feature of the first-century pharmacological text *Materia medica,* by Dioscorides. Finally, scattered recipes of various other types have been discovered on papyral fragments. The third-century *Leyden* and *Stockholm Papyri* comprise collections of prior collections, providing evidence of the prior tradition.[26]

Very little is known about the specific contexts within which recipe writings were created and disseminated. Yet Oppenheim's remark concerning Mesopotamian glass-recipe tablets, namely, that they cannot be taken simply as technical instructions from one artisan to another but must be considered "as literary creations within a complex literary tradition," is relevant for later texts as well. Poised somewhere between workshop and the scriptorium, recipes may have served as reminders of ingredients or processes for artisans who could read or have them read to them. Yet it is clear that they also were copied by scribes who were not connected to the workshop. Such scribes may have been separated by decades or even centuries from the original workshop context and may have understood little about the procedures and ingredients they recorded.[27]

The earliest examples of Latin craft recipes in the West contain evidence that they had been copied for centuries. The oldest extant collection of such recipes is an early medieval collection from the eighth century C.E. Known as the *Compositiones variae,* the collection is found in Codex Lucensis 490 in a library in Lucca, Italy. It is thought to have been compiled and translated into Latin during the eighth century from a number of earlier Greek sources, and it is closely related to the third-century *Leyden* and *Stockholm Papyri.* The *Compositiones variae* contains many words in Greek, including an entire recipe. Shirley M. Alexander points out that it is also significantly different from earlier collections in that it includes far more detail and elaboration. The collection begins with instructions for building in water and includes recipes for dyeing, mosaics, metallurgy, pigments, and the metallic decoration of manuscripts. The recipes appear to be straightforward and contain no hint of craft secrecy.[28]

A larger collection of craft instructions known as the *Mappae clavicula* includes all of the Lucca recipes and many others besides. Although the relationship between the two collections is not entirely clear, both were in existence at the beginning of the ninth century and both appear to derive from earlier collections of recipes. The Lucca manuscript has a provenance south of the Alps, whereas the *Mappae clavicula* comes from the north. The earliest known reference to the latter is a catalogue entry in the library of the Benedictine monastery of Reichenau at Lake Constance that reads, "Mappae Clavicula de efficiendo auro volumen 1" (Little key of the world on making gold, 1 volume). The manuscript to which this catalogue entry refers no longer exists, yet the description is consonant with a group of recipes on gold and imitation gold contained within the existing manuscripts.[29]

The *Mappae clavicula* contains an intriguing prologue that precedes the recipes for making gold color and imitation gold. The author, indicating his reasons for writing and his intended audience, notes that "as many admirable things are written in the books of Hermes," he has written a commentary. However, he has not written to encroach upon the sacred books, thus working hard and accomplishing nothing. Rather, avoiding such "heresy," he will disclose to those who want to understand all coloring and the work and processes used in it. He has called the collection the little key of the world *(mappae clavicula)*. He explains that "in a closed house it is impossible without the key to possess what is easily possessed by those who are in the house. Thus, without this commentary all writing consigned to the sacred books remains closed, and the sense obscure to those who read it." Moreover, in the name of the "Great God" the author vows to give his book to no one other than his son, and this only when he has "the piety and just sense . . . to conserve these things." These sentiments are strikingly similar to those found in some of the alchemical and Hermetic texts discussed in the previous chapter. In a convincing study, Robert Halleux and Paul Meyvaert demonstrate that the *Mappae clavicula* derives from Greek alchemical texts, including the *Leyden* and *Stockholm Papyri*, writings of Zosimos, and the Hermetic corpus.[30]

Three of the recipes that follow this prologue explicitly exhort secrecy. Recipe 11 concludes that if you take certain amounts of silver and "certified gold" (i.e., the imitation gold described in the recipe) and melt it, "you will find out how it behaves, a sacred and praiseworthy secret."[31] Recipe 14, on gold coloring, ends by admonishing, "Keep this as a sacred thing, a secret not to be transmitted to any one, and you will not as a prophet have given it away."[32] Finally, at the end of recipe 52, for "a seal of gold better than the real thing," appears this admonition: "To avoid being called dishonest, keep the recipe secret."[33] All of the recipes that exhort secrecy concern the fabrication of imitation or fake gold.

The *Mappae clavicula* is a compilation that comes out of a variety of sources; as Cyril Stanley Smith and John G. Hawthorne emphasize, it is a "compilation of compilations." Halleux and Meyvaert add, however, that the collection can no longer be viewed in the context of early medieval workshop practice.[34] The references to secrecy in the prologue do not refer to early medieval craft workshop secrecy but rather to alchemical secrecy and esoteric transmission that derives from late antique alchemical and Hermetic texts. It is consistent with that esoteric tradition that the author writes for his "son" and will give the book to him only after the son's character is prepared (in terms of piety and the ability to conceal information). Yet, as the prologue and the recipes that follow tell us, the content of the writings reveal physical operations or processes, not the spiritual content of the "sacred books," the revelation of which would seem to be heretical. Such physical processes are the key that will lead ultimately to the spiritual secrets of the Hermetic corpus.

We come much closer to medieval workshop traditions in a third Latin collection of recipes, *De coloribus et artibus Romanorum*, by one Eraclius. It consists of twenty-one stanzas or chapters in two books, written in metric verse and dating from the tenth century. (A third book, written in prose under the same author's name, actually dates from the twelfth century.) Eraclius's verses offer instruction on such topics as pigmentation, polishing gems, cutting crystal, and copper gilding.[35]

Two aspects of *De coloribus* are notable. First, Eraclius exhibits a self-conscious view of the great distance between his own time and the era of the ancient Romans, and he assumes the superiority of the latter. Second, he appreciates the value of actual hands-on practice and emphasizes his own practice. The prologue is addressed to "brother," which, together with the Latin of the text, points to a monastic context: "I have described brother, various flowers for your use, as I best could." Eraclius notes that he has added "flowers," that is, recipes, that relate to writing and are true to practice. He emphasizes his own practice: "I indeed write nothing to you, which I have not first tried myself." He believes that "the greatness of intellect" once possessed by the Romans has faded and that "the care of the wise senate has perished." He asks who now can investigate these arts, which the Romans discovered by themselves, "powerful by their immense intellect." Now, he answers, it is God: "He who, by his powerful virtue, holds the keys of the mind, divides the pious hearts of men among various arts."[36]

Eraclius, who was probably from Italy, refers to pigments derived from ground flowers and to pigments for writing. The brother to whom the verses are addressed is unknown, as are any biographical details concerning Eraclius himself. He is described in the title as "a very wise man," probably by a later editor of the verses. His debts to the ancients include direct quotations from Pliny's *Naturalis historia* concerning stones.[37] Yet he emphasizes his own practice. He suggests further that in his own day knowledge of the arts comes not from human discoveries (as he assumes was the case for the ancient Romans) but as gifts from God.

The craft recipes collected by Eraclius and the *Mappae clavicula* are similar in some ways, and both emerge from a monastic context. There is at least one significant difference between the two collections, however. The *Mappae clavicula* is a product of the monastic scriptoria. It was copied over and over by monks who worked generations and even centuries after the initial (and historically unretrievable) link between scribe and workshop was broken. Part of the *Mappae clavicula* is derived from the *Stockholm* and *Leyden Papyri*, which themselves were made up from collections of copies probably by scribes in an Egyptian temple. We can imagine a kind of "paper trail" from clay to papyrus to parchment and from the Mesopotamian palace to the Egyptian temple to the Christian monastery in the West. Eraclius undoubtedly also composed or compiled his work, which gives recipes for pigments and inks, in a monastic scriptorium, and he also utilizes some ancient writings and traditional recipes. Yet his collection seems to consist of recipes used in his own practice of, perhaps, manuscript illumination. The immediate reciprocity between scriptorium and workshop that is evident in his collection separates it from those that preceded it.

The context of a monastic workshop is also apparent in *De diversis artibus*, a twelfth-century treatise on painting, glass, and metalwork by one Theophilus, "lover of God." Theophilus was a Benedictine monk and metalworker, probably Roger of Helmarshausen. Three highly decorated metal objects by Roger have been identified—a jewel-studded book cover in Nuremberg and two portable altars in the cathedral treasury of Paderborn. Cyril Stanley Smith, a metallurgist and an editor of Theophilus's work, underscores that "Theophilus unmistakably wrote directly from his own experience in Book III [on metalwork] and from intimate observation and critical inquiry of his fellow artisans in the first two books." He wrote, Smith continues, "for the express purpose of disseminating information and instructing youth." There is no doubt "that Theophilus was himself a practicing worker in metal." Many of his descriptions "convey vividly the feel of the material and the appearance of the activity of the workshop."[38]

The presence of the artisanal workshop is striking, as is the context of Benedictine monasticism, which influenced the author's view of handwork and of authorship itself. Theophilus, who writes anonymously, describes himself as "humble priest, servant of the servants of God, unworthy of the name and profession of monk." In accord with a long Benedictine tradition, he saw a positive value in work and its role in disciplining the human soul. John Van Engen cogently argues that Theophilus in his own way also utilized more recent Benedictine theological ideas derived from the Benedictine Rupert of Deutz and others. This theology emphasized that man was made in the image and likeness of God and that the resultant dignity was reflected in human abilities in the arts. Probably in response to recent attacks on Benedictine luxury by Bernard of Clairvaux, both Rupert and Theophilus defended the Benedictines' fabrication and use of fine religious objects and garments.[39]

Theophilus addresses his book to those "who are willing to avoid and spurn idleness and the shiftlessness of the mind" by occupying their hands and contemplating new things. He wishes heavenly rewards for them. Describing God's creation of Adam and Eve and their fall, he emphasizes that although they lost their immortality, they transmitted to posterity their knowledge and intelligence (a consequence of their creation in the image of God). As a result, "whoever will contribute both care and concern is able to attain a capacity for all arts and skills, as if by hereditary right."[40]

In each of three prefaces, Theophilus stresses the intrinsic relationship between the craftsman's virtue and his handwork. In the second book, on glasswork, he emphasizes how delightful it is "to give one's attention to the practice of the various useful arts." In contrast, those who are idle and irresponsible also indulge "in empty chatter and scurrility, inquisitiveness, drinking, orgies, brawls, fighting, murder, fornication, theft, sacrilege, perjury and other things of this kind which are repugnant in the sight of God." Citing Paul on the value of working with the hands (Eph. 4:28), Theophilus emphasizes his own careful experimentation and testing of methods of painting on glass and other aspects of glasswork. Addressing the reader ("dearest brother"), he assures him that his own knowledge of these matters is sufficient and that "without envy" he has clearly set them forth for study.[41]

In his third book, on metalwork, Theophilus again stresses the intrinsic closeness of virtue and handwork. The book includes detailed instructions concerning how to make religious objects such as chalices and censers. Pointing to Moses's instructions for building a tabernacle (Ex. 31:3–5) and to the temple of Solomon, he stipulates that the spirit of God, a sevenfold spirit, should fill the craftsman's heart as he decorates the house of God; thereby he will create a likeness of paradise.[42]

Theophilus catalogues for the craftsman the seven gifts of the spirit and their accompanying virtues. Through the "spirit of wisdom" you know that "all created things proceed from God," without whom nothing exists. Through the "spirit of understanding" you have received "the capacity for skill—the order, variety and measure" with which you pursue various kinds of work. Through the "spirit of counsel" you do not bury your God-given talent but "by openly working and teaching in all humility, you display it faithfully to those wishing to understand." Through the "spirit of fortitude" you drive away sloth, and whatever you attempt, you vigorously complete. Through the "spirit of knowledge" you are the master of your skill. With the confidence of a full mind, you use "that abundance for the public good." Through the "spirit of godliness" you regulate the work "with pious care" to prevent "the vice of avarice or cupidity." Through the "spirit of fear of the Lord" you realize that you do nothing yourself and have nothing except as a gift of God. By acknowledging this and giving thanks you ascribe whatever you have or are to divine compassion.[43]

The Benedictine ideals of Theophilus are evident both in the humility that motivates him to write anonymously and in his explicit affirmation of openness. He exhorts faithful artisans not to neglect their inheritance of technical skill and knowledge but to work hard to acquire it, not for personal glory, but in thanks to God: "nor let him conceal what has been given in the cloak of envy, or hide it in the closet of a grasping heart." Rather, without boasting, let him "with a joyful heart and with simplicity dispense to all who seek." Theophilus offers himself as an example of the way to proceed: "I, an unworthy and frail mortal of little consequence, freely offer to all, who wish to learn with humility, what has freely been given me by the Divine condescension, which gives to all in abundance and holds it against no man."[44] Theophilus openly and in humility communicates his knowledge and God-given craft skill for the benefit of all.

By fusing Christian piety with an appreciation of handwork Theophilus provides dramatic evidence for a positive view of artisanal crafts. Lynn White Jr., George Ovitt, and Elspeth Whitney, among others, point to the medieval rise in the status of work in general and of handwork in particular, whereas recently Birgit van Den Hoven has suggested that the difference between medieval attitudes and those in the ancient world was not as radical as these scholars suggest. Whether or not Theophilus's views are representative of his age, his view of the Christian calling of the artisan is unambiguously positive. John Van Engen enlarges our understanding of the more immediate theological context of Theophilus's work, tying his views to writings and debates of the mid- or late 1120s. Van Engen notes that Theophilus "stood at the end of a long tradition." In the decades after he wrote "lay professionals" took over craftsmen's arts, so that the monastic artisan came to be very much the exception rather than the rule.[45]

Theophilus's essential context involves his recognition of the rising merchant culture that was developing in urban centers around him. The defense of Benedictine luxury, in which he took part, was an aspect of this same prosperous culture. I suggest that Theophilus's stated ideal of openness, although tied to theological issues, may also have been a negative response to the rise of the urban artisan and to the practice of craft secrecy that developed in tandem with the expansion of medieval urban craft production. This can only remain a speculation.

There are few sources for the early development of the medieval cities, for the initial establishment of merchant and artisanal guilds that arose within those cities, or for the craft secrecy that is documented as habitual in guild records of the late medieval period. Most existing documentation, especially for the guilds and for craft secrecy, comes from the thirteenth century and later. Yet we know that in the 1120s, when Theophilus wrote his book, many cities were flourishing, as were urban merchant and artisanal activities. It is reasonable to suppose that craft secrecy and proprietary attitudes so abundantly evident in later documents were already taking root and that Theophilus, the artisan-monk, was reacting against them.

Craft Secrecy and Proprietary Attitudes toward Intangible Craft Knowledge

The development of urbanism in the eleventh and twelfth centuries brought with it the rise of the merchant class and the great expansion of artisanal crafts and trades. A great variety of merchant and craft guilds flourished in these urban centers during the thirteenth and fourteenth centuries, as well as later. The relationship between town and guilds differed from town to town. The particular crafts that were organized into guilds also differed from one location to another. Steven Epstein emphasizes the obscure origin of the guilds and suggests that in some cases the towns may have created guilds, whereas in others they arose independently. Some communes controlled the guilds; others were controlled by them. Gervase Rosser argues that too exclusive a reliance on the statutes of formal craft organizations has led to an impression of coherence that was not there. Rosser emphasizes the many different organizations that flourished alongside the craft guilds in the medieval communes. Workers of many types were members of guilds but also of other fraternities and clubs; membership helped them to gain the needed reputation, credit, credibility, and patronage that was necessary for them to succeed in the fluid work environment of the medieval cities.[46]

Medieval guilds and the confraternities associated with them served social needs and carried out charitable, religious, and funerary functions; they also participated in political life in a variety of ways. Nevertheless, medieval guilds, in contrast to their ancient counterparts, functioned primarily in an economic sphere that involved craft production and the regulation of trade. Craft regulations were often aimed at ensuring quality control of products. Regulations regarding work hours and days, training and apprenticeship, the steps from apprentice to journeymen to master, and who could ply the trade (usually ensuring a monopoly on the production and sale of craft products) all were usual provisions of medieval guild statutes. The variability of the guilds from one urban center to the next and their diverse relationships with the towns and one another should not obscure the centrality of their economic functions.[47]

Variations among the guilds were determined by the nature of the craft or crafts included within the guilds and by the relationship of the guilds to the polity. Much of what we know about the guilds comes from guild regulations or capitularies recorded in various urban centers beginning in the mid-thirteenth century, as well as from other archival materials. Some crafts were not organized into guilds, whereas some communes (most importantly Nuremberg, in southern Germany) prohibited guilds but incorporated craft regulation into communal law.[48]

Medieval urbanism, the rise of merchant culture, the expansion of artisanal trades, and the development of merchant and artisanal guilds—all intrinsically related phenomena—provide the context for the development of proprietary attitudes toward craft knowledge. In the medieval urban context both knowledge of craft processes and mechanical inventions came to be considered intangible property separate from craft products and from the labor required to produce them. Such proprietary attitudes are manifest in two separate phenomena, the burgeoning of craft secrecy to protect craft knowledge from theft and the development of the privilege or patent as a limited monopoly on inventions and craft processes. Proprietary attitudes are far from being universal attributes of human culture in general or of craft production in particular. I suggest rather that craft secrecy and patents, both manifestations of proprietary attitudes toward craft knowledge, developed rapidly in the specific historical context of medieval urbanism.

Proprietary attitudes can be demonstrated with the example of glassmakers of Venice, famous for making the finest glass in Europe and also for the secrecy of their techniques. Venetian guilds, as Richard Mackenney notes, "fulfilled a dual function as administrative units for the government, and as associations of craftsmen." Like other Venetian guilds, that of the glassmakers, or *fieroli*, maintained a corporate organization that included the guild itself, the *arte*, and an associated religious confraternity, the *scuola*. The guild was governed by the Giustizia Vecchia, a magistracy chosen by and representative of the commune of Venice. Direct oversight was provided by a *gastaldus* and other officers, *judices*, who were elected by guild members but beholden to the Giustizia Vecchia and to the regulations approved by it. Thus Venetian glassmakers, like other Venetian artisans, were controlled by the commune.[49]

The capitulary of the glassmakers concerns governance of the guild members and the regulation of the guild itself. Glassworkers included furnace owners *(padroni di fornace)*, masters, and workers, of whom some were apprentices and others were low-level laborers with no chance of guild membership.[50] Glassmaking in Venice was restricted to guild members who had taken the oath. They were required to join the scuola of the guild and to hear the regulations read once a year.[51] Guild regulations prescribed such matters as legal work days for glassmaking,[52] instructions for the election of guild officials,[53] judicial procedures,[54] rules governing apprenticeships, and the relationship of masters and owners.[55] Handling stolen goods and selling defective or non-Venetian glass products were forbidden.[56] Only a few regulations dealt with the technical processes of glassmaking, although the type of wood to be used in the furnaces was specified, as was the number of openings, or "mouths" *(boche)*, allowed in the main furnace, both of which affected the quality of the glass.[57]

Glassmaking was a complex process that required three furnaces (sometimes combined into one or two structures). The clarity, strength, and quality of the final product depended on the nature, purity, and mixture of the original ingredients, on a detailed knowledge of the proportions and ways of combining them, and on the effect of furnace temperatures on the materials at various stages of the process. Requisite knowledge of coloration formulas and techniques and of glassblowing added to the specialized nature of the craft. Venetian glass was superior to the glasses of northern Europe in part because the Venetians had access to superior materials, including Syrian soda ash (sodium carbonate), made by burning seaweed that provided ballast for Venetian ships from the East. Soda ash produced a clearer glass than the potash (potassium carbonate, obtained from burning hardwoods) used by northern glassmakers. Yet the skill and knowledge of Venetian glassmakers were also intrinsic to their success. And as the fame of Venetian glassware increased, the specialized knowledge of Venetian glassmakers acquired ever greater value.[58]

The Venetian senate considered the craft knowledge of the glassmakers communal property, to be used for the benefit of Venice and the guild. By means of lucrative sales Venetian glass products spread across Europe, but export of the craft itself, that is, of information concerning the craft processes and the practice of the craft, was strictly forbidden. Glassworkers were prohibited from plying their craft outside of Venice. Guild capitularies of 1271 specify that "anyone of the aforementioned art who will have gone outside Venice for the occasion of practicing the said art" would pay a fine and that the "gastaldus must not accept the oath of guild membership from men who will have gone out beyond Venice with the reason [*causa*] of this art without the permission of the justices."[59] In 1281 the regulation was repeated, with higher fines levied as penalties.[60] In 1295 the gastaldus and other officials of the guild petitioned the council to punish glassworkers who worked outside by banishing them from the guild and preventing them from working in Venice for a prescribed time. They complained that the small fine was ineffective and that in just a short time furnaces had appeared in Vicenza, Padua, Mantua, Ferrara, Ravenna, Ancona, and Bononia. The petition also requested that documents necessary for exporting glassworking materials (e.g., sand and alum) be withheld from such men, that these men "may be forsaken and moreover may be disparaged," and that guild members be required to inform officials about members that they knew of or discovered working outside the city. The petition was granted.[61]

Yet effective control was elusive. Guild capitularies from 1303 note that the law was shouted out in San Marco and the Rialto: any glassworkers who had worked outside of Venice had to come before the Giustizia Vecchia to make their excuses and might be banned perpetually from the guild.[62] In 1313 it was recorded that a glassworker named Petrus Caldera had indeed been banned from the guild for working outside the city and that he could only be restored to grace after paying a fine of two hundred libras and promising not to repeat the infraction. His two glassmaker sons-in-law posted the deposit for him, and it was noted that if he worked outside again he must pay the fine again and that they also would be fined. The strain on family relationships must have been great, for in 1315 one son-in-law, Donatus, renounced his security pledge.[63] In the same year the Giustizia Vecchia tried another tactic. It had a crier shout in the Rialto and San Marco that all glassworkers working outside of Venice must return within two months and could do so without penalty; however, those who stayed away would be penalized.[64]

These Venetian efforts to prevent glassworker mobility do not represent an attempt to keep glassworkers in Venice for the sake of their labor per se. Rather, Venice and the fieroli wanted to keep knowledge of the fabrication of Venetian glass in Venice. Glassworkers who went to, say, Ferrara to make glass set up furnaces there and trained apprentices. The valuable commodity being transported was *knowledge* of the process of Venetian glassmaking. Venetian glass was superior to other glasses. The Venetian state and Venetian glassworkers did not want to compete in the market with what they considered their own wares manufactured outside Venice. They viewed the knowledge of the process of making Venetian glass as their own property and instituted punishments for those who took it elsewhere.

Proprietary attitudes toward craft processes also are evidenced by Venetian glassworking families in competition with one another. The career of one fifteenth-century glassmaker, Giorgio Ballarin, illustrates the importance of the craft secrecy that protected the intangible property of craft knowledge. The fifteenth-century account was written by a monk, Gian Antonio, as part of an elegy for his master, Paolo de Pergola. Gian Antonio claims that Paolo was "the inventor and author of various colors mixed in glass" and that he had taught the recipes to Angelo Barovier, one of his pupils. (Angelo Barovier was from a well-established glassmaking family and is famous as the inventor of crystal.) According to Gian Antonio, Angelo guarded the recipes carefully and entrusted them to only one of his own children, his exceptionally pious and chaste daughter, Marietta. One day Marietta inadvertently left the recipes out, and Giorgio quickly transcribed them. Giorgio was not a Venetian by birth, and thus the law forbade him to set up his own glassmaking furnace. However, he subsequently gave the recipes to a rival glassmaker who had an eligible daughter, married the daughter, and was thus able to acquire a furnace of his own. He became one of the leading glassmakers of Murano and served as an officer in the guild.[65] Although all the details of this story cannot be verified by other sources, it reflects the well-documented practice of secrecy within the Venetian glassmaking culture.

The story demonstrates competition among the glassmakers themselves. Some aspects of the glassmaking craft were kept secret within particular Venetian glassmaking families and could provide significant competitive advantages. The Barovier family apparently concealed craft recipes in order to produce unique glassware that evidently possessed considerable commercial value. The glassmaking recipes represent intangible property—knowledge of craft processes—that was protected by secrecy. The Venetian state, the glassmaker's guild, individual glassmakers disappearing over the border, and those who stayed at home all recognized that knowledge of glassmaking processes was a precious commodity.

The city-state of Venice and individual Venetian glassmakers were significant players in the development of the concept of craft knowledge as intangible property. Yet I am not making an exclusive claim for them. Many urban centers and the guilds associated with them claimed proprietary rights over certain crafts and industries and over the processes, techniques, and recipes that constituted the knowledge base of particular crafts.

Such claims for the intangible property of the craft—what in a later era would be called "intellectual property"—are particularly evident in highly technical, complex crafts. Yet this is an area in which more research is needed. The requirement of craft secrecy involving the protection of intangible property of craft knowledge would have been just one part of a town's, guild's, or artisan's attempt to monopolize the craft, and it would have been just one of a number of other goals involving, for instance, quality control over craft products and organizational, political, and economic regulation. Given the diversity of medieval cities, guilds, and crafts, a uniform picture should not be anticipated. Yet it is clear that within medieval cities the attitude developed that craft processes constituted intangible property with commercial value subject to conditions of ownership.

In addition to craft secrecy, another sign of proprietary attitudes toward craft knowledge was the emergence of patents for both craft processes and mechanical inventions. Privileges or patents and patent laws developed as an aspect of urban economic policies concerned with maintaining control over the crafts, maintaining the benefits of craft knowledge, and encouraging and compensating strangers who brought new techniques and processes from other localities. Evidence for patents exists from the thirteenth century. By the fifteenth century there was a steady flow of patent awards from Venice and elsewhere. Patents represented time-limited monopolies (the time varied considerably, but ten years was usual) awarded for (locally) novel craft processes or the invention of new machines and devices.[66]

Patents were awarded to individuals who possessed knowledge or inventions that were novel and seemingly beneficial to the state awarding the patent. Venetian glassmakers, for example, knew that they could leave Venice and be awarded a monopoly for making Venetian glass in other locations by obtaining a patent. Early patents were issued to Italian glassmakers in Antwerp, Holland, England, France, Germany, and Austria.[67] A developing patent system gave glassmakers good reasons for leaving Venice to ply their trade elsewhere despite the many prohibitions and penalties promulgated against such a move by the Venetian state and the guild of the fieroli.

Early patents were usually granted for the possession of craft knowledge or inventions novel to the town or region granting the patent. Original authorship or invention was not a requirement. Since the holder of the patent, even if from another region, would normally train local apprentices, the skill and knowledge embodied in the patent ultimately would transfer to local artisans and the local economy. The granting of limited monopolies enabled the state or city to possess the craft processes or inventions thus protected.[68]

Usually the possession of a novel craft process or mechanical invention rather than original authorship or invention was the key to obtaining a patent. Yet at times patents were granted as rewards for original authorship and innovation. A very early example is a law passed in 1297 by the Great Council of Venice. The law concerned the manufacture and sale of medicines, which could be sold only in shops organized as public firms that were subject to the strict supervision and control of the Giustizia. Yet the invention of new medicines was encouraged by the provision that "if any physician wishes to make any of his own medicine in secret, he may be empowered to make it, if only, of course, of the best materials, and all may hold in confidence, and all guild members may swear not to interject themselves into the abovementioned."[69] In effect the council gave the physician monopoly rights over his own medicinal invention by allowing him to protect his secret—and sell it—as long as he used high-quality ingredients.

Venetian interest in actively promoting the economy is evidenced by thirteenth-century licenses. The historian of Venetian patents Giulio Mandich pointed to a number of documents from the years 1281–96 in which the Great Council gave licenses for building various kinds of mills. One from 1281 seems to refer to a new invention involving a windmill, although the language is imprecise. In 1323 the council agreed to support one Joannes Teuthonicus, "inzenerius molendinorum" (builder of mills) as he made machines sufficient for the needs of Venice. It would support the cost of the "experientia" up to eighty ducats. Then, the doge and the city council, the Signoria, with the help of two experts, would establish a just compensation for the inventor (or importer) of the new mills "if said work is good and useful for the commune." In the same century, licenses were given for new dredging machines.[70] Such licenses indicate that Venice actively encouraged experimentation and innovation in the development of certain kinds of machines.

To encourage technological invention Venice at times granted monopolies to outsiders who were competing with local artisans. For example, in 1416 one Franciscus Petri from the island of Rhodes petitioned the Great Council for a privilege for "structures with pestles for fulling fabrics." The council granted him the patent "because they full thus perfectly, better than the fullers existing in the waters surrounding" the Rialto. For fifty years no one but Petri and his heirs was permitted to build or have built similar fulling devices, nor could any additions or reconstructions be made of any part other than what he provided. The council emphasized that "his fulling is better than that of the usual devices and different therefrom and his device is superior to the usual fulling devices and better than the same."[71] It is striking that Venice awarded a patent to a foreigner for a purportedly superior fulling device without original authorship being an issue. Venice gave Petri the patent because he possessed the new invention. Whether he invented it himself or got it from someone in Rhodes or elsewhere is unknown. This was not an isolated case; numerous patents were issued for various devices in Venice and in other cities and states as well.[72]

On 19 March 1447 the Venetian senate passed the first general patent law, distinct from individual grants of patent. The law explained that the greatness of the city had drawn men of diverse origins who possessed "very sharp ingenuity" and were "skilled in devising and discovering various ingenious devices." These men would exercise their ingenuity and discover and make things useful and beneficial to the state provided that others did not increase their own honor by taking their works. Thus, the law required that each person who invented an ingenious device that had not been made before in the Venetian domain must give notice of it to the Provveditori di Comun. No one within the Venetian domain could make a similar device for ten years. If anyone violated the law, the "author and inventor" could report it to city officials. The infringer could be required to pay one hundred ducats, and the device might immediately be destroyed. However, the Venetian government could itself take and use the device for its own needs as long as others did not.[73] This law did not embody new principles but was a codification of prior practice. It was followed by a steady stream of Venetian patent awards from the late fifteenth century through the sixteenth.[74]

Several features of the Venetian patent law are noteworthy. First is the conspicuous interest of the state in new inventions. In a provision that seems clearly unenforceable inventors were required to report their inventions to the Venetian magistery. The benefit of the state is invoked as the rationale for the law, and although a patent protected the inventor from theft by private individuals, the state itself could appropriate the invention for its own needs at any time. Novelty and original authorship as a prerequisite for a patent are circumscribed: the invention had to be novel for the Venetian domain only. There seems to have been no interest in discovering whether the claimant for a patent actually invented the device. The requirement found in most modern patent laws for both novelty and original authorship is absent from the Venetian law. Yet it is significant that authorship and ingenuity are mentioned as desirable characteristics that the law will encourage.

Venetian records provide insight into attitudes toward intangible craft property from the point of view of the guild and of the state. Venetians, particularly the Venetian glassworkers, may have been precocious in their view that craft knowledge and inventions constituted property, but they were by no means unique. During these centuries many different crafts and guilds throughout Europe protected craft secrets. The awarding of patents for inventions came to be a widespread practice. Craft secrecy and patents developed as different ways of either protecting craft property or acquiring it. Patents were usually awarded by cities or rulers to people who possessed novel inventions whether or not they themselves had invented them. In exchange, the patentee usually was required to teach the craft to apprentices from that locale. Thus, patents were a means by which cities or states acquired new technologies and new craft processes.

Both craft secrecy and the emergence of patents suggests rivalry among skilled artisans. Yet other modes of artisanal practice also emerged in the fifteenth century.

Beginning early in the century a few exceptional practitioners were able to move away from the guild context altogether. Most turned to patronage as a substitute; many, as we shall see, undertook authorship as an aid to obtaining patronage. The Florentine Filippo Brunelleschi, in certain ways a transitional figure, achieved fame and independence without writing books and without prominent patronage (although he himself was involved in Florentine political processes). Brunelleschi was exceptional in many ways, yet he was also fundamentally a product of his own society and culture, a culture that particularly valued craft knowledge and skill in the mechanical arts.

The Rise of the Architect-Engineer: Filippo Brunelleschi

Brunelleschi's reputation was based on his invention of artist's perspective and on his classically proportionate architectural designs but most of all on his design and construction of the massive dome of the Florentine cathedral, which dominates the city to this day. The dome was constructed with a two-layer shell; the herringbone pattern of the brickwork allowed it to be built without centering armature. Brunelleschi invented specialized cranes and other machines that were used in the construction. Documents confirm that Brunelleschi was paid for designing a complex hoisting machine in 1421 and that he periodically received money for expenses as well as prizes for new machines that he had invented and built. Fifteenth-century notebooks record Brunelleschian crane hoists, elevated cranes, elevated load positioners, and smaller machine parts and devices.[75]

Brunelleschi's headstrong personality and confidence in his own skill as an engineer is well documented. Trained as a goldsmith, in 1398 he applied for membership in the Arte della Seta, the silk guild, which also included the goldsmiths. He entered the competition to design bronze relief panels for the doors of the Florentine baptistery, losing to Lorenzo Ghiberti. Matriculating in the silk guild as a master in 1404, Brunelleschi also became a member of the overseer's committee for the Florentine cathedral around this time and gave advice on one of the buttresses. His model for the cupola of the cathedral was accepted in 1418 after strenuous efforts of persuasion on his part. Thereafter, his biographer Manetti writes, he was involved in every detail of the construction. Brunelleschi bitterly resented the appointment of cosupervisors for the construction of the cathedral dome, and his rivalry with one of them, his old nemesis Ghiberti, was notorious. On one occasion Brunelleschi feigned illness so that Ghiberti's "incompetence" could be demonstrated. On another, he divided the work completely so that Ghiberti's mistakes would become obvious. Brunelleschi clashed with other workers as well. In the course of a dispute with the woodworkers and stoneworker's guild he was jailed for eleven days. (The dispute concerned

Brunelleschi's refusal to join their guild when he was appointed head of the cathedral construction.)[76]

Brunelleschi's own valuation of his skills and knowledge of the mechanical arts is vividly exemplified by a petition to the Signoria of Florence on his behalf for a privilege on a large cargo boat that he had purportedly invented. In 1421 the Signoria awarded him the privilege for a ship to haul loads (undoubtedly materials for the dome construction) up the Arno River. The petition describes him as "a man of the most perspicacious intellect, both of industry, and of admirable invention." It claims that the newly invented ship could haul loads more cheaply and that it would provide further benefits to merchants and others. However, Brunelleschi "refuses to make such machine available to the public in order that the fruit of his genius and skill may not be reaped by another without his will and consent." If he could enjoy "some prerogative" concerning his invention, he "would open up what he is hiding and would disclose it to all." A patent would allow the matter to be brought to light for the benefit of Brunelleschi and everyone else. Further, Brunelleschi would be motivated "to higher pursuits, and would ascend to more subtle investigations."[77]

The petition was successful. The Signoria granted a three-year patent specifying that no one except Brunelleschi could have or use on Florentine waters any new ship or machine designed for transporting goods on water. Only familiar and usual ships could be used by others. Any "new or newly shaped machine would be burned." In addition, all new ships must be made either by Brunelleschi or "with his will and consent." All goods transported on the newly invented ship would be free of levies and taxes not previously imposed for three years.[78]

This patent, which was phenomenally advantageous for Brunelleschi, never brought results because Brunelleschi's cargo boat, whose technical details are unknown, was a fiasco. The *baldone*, as it was called, had not yet been built when the three-year term of the patent expired. One member of the Opera of the cathedral, the committee supervising the construction, was Giovanni Acquettini, a professor from Prato who taught at the Florentine *studio*, or university. Acquettini wrote an invective sonnet against Brunelleschi and his ship:

> O you deep fountain, pit of ignorance,
> You miserable beast and imbecile,
> Who thinks uncertain things can be made visible:
> There is no substance to your alchemy.

Acquettini scorns Brunelleschi's ship:

> So if the Baldalon, your water bird,
> Were ever finished—which can never be—

> I would no longer read on Dante at school
> But would finish my existence with my hand.
> For surely you are mad. . . .

Acquettini favors letters (reading Dante in school) above handwork. In his reply, Brunelleschi sharply defends the artisan against foolish scholars:

> For wise men nothing that exists
> Remains unseen; they do not share
> The idle dreams of would-be scholars.
> Only the artist, not the fool
> Discovers that which nature hides.[79]

He praises the special knowledge of the artisan, whose activities of construction allow him to understand visible things—nature itself—in contrast to the phantasms of the scholar.

Brunelleschi's confidence in his own technical expertise is vividly revealed in a conversation recorded by the Sienese engineer Mariano Taccola. Taccola describes Brunelleschi as "a singularly honored man famous in several arts, gifted by God especially in architecture, a most learned inventor of devices in mechanics." Taccola reports that Brunelleschi was "kind enough to speak to me in Siena," when he admonished Taccola: "Do not share your inventions with many persons, share them only with men who understand and love knowledge." He explained to Taccola that if he disclosed too much about his inventions, he would give away the fruit of his "genius." Many people belittle and deny the achievements of the inventor, Brunelleschi told him, so that honorable people do not listen to him. Then after some time "these persons use the inventor's words, in speech or writing or design," boldly calling themselves the inventors of things that they previously condemned. "They take for themselves the glory that rightly belongs to the inventor." Other kinds of people who abuse inventors include "the big ingenious fellow" who hears of an invention and then calls it surprising and ridiculous, telling the inventor to go away and to conceal such inventions lest people "call him a beast."[80] As we know from Giovanni Acquettini's sonnet, Brunelleschi was speaking from experience.

In the context of giving Taccola advice on mechanical devices that control flowing waters, Brunelleschi recommends who should and should not hear about inventions: "God's gifts to us must not be divulged to envious and ignorant people who ridicule them. . . . We must not show the crowd our secrets." He suggests that decisions be made by a council, that is, "an assembly of experts and masters in mechanical art to discuss plans and the construction of the work."

What usually happens instead, he says, is that everyone, both educated people and "morons," wishes to hear about the plans. The intelligent understand it, or at least some of it, whereas "morons and inexperienced men understand nothing," not even when things are explained to them. Their ignorance moves them to anger, and although they remain ignorant, they want to seem intelligent. They then persuade morons to think as they do and to scorn intelligent men. Brunelleschi emphasizes that "blockheads and morons can do much harm in questions" about aqueducts and other problems concerning water. A council should be formed of people who know about these things, he emphasizes, while "the headstrong charlatan should be sent to war."[81]

In this conversation as it was recorded by Taccola, Brunelleschi expresses two separate concerns. First, he condemns the theft of ingenious inventions by people who ridicule novel inventions and later claim them as their own; and he advises secrecy as a remedy. His second concern is quite different. It involves his impatience with citizen committees, on which sat men who did not really understand the technical details of problems but obstructed plans proposed by others. Clearly, he is speaking from his own experience, which involved a contentious relationship with the Opera del Duomo, the committee that supervised the construction of the Florentine cathedral.

Brunelleschi possessed the confidence and the ability to spar with both scholars and guildsmen. A combination of social and cultural circumstances gave him a particularly wide berth for maneuvering in early-fifteenth-century Florence. He was from a well-placed family and enjoyed a higher social status than most artisans. His father was a respected notary whose work centered on finding and provisioning soldiers, usually mercenaries from the north, for the Florentine army.[82]

Florentine pride and the legitimacy of its ruling oligarchy came to be closely linked to the visual and constructive arts—painting, sculpture, and building construction. The relief panels on the doors of the Florentine baptistery executed by Lorenzo Ghiberti and his workshop, the sculptures of Donatello and others that stood in the public niches of guild halls and of the cathedral itself, the paintings of Masaccio, the buildings of Brunelleschi and Michelozzo, all exhibited a classicizing style and all contributed to the pride of Florence. Overshadowing the city as a whole was the great dome of the Florentine cathedral, designed and executed by Brunelleschi. When he died in 1444, he was buried in the cathedral; the epitaph under his portrait bust expressed the views of his city: "Not only this celebrated temple with its marvelous shell but also the many machines his divine genius invented can document how Filippo the architect excelled in the Daedalian art."[83]

Brunelleschi lived at a time when political power and legitimacy came to be closely connected to the constructive arts. This cultural and political fact gave him wide room to maneuver in his lifetime. It also gave him personal fame both while he lived and after his death. Christine Smith emphasizes that this fame was specifically tied to his abilities as an engineer and inventor, as one who with "divine genius" invented machines. Martin Kemp underscores the development of the notion of inventive imagination in the visual arts as a specifically fifteenth-century occurrence identified with three individuals—Filippo Brunelleschi, Leonardo da Vinci, and the Sienese architect-engineer Francesco di Giorgio.[84] The first of these, Brunelleschi, achieved remarkable feats in architecture, engineering, and other visual and mechanical arts. Yet he became famous in Florence and beyond not solely because of those achievements but because he lived in a culture in which political praxis and the mechanical arts had become far more closely associated with each other than they had been before.

The social meaning of artisanal work depended upon the status of labor and handwork within the larger society. Organizations of artisans functioned very diversely from one era to the next. Similarly, although craft-recipe collections exhibit some continuity in terms of procedures and ingredients, they were discontinuous, sometimes separated by hundreds of years. Collections of craft recipes were created and transmitted in both the scriptorium and the workshop. Where specific recipes show evidence of secrecy, the import of that secret depends on the specific circumstances of their creation.

Craft production was an activity intrinsic to past societies; its history includes the specific details of making things but extends beyond that to encompass social and cultural considerations. Generalizations that assume uniformity across historical eras, such as the one that proposes that most craft production involved craft secrecy, are untenable.

Proprietary attitudes toward craft knowledge arose in the context of medieval urbanism. Such attitudes are manifest in two related phenomena—the emergence of widespread craft secrecy and the development of patents. Both involved the view that intangible property, whether craft processes or ideas for inventions, could be protected by some means. Patents became a means not only for profiting from craft processes and inventions but also for transmitting craft knowledge as artisans migrated across Europe, bringing their craft knowledge or inventions with them and gaining limited monopolies and apprentices in localities that sought new crafts and inventions.

From another point of view, medieval commercial capitalism and urbanism brought economic and political power to a new group made up primarily of merchants and bankers. The members of this new urban class validated their power in part by reconstructing and ornamenting the cities that they ruled. They transformed many cities by means of architecture, engineering, painting, sculpture, and the other visual arts. The reputation and career of an architect such as Brunelleschi was based on the civic pride of Florence and the relationship of that pride to the arts of construction.

The cities provided new arenas for an expanded practice of engineering, architecture, and other mechanical arts. Urban centers constituted cultural spaces where new styles of painting, sculpture, and architecture developed. In addition, an increased appreciation for the technical arts encouraged the writing of treatises, not only on visual arts such as painting but on topics such as gunpowder artillery and on machines of various kinds. Authorship on the mechanical arts initially developed within the context of patronage. Such authorship, which was particularly significant from the early fifteenth century, transformed some of the arts as it also led to openness—both in practice and as an explicitly stated ideal.

Chapter 4

Authorship on the Mechanical Arts in the Last Scribal Age

WRITINGS ON THE mechanical arts expanded greatly in the fifteenth century, especially in northern and central Italy and southern Germany. Throughout the century, authors created pictorial and textual books on machines of various kinds, gunpowder artillery, fountains and pumps, hydraulic works, painting, sculpture, architecture, and fortification—an array of disciplines classified as mechanical arts. This activity of authorship emerged from within manuscript culture, either before the invention of the printing press (ca. 1450) or outside of its influence. The printing press cannot be claimed as a cause, although eventually printed books on the technical arts proliferated.[1]

Scholars investigate these books mostly in relation to specific topics, such as gunpowder artillery or architecture. Here I explore them in terms of the cultural phenomenon of authorship. My thesis is twofold. First, authorship in the mechanical arts expanded because of a changing political culture in which the legitimacy of rulership was increasingly supported by the constructive arts. A new alliance of technē and praxis led to openly purveyed treatises on the mechanical arts. Second, the expanded production of writings on the mechanical arts, beginning in the fifteenth century but increasing exponentially in the sixteenth century and beyond, significantly influenced the culture of knowledge.

In the early fifteenth century the legitimation of political and military power came to be closely associated with building construction and technical arts such as painting and sculpture. Rulers, princes, and military captains who wanted to consolidate their power achieved legitimation through the remodeling of urban space and the creation of material artifacts such as buildings, paintings, and sculpture. As Christine Smith put it, such projects were indicative of the "use of the built environment as evidence for the authority of state."[2] The architecture of the Renaissance city-state, including palaces, loggias, churches and cathedrals, and urban design itself, constituted legitimating modalities for the political authority of new urban elites. In addition, the ornamentation of this built environment—painting, sculpture, and the other decorative arts—served to display the power of rulers and reinforce their authority.

In oligarchic republics such as Florence, Venice, Nuremberg, and Augsburg banking, commercial activity, and wealth did not ensure political legitimacy, which traditionally had been conferred by kinship ties to the nobility. Nevertheless, the German emperors relied on the patrician oligarchies of the free imperial cities in their struggle against recalcitrant German nobles. In Italy, men who started their careers as *condottiere*, military captains leading mercenary armies, often achieved autonomous power as princes of cities and territories. Initially they established control through military force, a consequence of their ability to command and lead soldiers. Thereafter, like the merchant oligarchs of the republics, they sought to augment their power and authority by additional means, such as the creation of visible manifestations of power, including the reordering of urban space, the construction and decoration of great palaces and churches, and the staging of elaborate festivals.[3]

The cultural value of many mechanical arts was enhanced by their explication in treatises dedicated to patrons. No longer the result of mere craft know-how, the built environment came to be seen as the material manifestation of rational and mathematical principles appropriately elaborated in books. Construction projects were carried out by machines that themselves became illustrated and discussed in codices. Similarly, the praxis of military leadership came to be closely associated with armaments and techniques, in contrast to most ancient models, in which generalship was perceived to rest on character rather than technology. Along with a new image of princes and rulers who controlled military technologies came a new technology—gunpowder artillery. Several developments, then, were intrinsically related: the legitimization of rulership; the elevation of the technical arts, in part achieved through textual and pictorial authorship; the rising status of some categories of workers involved in material production, such as architects, engineers, painters, and sculptors; and finally, the proliferation of writings on the mechanical arts.

To suggest an expansion of such authorship in the fifteenth century is not to deny prior traditions of writing in the mechanical arts, both those discussed in earlier chapters of this study and others, especially medieval Islamic writings. Yet fifteenth-century authorship represents an expansion both in the number of treatises written and in the range of topics treated. Artisan practitioners wrote some of these books; university-educated men, including physicians and learned humanists, such as Alberti, wrote others. Most authors, whether trained in a workshop or in a university lecture hall, dedicated their books to potential or actual patrons, many of whom were rulers or members of the urban patriciate. One long-term result was that certain arts, most importantly painting, sculpture, and architecture, eventually became liberal, or "fine," arts. Another consequence was that particular crafts and constructive arts, having been transformed into written, discursive disciplines, came to be treated as forms of "knowledge," characterized by rational and sometimes mathematical principles.

Both university-educated men and workshop-trained artisans wrote treatises on one or another of the diverse mechanical arts. Yet in the fifteenth century a university education and artisanal practice represented very separate domains. University education, carried out in Latin, required a facility in that language that could only be acquired through long study, either in Latin school or through private tutoring. University education was based primarily on lectures and commentaries on authoritative books, as well as disputations requiring training in rhetoric and logic and knowledge of the required texts. In contrast, workshop apprenticeships usually began at an early age (8–14 years), often after an elementary education involving vernacular reading, writing, and arithmetic. Workshop apprentices undertook orally transmitted, hands-on training.[4] The expansion of authorship on the mechanical arts in the fifteenth century shows that at least a few learned men turned their hand to technical matters, whereas some artisans took pen in hand to write treatises. Authorship on the mechanical arts began to mediate the wide gap between the cultures of learning and artisanal craft production.

Early Latin Technical Authorship

Physicians authored many of the Latin treatises on the technical arts in the fourteenth and early fifteenth centuries. Guido da Vigevano, court physician to the French queen Jeanne of Burgundy, wrote one such treatise in 1335 in an effort to help the French king conquer the Holy Land. He filled its pages with medical advice, ideas for siege machines, and illustrations of those machines. Later in the fourteenth century Giovanni de' Dondi, a physician who taught medicine at the University of Padua, constructed a famous astronomical clock, or *astrarium*, and wrote an illustrated treatise about it. In addition to marking the time, the astrarium displayed the motions of the sun, moon, and planets, feast days, nodes or points of intersection of solar and lunar orbits, and the times of Paduan sunrises and sunsets. De' Dondi's treatise describes and illustrates many of the complex parts of the instrument, especially gearing. Bert Hall remarks that these treatises "represent a more open and discursive phase" of the technical arts.

Pointing to the apparent existence of an "open audience" and to the fact that such books were patronized and collected by the wellborn, Hall suggests "a fundamental realignment of attitudes about technical matters."[5]

Physician authorship on the mechanical arts continued into the early fifteenth century. Especially notable are the Latin treatises of Conrad Kyeser and Giovanni Fontana. Kyeser, from Eichstätt in Bavaria, completed *Bellifortis* (Strong war) about 1405. Fontana, from the Veneto, wrote his partially encrypted *Bellicorum instrumentorum liber* (Book on the instruments of war) about 1420. Both Kyeser and Fontana worked as military physicians, the former in the imperial army and the latter in the Venetian army. Physicians were perhaps prime candidates among the learned for writing treatises on the mechanical arts, since medical practice involved them in significant areas of practical application involving the treatment of patients. Military service led to a technical practice in the surgery of wounds (perhaps abrogating the traditional separation of physicians and surgeons) and in addition offered plenty of opportunity to those so inclined to think about military technologies.[6]

Kyeser wrote *Bellifortis* after being exiled from Prague to his hometown of Eichstätt. A victim of imperial warfare, he suffered exile when Sigismund, the king of Hungary, conquered Prague and imprisoned his half brother, Wenceslas, king of Bohemia, and exiled Wenceslas's retainers. Kyeser, who probably was one of them and who hated Sigismund, had been present at Nicopolis on the "impious day" in 1396 when the Ottoman Turks defeated Sigismund's imperial army. He described the rout "with enormous bitterness of spirit." He had witnessed at a distance "the unexpected flight with unheard-of impudence of the Prince Sigismund, hermaphroditic king of Hungary, and of his lords and subjects." Later, Kyeser calls Sigismund "a deserter and madman, deceitful good-for-nothing."[7]

Kyeser wrote *Bellifortis* in bitter response to his exile, dedicating it to the Holy Roman Emperor, the weak Ruprecht III (ruled 1400–1410). The treatise, in Latin verse, contains magnificent illustrations of weaponry and warfare created by German illuminators who passed through Eichstätt after their own ouster from the scriptorium in Prague. Kyeser believed that effective warfare was based solidly on technique and technology in the broadest sense, whether it involved astrology, sorcery, incendiary devices, or siege machines. The status of this technologically based art of war was enhanced by its association with Latin letters and learning and by its frequent allusions to antiquity. The treatise depicts and discusses cannon, rockets, chariots, trebuchets, battering rams, mobile bridges, ships, mills, scaling ladders, incendiary devices, crossbows, and instruments of torture—weapons, machines, and devices, both old and new. The large, expensive format, the striking illustrations, and the number and variety of devices and machines depicted all suggest that the emperor Ruprecht was also representative of the intended audience. Kyeser wrote a prince's book, not an engineer's.[8]

In his dedication to the emperor, Kyeser emphasizes the alliance of technical skill and learning. Praising the robust soldiers of Germany, he adds: "Just as the sky shines with stars, Germany shines forth with liberal disciplines, is embellished with mechanics, and adorned with diverse arts." Kyeser reinforces the impression of learning with references to the ancients, particularly Alexander the Great. He depicts Alexander as a young nobleman whose conquests involved technological expertise. In one illustration Alexander is shown standing with a rocketlike weapon inscribed with the mysterious letters MEUFATON; the accompanying verses explain the efficacy of the weapon against all enemies (Fig. 1). Elsewhere Kyeser describes Alexander as the inventor of a huge war carriage (Fig. 2). Alexander is portrayed as a military leader heavily armed with magical, mechanical, and incendiary devices who not only handled weapons with his own hands but actually invented them.[9] Kyeser's portrayal of Alexander's technological expertise stands in striking contrast to the ancient separation of technē from praxis.

Kyeser concludes his treatise with a remarkably bizarre self-presentation. He reveals himself to be a dying man and provides his own brooding portrait, described by Lynn White Jr. as "the first realistic portrait of an author since antiquity" (Fig. 3). He includes his own funeral ode and an epitaph, including a prayer for his own soul: "May my soul be joined to your very high one." He writes a verse for each of the arts that he claims have informed his treatise—the seven liberal arts, namely, grammar, logic, rhetoric, arithmetic, geometry, music, and astronomy, as well as geomancy, theology, philosophy, law, canon law, physics, alchemy, the theurgical arts, and finally the military arts.[10] He emphatically joins traditional learning to the technical arts, including magical ones, which he views as effecting change in the material world.

In his epitaph Kyeser leaves a blank (never to be filled in) for the date of his death. He lists the great rulers and princes with whom he has been associated—Wenceslas, Sigismund, Francis of Carrara in Padua, and others—and the places he has been, by his account practically all of Europe, including Norway, Sweden, Russia, Poland, France, and many parts of Italy. He depicts himself as famous and loved by all, an expert on military matters whose *Bellifortis* will scatter armies. He will be mourned by all—nobles, the wealthy, the poor, even animals (quadrupeds, tripeds, birds, even vermin).[11]

While nothing can compare to this image of worms and such grieving for their departed hero, the career of Giovanni Fontana was in some ways parallel to

Fig. 1. Alexander of Macedon holding a rocket, from Conrad Kyeser's *Bellifortis*, fol. 11b, 2 Cod. ms. philos. 63 cim., Universitätsbibliothek, Göttingen. Alexander's powerful rocket is inscribed with the mysterious letters MEUFATON. Kyeser ties Alexander's power to both weaponry and magic.
Courtesy Niedersächsische Staats- und Universitätsbibliothek, Göttingen.

Fig. 2. A giant tank invented by Alexander of Macedon, from Conrad Kyeser's *Bellifortis*, fol. 16a (accompanying text on 15b), 2 Cod. ms. philos. 63 cim., Universitätsbibliothek, Gottingen. Kyeser reports that this gigantic troop carrier was invented by Alexander the Great. The soldiers entered by the door in front. The vehicle's defenses included a combination of scythes and cannon placed in alternating positions between the wheels, which we are told were hidden. Thus could the army safely approach the enemy and destroy it. The Alexander tradition here entered a new phase as the Macedonian king became the inventor of large-scale military hardware.
Courtesy Niedersachsische Staats- und Universitätsbibliothek, Göttingen.

Fig. 3. Portrait of Conrad Kyeser, from his *Bellifortis*, fol. 139a, 2 Cod. ms. philos 63 cim , Universitatsbibliothek, Gottingen. Kyeser wears the green of physicians and those born under the sign of Jupiter (including Alexander of Macedon). He presents his *Bellifortis*, with its many illustrated technical devices and allusions to magical powers, as a new light that will vanquish the unfaithful. In contrast, his portrait shows his deeply scarred face, presumably in battle, and an agonized expression. The portrait is placed above his own self-composed epitaph and depicts a man about to die. The blank space that he left for his date of death was never filled in.
Courtesy Niedersächsische Staats- und Universitätsbibliothek, Göttingen.

Kyeser's. Fontana attended the University of Padua and probably spent most of his professional career as a physician to the Venetian army. His writings include treatises on clocks, measuring devices, instruments, and machines, including war machines. He drew inspiration from the Hellenistic mechanical writings of Philo of Byzantium and Hero of Alexandria and from Arabic mechanical treatises, including a lost work on fountains by the tenth-century Arabic writer Al-kindi. His early tracts on clocks and measurement, dating from his years at the University of Padua, shortly before 1420, were influenced by Giovanni de' Dondi's treatise on the astrarium.[12]

Fontana's treatise on mechanical devices, the *Bellicorum instrumentorum liber*, is somewhat mistitled in that military machines do not predominate. The manuscript consists of 70 folio sheets and close to 140 drawings of numerous kinds of machines and devices: fountains and pumps, magic lanterns, machines for lifting and transporting heavy weights, clocks, defensive devices to protect forts and castles, battering rams, incendiary devices and other offensive weapons, alchemical furnaces, measuring instruments, a mechanical devil and witch, defensive towers, locks and keys, scaling ladders, and masks, to name a few. The pictorial representations are often self-explanatory. Most sheets also contain from several lines to several paragraphs of text, which usually begins with normal alphabetic writing and then shifts to a cryptographic code.[13]

Encryption might suggest the appearance of craft secrecy within traditions of authorship on the mechanical arts. Yet the modern editors of the manuscript, Eugenio Battisti and Giuseppa Saccaro Battisti, suggest otherwise. They compare it with another of Fontana's treatises, the *Secretum de thesauro experimentorum ymaginationis hominum*, a book concerning experiments and natural philosophy, in which the title and contents are given in alphabetic Latin and the remaining text is given in the same cryptographic script that is used in the *Bellicorum instrumentorum liber.* In a cogent analysis they argue that scribes produced both of the manuscripts. They suggest that Fontana developed the idea of cryptic writing and the code itself when he wrote the *Secretum* and that he then instructed the scribes to use the same code to copy his earlier *Bellicorum instrumentorum liber.* They describe Fontana's cryptographic system as a simple, rational cipher "based on signs without letters or numbers." Fontana devised a "completely organic system of signs" that was equivalent to the conventional alphabetic system and could substitute for it. The editors conclude that Fontana's encryption responded "not so much to the criterion of secrecy as to an awareness of a conscious symbolism of signs and their systems." He was interested in "a system of significance" rather than secrecy per se.[14] Rather than protecting the technical content of the manuscript, he was interested in techniques of encryption.

Battisti and Battisti's interpretation is supported by examination of two facing pages showing a fountain and its parts (Fig. 4). Here Fontana writes in standard alphabetic form: "Of fountains, perhaps none have been invented more artfully and more durably and also by my own imagination, because I, Johannes Fontana, always have been pleased to study these things." Then he continues in cryptographic code: "All parts of the Fountain, and the complete figure of this thus are depicted clearly so that you can understand with facility."[15] As on the other pages of the treatise, the scribe began the page in alphabetic Latin and finished it in code. The irony seems to have been unintended.

In the same decade that Fontana wrote the *Bellicorum instrumentorum liber*, the Sienese notary Mariano Taccola created a machine book that he called *De ingeneis*. Taccola worked in various positions in Siena and also seems to have occupied himself with the hydraulic, military, and engineering problems of his city. In addition, he attempted to acquire the patronage of Kyeser's nemesis, Sigismund, crowned Holy Roman Emperor in 1433. Later he spent time in the Veneto. Taccola's background is unusual in that it apparently included both craft and notarial training. Craft training is suggested by his commissions to carve heads for wooden choir stalls and to produce sculptures for the Sienese cathedral. His notarial training is indicated by his nomination in 1417 for admission to the guild of judges and notaries. Taccola served as secretary of the Domus Sapientiae, a hospital and student lodging that served as a temporary residence for many foreign visitors. Some of Sigismund's entourage stayed there during the emperor's nine-month visit in 1432. (Sigismund waited in Siena while his passage to Rome, where he was to be crowned emperor, was being negotiated.) When Taccola's position at the Domus ended in 1434, he became a *stimatore*, one who tallied completed work at building sites, assigned wages, and perhaps estimated the amounts of materials needed, for Siena. In 1441 he was appointed *viaio*, superintendent of the streets, fountains, and bridges.[16]

While carrying out his various duties from the mid-1420s until 1449 Taccola created two pictorial machine books. In both works the sheets are filled with drawings of constructions and machines—cannon, trebuchets, winches, pumps, mills, gears, cranes, and ships, among many others. Most sheets also contain explanatory text. Taccola's pictorial treatises reflect the concerns and problems of the Sienese state. He illustrates ships, harbors, and defensive and offensive technologies of naval warfare, reflecting Siena's long-term efforts to control Tuscan rivers and to develop a commercial port at Talamone, on the Tuscan Sea. He depicts pumps, dams, siphons, aqueducts, and other water-controlling devices—a response to the perennial Sienese problem of water supply, which was insufficient for the needs of the population and local industries. Finally, he provides

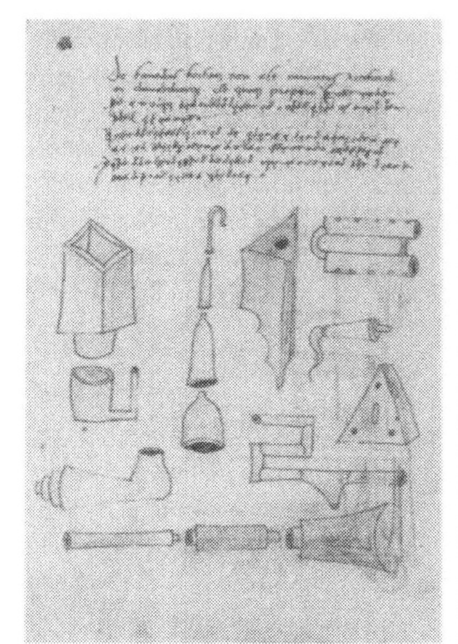

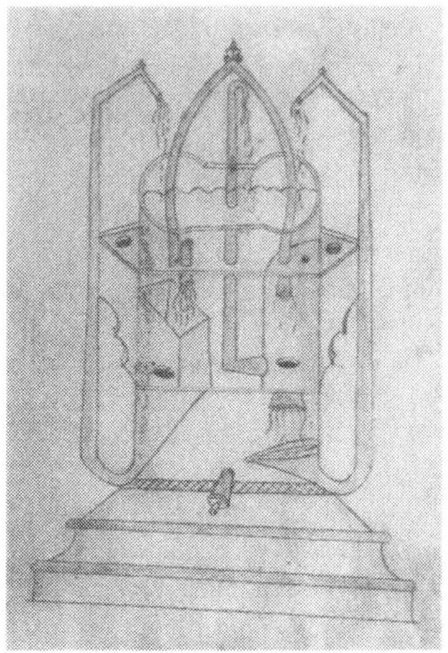

Fig. 4. Giovanni Fontana, parts of a fountain *(left)* and a transparent view of the assembled fountain *(right)*, from his *Bellicorum instrumentorum liber*, fols. 62v and 63r, Cod. icon 242, Bayerische Staatsbibliothek, Munich. The text, in alphabetic Latin, discloses Fontana's delight in fountains, whereas the cryptographic writing assures the reader that everything has been explained clearly. This example suggests that Fontana did not use encryption to hide technical information. Courtesy Bayerische Staatsbibliothek, Munich.

multifarious images and descriptions of cannon, firearms, cannon lifting and transport devices, and weaponry of various sorts. Such images suggest a response to Sienese military problems that included the frequent ravaging of the surrounding countryside by marauding mercenary companies.[17]

Taccola's authorship developed out of Sienese concerns, but he also used it in his own quest for patronage. He took advantage of Sigismund's Sienese sojourn of 1432 by presenting the emperor with books 3 and 4 of the *De ingeneis*, the presentation copy of which contains two portraits of the emperor. Desiring to obtain a position in the Hungarian court, Taccola used Saint Dorothy as his mouthpiece (Fig. 5). After addressing Sigismund in his own voice—"for you, Sigismund, most exalted prince"—he has Dorothy speak for him. Dorothy notes that Mariano has composed the book and drawn the emperor's picture. She asks that Sigismund accept him in the Hungarian court and make him "master of machines for waterworks." Mariano wants to finish his days in Hungary. He wants to attend to waterworks and to compose books describing everything that Sigismund and his predecessors have done. He desires "to compile [what is known of] every place; and in said books in margins of the pages to design and illustrate stories."[18] In his bid to go to Hungary as a master of waterworks Taccola included authorship among the services he would perform. He seems to have had in mind a book like the one he was presenting, but with Sigismund's Hungarian machines and waterworks, rather than those of Siena drawn on the pages.

Although Taccola probably never made it to Hungary, there is indirect evidence that he may have worked in the Veneto and northern Italy in the late 1430s and 1440s and that he may have written his treatise *De machinis*, completed in 1449, there as well. A manuscript of this treatise written in Taccola's own hand, Latinus Monacensis 28,000, now in the Bayerische Staatsbibliothek, in Munich, is of Venetian provenance, as are all of the copies. One copy was made for the Venetian captain-general Bartolomeo Colleoni (well known today because of Verrochio's famous equestrian statue in Venice); another made its way from the Veneto to the Ottoman empire as a gift to the sultan of Turkey. Although the extant copies do not contain dedications, Taccola's authorship again may have been influenced by renewed hope for imperial patronage. The Roman coronation of the first Hapsburg emperor, Frederich III, was being discussed in Italy in 1447. Taccola may well have intended to dedicate the *De machinis* to him.[19]

The *De machinis* is a pictorial treatise divided into ten books. The table of contents represents a rationalization of types of machines into coherent groups, such as: "Bridges and structures on water," "Bombards and guns," and "On weights raised higher and lowered down." The work contains more than two hundred full-page illustrations, many of which are accompanied by explanatory text. The introduction reflects overlapping imperial and northern Italian concerns: the struggle of the empire and the papacy against the Ottoman Turks, as well as Siena's difficulties with neighboring states, such as Florence. Taccola does not want to aid fighting among Christians: "I, Ser Marianus Taccola, also called Archimedes of great and magnificent Siena, have not designed these engines, machines and weapons that they may operate against Christians, but have invented, composed and designed them that they may go against the infidel and barbarian people."[20]

In his self-appellation, the Sienese Archimedes, Taccola invokes the authority of antiquity. Archimedes was famous for inventing ingenious catapults and

Fig. 5. Saint Dorothy recommending Mariano Taccola to the emperor Sigismund, from Mariano Taccola's *De ingeneis*, fol. 69r, Palat. 766, Biblioteca Nazionale Centrale, Florence. Saint Dorothy notifies the emperor that Taccola would like to be master of waterworks in Hungary: "I always recommend Ser Marianus Jacobi, secretary of the Domus Sapientiae in Siena who composed this little book and drew my picture. May you be pleased to accept him as one of your family and court."
Courtesy Ministero per i Beni e le Attività Culturali.

defensive weapons used to fight the Roman siege of Syracuse. Yet, in Plutarch's account the ancient Archimedes refuses to write about his inventions, regarding "the work of an engineer and every art that ministers to the needs of life as ignoble or vulgar."[21] In vivid contrast, Mariano Taccola, the Sienese Archimedes, embraces technical authorship with enthusiasm. He presents a great array of machines in both pictorial and textual explication and leaves no doubt that they are both the proper subject of rational thought and worthy of written treatment.

Taccola's treatises openly display machines, constructed objects, and mechanical devices. Yet there are occasional instances when he explicitly states that he is withholding information. Perhaps the most striking example occurs in the *De ingeneis* with reference to an underwater apparatus in the shape of a fish (Fig. 6). Under the illustration, Taccola writes:

> I know what I am doing on the swimming fish. I feed it oil from the sponge so that he who rides is carried by the fish. It has inside itself what supports it, and what is self-supporting. Nothing in the Chapters on Weapon Throwers and Artillery can be brought to perfection without me. My speech has been veiled. What I have acquired during a rather long time with labor shall not be known at once.[22]

Taccola emphasizes that the reader cannot know, without being told, the exact nature of the apparatus inside the fish that holds it up (and presumably allows it to change depth). Yet it seems unlikely that the fish actually worked as a kind of submarine or that its interior mechanism was anything more than an idea in Taccola's mind. Taccola's suggestion that there are missing elements in his presentation of artillery weapons that only he can supply is also notable. He holds back essential technological information in order to entice potential patrons to hire him.

In sum, Latin books on machines and devices reveal a fascination and even delight in technological apparatus of all kinds. Military technology is of particular but not exclusive interest. Latin authors for the most part openly explain and illustrate their material. Where there is evidence for secrecy, it suggests very particular meanings. Kyeser's placement of the mysterious letters MEUFATON on the rocket of Alexander suggests the rocket's magical powers. Fontana's cryptographic treatise reveals his interest in the techniques of encryption rather than a desire to conceal technical details. Amid a plethora of openly illustrated and explained devices Taccola occasionally alludes to secrets that, he intimates, he will reveal upon being employed by a patron. These specific instances of secrecy exist within the context of the open display of machines and devices in lavishly illustrated manuscript books.

Fig. 6. Man in fish submarine, from Mariano Taccola's *De ingeneis*, fol. 31r, Codex Latinus Monacensis 197, pt. 2, Bayerische Staatsbibliothek, Munich. Taccola tells us that we don't know all there is to know about how this fish works, that his "speech has been veiled" and that what he has taken a long time to acquire, he will not reveal quickly.
Courtesy Bayerische Staatsbibliothek, Munich.

German-language Writings on Gunpowder Artillery and Machines

Alongside the Latin books discussed above, numerous German-language treatises, primarily concerning gunpowder artillery and machines, appeared beginning in the early decades of the fifteenth century and continuing until its end. Although the authors of some of these German books are unknown, clearly many of them were practitioners in the burgeoning field of gunpowder artillery. Such practitioners included the *feuerwerker*, who produced gunpowder; gun founders, who manufactured cannon and other firearms; and gunners, or *büchsenmeistern*, who supervised such activities and in addition transported and discharged artillery. The books contain a wealth of technical information, such as how to make gunpowder and how to load cannon. They also attempt to rationalize practice by treating the causes of particular phenomena, and they discuss desirable attitudes and deportment for both practitioners and patrons.[23]

An example from the early fifteenth century, Cod. germ. 600, in the Bayerische Staatsbibliothek, in Munich, is a small manuscript of twenty-two sheets about gunpowder and gunpowder artillery. Woodblock images dominate the pages, with text at the bottom. On one sheet the author explains how to load the barrel, while another shows a gunner igniting a cannon (Fig. 7). The manuscript shows individuals carrying out technical processes and includes explanatory text. Cod. germ. 600 represents and validates a new group of practitioners. Each of twenty-eight human figures in the woodcuts is distinguishable from the others—each has a distinct physiognomy and wears his own particular clothes and hat or hairstyle. Some work in pairs, some alone. All are dressed rather plainly and clearly belong to the artisanal class. All are busy at work, with serious demeanor, carrying out the process described in the text below. The pamphlet explains technical practices and displays practitioners at their work, their individuality revealed along with their virtues of industriousness and care.[24]

The practitioner also emerges as a central figure in the *Feuerwerkbuch*, an anonymous treatise written about 1420 and widely distributed in manuscript copies before its first printing in 1529. The book treats guns, missiles, gunpowder, and fireworks, by which princes, counts, lords, knights, and cities protected themselves when besieged in a city, stronghold, or castle. The author advises princes and cities to have "servants who are godly and steady people" and who will give their own lives and property to enemies before they surrender. They should also be "wise people" who know how to bring guns and missiles in an emergency, how to attack, how to build walls and bulwarks, and how to dispatch themselves with their arms effectively. Finally, they need to get along with one another, not

Fig. 7. Loading a cannon barrel and discharging a cannon, from Cod. germ. 600, Bayerische Staatsbibliothek, Munich. In the woodcut on the left, two gunners are loading a cannon barrel. The text explains that one is hammering in wooden wedges between the stone ball and the interior surface of the barrel, while the other holds containers with other substances to be poured into the barrel—loam, oakum, or hay. On the right, a gunner ignites the powder in the cannon. The text warns of the danger that the barrel might explode. The gunner should stand back and ignite the powder with a long rod. The clamps to secure the barrel to the carriage are shown in the illustration but are not mentioned in the text. The gunner is depicted as an artisan wearing simple clothing who carefully carries out a dangerous task. Notice the serious expressions, the industrious demeanor, and the simple clothing of the artisans, all of whom are depicted as particular individuals. Courtesy Bayerische Staatsbibliothek, Munich.

quarrel, and settle their disputes according to the best advice. Princes and rulers need good smiths, bricklayers, carpenters, cobblers, and also good gunners, and they must be good masters to their gunners by giving them all the supplies and equipment that they need.[25]

The author claims to have written the book because the technical details of gunpowder manufacture are too complex to remember without the help of writing: "And thereupon since the subjects belonging to it [gunnery] are so many, which every good gunner should know, and which a master without writing cannot remember in his mind," all the necessary details are provided. The transmission of technical information is an important aspect of the treatise; it is filled with details concerning how to treat the ingredients of gunpowder (sulfur, saltpeter, and charcoal), how to mix them together, how to load the barrel, and how to discharge the cannon in differing situations and with various kinds of ammunition, such as flaming bolts and stone balls.[26]

The treatise also treats issues involving the causes of gunpowder explosions. It begins with twelve questions, reminiscent of the university exercise known as the *quaestio*. The first question asks whether it is the fire, or the force of the vapor produced by the fire, that drives the stone forth from the cannon. The author argues that it is the vapor and proposes an experiment to prove it: Take a pound of good powder, put it in a thick wine cask, and close it well so that no vapor may come out. Then ignite the powder at the hole; the powder will soon be burned, and the vapor will destroy the cask. The second question also concerns the nature of the process, rather than technique: Is it the saltpeter or the sulfur that pushes the stone? It is both. When the powder is kindled, the sulfur is hot and the saltpeter cold. Since heat will not admit cold and cold will not admit heat, they will not tolerate each other. Other questions concern technique: How much powder should be used? Should the wedges pushed against the stone ball in the barrel be made of soft wood like linden or hard wood like oak or beech? Is the gunpowder better corned or in the form of a fine powder?[27]

The treatise provides a history of the invention of gunpowder by the legendary alchemist "Niger Berchtoldus" and presents a summary of character traits and training appropriate to the gunner: "Above all he should honor and hold God before his eyes." In addition, he should exercise great care when handling guns and powder. He should "rest content with the world in which he travels" and should be a "strong, undaunted man." In war he should "hold himself pleasantly," for such people are greatly trusted. "The master should also know how to write and read" since all the elements of the art cannot be held in the head but can be read in the book at hand. He should know something about purifying and distilling and about defensive operations, and he should understand weights and measures. He should teach with honesty and friendliness in both words and works, and he should be thoughtful and avoid drunkenness at all times.[28]

Just as the author offers advice concerning the character of the gunner and what he should know, he also counsels the gunner's patrons—nobles, princes, lords, and cities. Many have found themselves besieged and have not taken proper care for defense, nor have they brought people to them "who through their technical wisdom [*Kunst Weissheit*] may bring advice and help to withstand their enemies." As a result, many god-fearing princes and lords of the Holy Roman Empire have come to destruction. Princes as well as other nobles and cities should provide themselves with sober, skillful gunners; they should give these gunners plenty of food, clothing, and accommodations and provide them with supplies such as saltpeter, sulfur, and suitable wood.[29]

The *Feuerwerkbuch* thus addresses a double readership—gunners and the princes and nobles who supported them. The author offers advice concerning how members of each group should comport themselves to their mutual advantage. Undoubtedly this anonymous author was himself a gunner seeking patronage. He offered a written text that could enhance an image of learning and technical competence for gunner and patron alike.

Manuscript treatises on weaponry and other kinds of machines and devices proliferated in the fifteenth-century German empire, particularly in Bavaria, Bohemia, and other areas of southern Germany.[30] Many are lavishly illustrated. Some are specialized. For example, HS 25,801, at the Germanisches Nationalmuseum, in Nuremberg, depicts the different ways in which gun carriages and their aiming mechanisms (such as the elevating arch and the screw) might be constructed (Fig. 8).[31] Others range much more widely. The manuscript known as the "Anonymous of the Hussite Wars" includes illustrations and textual descriptions of mills, cranes, aquatic devices, and gearing, among other mechanical devices and machines.[32] The *Mittelalterliche Hausbuch*, from about 1480, depicts cannon, weightlifting machines, mills, a spinning wheel, a wagon train, and mining activities and machines.[33]

In the last third of the fifteenth century several prominent gunners from southwestern Germany supplemented their military practice with authorship. One was Johann Formschneider, a gunner active in Nuremberg between 1460 and 1470, whose name is attached to a particular group of writings and drawings of cannon.[34] Not far from Nuremberg, in Pfalz or the Palatinate, two other prominent gunners undertook authorship. Martin Mercz served the energetic Duke Friedrich I of Siegreich, who reorganized his state administratively and expanded its territory and power. While in Friedrich's service, Mercz wrote a treatise on the

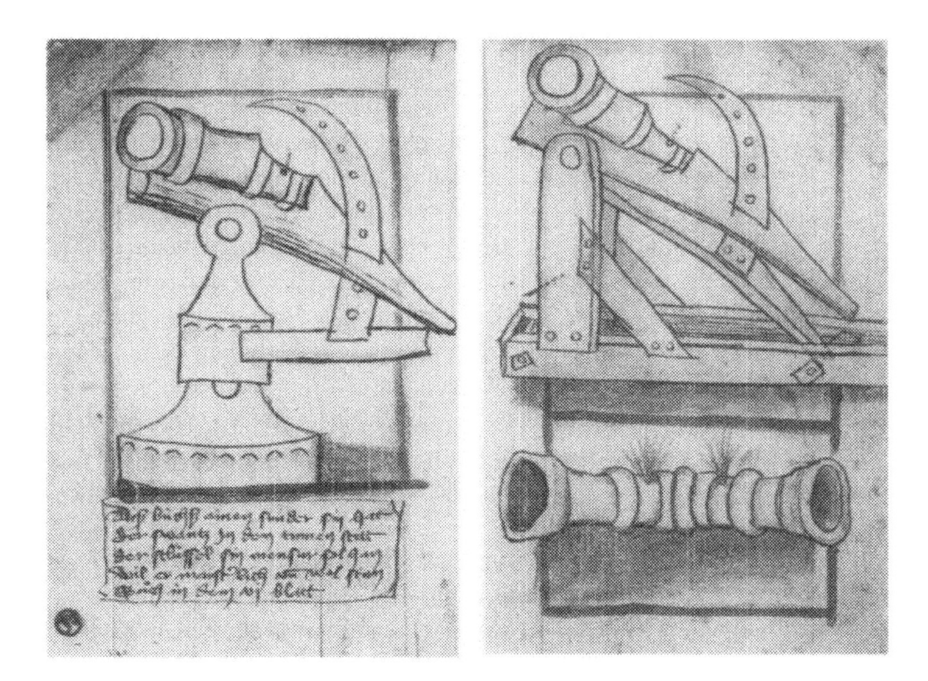

Fig 8. Two cannons mounted with elevated arches used for aiming and a double-barreled cannon, from HS 25,801, fols 2v and 3r, Germanisches Nationalmuseum, Nuremberg. These illustrations in an early-fifteenth-century military book show two different cannon with variations in the carriages and aiming arches, differences that would be very hard to explain verbally. Also illustrated is a double-barreled canon, probably to be mounted on a wall.
Courtesy Germanisches Nationalmuseum, Nuremberg.

mathematics of aiming cannon in 1471. Mercz's importance in his lifetime can be measured by his burial in the parish church of Amberg, his tomb covered by a full-length relief-sculpture portrait. In this, the oldest memorial of a European gunner, he is depicted as a somewhat portly man with one eye covered by a patch, his hands holding rosary beads in prayer, his feet resting on a cannon barrel. Inscriptions on the tomb tell that he was "famous before others in the mathematical art of gunning" and that the dukes of Pfalz always received his heart and work and were truly served by him.[35] Another Pfalz gunner, Philibs Mönch, following Mercz's example, wrote a military treatise near the end of the century. His portrait appears on the title page.[36]

The large, beautiful hand-painted *Zeugbücher* (Ordnance books), created for the emperor Maximilian, provide a final example of the German manuscript tradition of military authorship. Maximilian ordered the creation of spectacular inventory books as one part of his large-scale reorganization of the imperial army. The reform augmented the imperial arsenal by creating a new artillery-manufacturing center in the Tyrol and a new arsenal in Innsbruck. The inventory books record the content of the imperial arsenals both in images and in itemized lists. These codices are visually spectacular. One example, Cod. icon. 222, in the Bayerische Staatsbibliothek, Munich, is a large vellum codex of 296 folios. The folios display beautiful hand-painted illustrations of cannon with their stone or iron balls, culverins (light field guns), and other armaments at various sites in the empire. The illustrations were created by Bartholomeus Freysleben, a locksmith by trade who became a master gunner at Innsbruck.[37] Whatever their practical uses in inventory, such books displayed both the military power and the cultural authority of the prince.

The variety of the German codices treated here suggests a diversity of uses. Certainly the transmission of technical information from one practitioner to another was one. Also evident are the value of the artillery codex in patronage exchanges and their cultural utility as representative of the power of princes and emperors. A careful look at particular codices suggests another point. Take, for example, a codex in the New York Public Library—Spencer MS 104. The book retains its fifteenth-century binding of hand-tooled calf leather stretched over wooden boards and its iron clasp, lock, and key. Inside are careful copies of two treatises—the *Feuerwerkbuch* and a German translation of Kyeser's *Bellifortis*. The illustrations for the latter are only partially complete, and the provenance of the codex in the early modern period is unknown. Yet the two texts were bound together for a reader or collector who, the beautiful binding tells us, were far from lowborn. German writings on the mechanical arts seem to have enjoyed a readership of wellborn individuals who valued them and collected them for their libraries.

The Humanist Synthesis of Technē and Praxis: Alberti and Valturio

The close association of the mechanical arts and political praxis is particularly evident in the writings of two prominent fifteenth-century humanists, Leon Battista Alberti and Roberto Valturio. Humanism was a broadly based cultural and educational movement that favored rhetoric and ethics over logic and philosophy. In general, humanists advocated a life of active civic participation over contemplation. They attempted to reinstate classical, Ciceronian Latin in place of the "barbarisms" of medieval Latinity. Finally, they wrote histories, both ancient and contemporary, the latter often undertaken as part of their service in the courts and households of elite patrons.[38] Of the two humanists treated here, Alberti wrote prolifically on a wide range of topics, including painting, sculpture, and architecture, while Valturio composed only one treatise, *Elenchus et index rerum militarium*, which became the best-known military book of the fifteenth century. Both authors treated their topics as intrinsic to political and civic praxis.

Alberti's authorship was made possible by ecclesiastical benefices and princely patronage, including that bestowed by the Este of Ferrara, the Gonzaga of Mantua, and the Rucelli of Florence. The natural son of an exiled father, Alberti first traveled to Florence only sometime after the ban against his family was lifted in 1428. He was profoundly impressed by developments in the visual and constructive arts that had occurred there in the early decades of the century. Whatever the exact date of his initial visit, he was able to observe Brunelleschi's work on the dome of the cathedral of Florence and his classically proportionate architecture; the discovery and elaboration of painter's perspective by Brunelleschi, Masaccio, and others; Ghiberti's installment of classicizing relief panels on the doors of the Florentine baptistery; and Donatello's newly classical sculpture. Alberti, in sum, arrived in Florence just in time to see the stylistic revolution in the visual arts that was unfolding there in the 1420s and '30s, and he made his own significant contributions to that revolution, in part through his authorship.[39]

An extensive knowledge of classical Latin texts combined with an interest in practical mathematics allowed Alberti to construct a rational and mathematical as well as civic basis for disciplines such as painting, sculpture, and architecture. He conceived both painting and architecture as mathematical disciplines, but he also insisted on their practical, material, and moral foundations. Although a university-educated scholar, he took up painting and sculpture as an amateur and eventually made important contributions to architectural design. Although his exact role as an architect is tenuously documented, he was instrumental in the design of several major buildings in Ferrara, Rimini, Florence, and Mantua. He also contributed to Nicholas V's renewal of Rome in the mid-fifteenth century and perhaps helped to plan the town of Pienza, which was redesigned in the early 1460s at the behest of Aeneas Sylvius (Pope Pius II).[40]

In the 1430s Alberti produced two versions of his treatise on painting, one in Latin, dedicated to a potential patron, Giovan Francesco Gonzaga, and the other an Italian translation dedicated to Brunelleschi. Alberti insists that he is writing *De pictura* not as a mathematician but as a painter. "Mathematicians measure the shapes and forms of things in the mind alone and divorced entirely from nature." He, on the other hand, wants to speak in terms of Minerva, about visible things. The allusion is to Cicero's *De amicitia,* in which the ancient author refers to a more popular kind of knowledge.[41]

Alberti speaks as a painter, that is, as one who does not separate mathematical and theoretical concerns from the visible world. Yet he alludes frequently to ancient texts and ancient painters. He speaks not as an ordinary but as a learned painter. The *De pictura* treats the process of vision, the creation of the illusion of three-dimensional space on a flat surface by one-point perspective, and procedures of composition, light, and color. It also concerns the painter's creation of *istoria,* a dramatic effect in the painting that moves the soul of the viewer. Finally, it enumerates desirable moral attributes of the painter, of which the first is that he should "be a good man, well-versed in the liberal arts."[42]

Alberti dedicates the Italian translation of the *De pictura* to Filippo Brunelleschi, praising his great achievements in the arts, as well as those by "our great friend the sculptor Donatello," Ghiberti, Luca della Robbia, and Masaccio. He praises the achievements of these men as equal to or higher than those of the ancients. As Christine Smith has emphasized, Alberti commends Brunelleschi specifically for his "feat of engineering," namely, raising the cupola of the Florentine cathedral "without the aid of beams or elaborate wooden supports." And he asks that Brunelleschi the practitioner correct any aspect of his treatise that seems to need amendment.[43]

Alberti presented his Latin architectural treatise, *De re aedificatoria,* to Pope Nicholas V about 1452 while advising him on the renovation of Rome. John Oppel notes the usual assumption that Alberti "raised architecture to the status of one of the liberal arts" and remarks that "it would be at least as true and historically more appropriate to argue that he did just the reverse, that he brought the liberal arts down to the level of the mechanical ones." For Alberti, says Oppel, "the whole of architecture is based on mechanics."[44] Alberti did not remove design from construction but combined both in a Latin treatise that encompassed the design of buildings and projects of construction and engineering.

Alberti emphasized that the architect is not a carpenter but one "who by sure and wonderful reason and method" devises in his mind and realizes in construction whatever is needed "by the movement of weights and the joining and massing of bodies." Stressing the importance of architecture to human life and health, he includes within the discipline activities now subsumed under civil and military engineering. The architect constructs houses but also walks, swimming pools, and baths to keep men healthy, as well as vehicles, mills, timepieces, and other smaller inventions that play a vital role in everyday life. Architects have devised methods of drawing up vast quantities of water for many purposes and have produced buildings for divine worship. Further, by means of cutting through rock, tunneling through mountains, filling in valleys, restraining the waters of the sea and lakes, draining marshes, building ships, altering the course of rivers, dredging the mouths of rivers, constructing harbors and bridges, the architect has not only "met the temporary needs of man, but also opened up new gateways to all the provinces of the world," allowing nations to exchange food and goods as well as experience and knowledge. Architecture has provided ballistic engines and machines of war, fortresses, and other things that protect the liberty of "our country."[45] In contrast to ancient authors, Alberti posits the skill and ability of the architect-engineer as the most essential factor for achieving victory in most wars, more important than the command and foresight of any general.

Alberti indicates a personal interest in machines and practical problems of the mechanical arts. In 1447 he supervised the raising of an ancient ship from Lake Nemi, in central Italy, and wrote a small book about it, *De navis*, which is no longer extant. In his architectural treatise he mentions several of his own inventions that have to do with rigging the gangplanks of ships to prevent successful boarding by an enemy and devices for sinking and burning enemy ships. He also promises that he will deal with war machines at greater length elsewhere, perhaps implying that he was planning to write a treatise on the subject.[46]

Alberti stresses the material applications of architecture, but he also repeatedly emphasizes its ethical dimensions, involving intrinsic relationships between the work of the architect and civic life as a whole. He emphatically joins the mechanical art of architecture, which he transforms to a higher art by means of his treatise, and civic praxis. When you "erect a wall or portico of great elegance and adorn it with a door, columns, or roof," good citizens rejoice for their own sake because you have used your wealth "to increase greatly not only your own honor and glory, but also that of your family, your descendants, and the whole city." Alberti insists that "the security, dignity, and honor of the republic depend greatly on the architect."[47] The social hierarchies of the city and physical structures should be carefully correlated. It is the duty of patrons who commission buildings to take their own social standing into account. It is a sign of "a well-informed and judicious mind" to plan the building in accordance with both one's position in society and the requirements of use.[48] Architecture is at the center of civic society, contributing to the honor and glory of the city itself and to the social structures that exist within it.

Architecture requires empirical study, observation, and the gathering of information from the past. Alberti instructs that architects should collect, compare, and put into their own work "all the soundest and most useful advice" that learned ancestors have handed down in writing, as well as those principles that can be observed in their buildings. Architects should report things that they themselves contrive through their own invention, "by careful, painstaking investigation" that might be of some future use.[49] Architecture requires investigation of past forms, reporting of present-day inventions, and efforts toward innovation.

As both a body of knowledge and a civic praxis architecture must be openly and clearly communicated. Alberti wants "to be as limpid, clear, and expeditious as possible," for the subject has been "knotty, awkward, and for the most part thoroughly obscure." Alberti plans to explain the precise nature of architecture. He believes that "the very springs of our argument should be laid open," making the discussion that follows flow more easily. Elsewhere he laments the obscurity and poor Latinity of Vitruvius's text, stressing the difficulty of his task of writing on architecture. He would rather his speech "seemed lucid than appeared eloquent" and believes that what he has written is "in proper Latin, and in comprehensible form."[50] Openness and clear explanation are essential to a discipline that represents an accumulating body of knowledge and a kind of civic praxis.

Alberti's contemporary Roberto Valturio similarly elevated military science to the status of a learned discipline. He wrote the *Elenchus et index rei militarium*, also completed in the mid-fifteenth century, for Sigismund Malatesta of Rimini. At the age of fourteen Sigismund had seized Rimini from his uncle and his own brothers by decisive military action. Thereafter, he had obtained recognition of his status as lord of Rimini in 1433 from the emperor Sigismund, who knighted him on his way north from Rome, and in 1450 from Pope Nicholas V. Sigismund's vast building projects in Rimini exemplify the use of construction and urban redesign to legitimize political power gained by military conquest. These projects included a great fort, the Rocca Malatestiano, the ducal palace, and the redesign of the church of San Francisco. Sigismund also encouraged Valturio to write his learned treatise.[51]

Sigismund transformed the Gothic church of Rimini into the Tempio Malatestiano, a great classicized monument dedicated to his own glory. Individuals involved in the project included Roberto Valturio, who planned some of the decorative iconography; the painter Piero della Francesca, who painted a fresco depicting Sigismund Malatesta kneeling before a Saint Sigismund, who bears a striking resemblance to the emperor of the same name; and Alberti, who was responsible for the design of the facade (which suggests a Roman triumphal arch) and for the new exterior side walls. Evidence suggests that Sigismund himself was very involved in the Tempio's planning and construction.[52] Although the building was never finished, it stands as a monumental tribute to the glory of Sigismund's rule and also displays his symbolic connection to Roman antiquity.

The lord of Rimini encouraged authorship on the military arts with similar aims in mind. In *Elenchus et index rerum militarium*, completed between 1455 and 1460, Valturio cites hundreds of ancient authorities but also emphasizes the technological aspects of warfare. His book gained great popularity. It exists in at least twenty-two manuscript copies and was later published in numerous printed editions (the first published in Verona in 1472) and in many translations.[53]

Valturio's humanist education in both Greek and Latin is evident in his treatise, which demonstrates an extensive knowledge of ancient military literature. The manuscript copies also include numerous beautiful hand-painted illustrations of military machines. Valturio cites hundreds of ancient authorities but also emphasizes the technological aspect of warfare.[54] He is interested, not in the latest weaponry or military technology, but in weapons of all kinds, both ancient and modern. Yet weaponry holds a central place in his treatise, unlike in most ancient writings, which separated military engineering from strategy and tactics.

Valturio displays what might be identified as extreme anxiety concerning his own authorship. In his dedication to Sigismund he seems to ward off criticism with a vitriolic attack against critics. They include those who think nothing is done properly; those who, "like pigs grunting in the mud," aspire to glory by their own effort of letters and ingenuity and, having received little help, suppose that no one needs help; and those who, wanting fame for themselves, "just like drone bees and very perverse sycophants," must detract from everyone and carp at every single thing. Others may criticize Valturio because he lacks actual military experience. They will say he has gleaming skin and "soft body because of the shade, having never received a wound." How does he nevertheless dare to give advice about military matters to a leader like Sigismund, who was "brought up in a camp, and has led great armies always victorious." Other critics include the "difficult monstrous light-fleers," who boast of their own knowledge, "arrogating much to themselves" and "ineptly croaking."[55]

Valturio finds comfort in the knowledge that venerable ancient authors also suffered criticism. His examples mostly concern accusations of plagiarism. Homer was castigated by Zoilus a thousand years after his death. Vergil had many critics, including those who said that he was "not so much a true imitator indeed as a ravisher and plagiarist of the ancients and a public thief." Terrance, "the highest of the comedy writers," was accused of merely translating his plays from other ancient works, so that his achievement was "not by his own merit but by another, publishing for his own." Even Cicero, "of divine spirit and very powerful eloquence," endured the carping of the envious, who accused him of taking from Greek writings rather than using his own inventions, of speaking in a disorderly manner, of being verbose and a buffoon. Valturio shifts from his sympathetic portrayal of ancient authors who were criticized to a stern censure of ancient critics.

He is particularly incensed by Epicurus, who "with insufferable pride" criticized many authors and about whom "it is said that he attacked in all books those who differed in philosophy a little from himself and from his insane opinions." Valturio is equally censorious of "evil speaking . . . Zeno."[56]

Returning to his own authorship, Valturio concludes, "It must not seem amazing to you, Sigismund, nor to anyone, if certain very insignificant men, both certain unlearned students and itinerant ranters, bark and vent rage against me." He asks readers out of kindness to chase such critics "with invectives" and fill them with terror, and curses. Valturio asks further that his treatise be read thoroughly. Nothing can be written "entirely with new investigation," so his book should "not be considered something of new instruction and demonstration." Rather, it should be viewed as a restoration of things "from a lost . . . monument of illustrious men" and "with demonstration of the footprints of the ancients." He supports this statement with a long list of ancients whose military commentaries had been lost.[57] Sensitive to the accusation of plagiarism, he follows Pliny in providing a comprehensive list of the chapters of his treatise, each of which is followed by a long list of authors (primarily ancient) that constitute his sources.[58] Valturio's treatise in fact takes a very traditional form in that it is based overwhelmingly on examples gathered from ancient texts. His anxiety about that form and about his own lack of military experience reflects new concerns in his own time about both originality and the value of personal experience.

Much like the Tempio Malatestiano, Valturio's treatise is a monument to the glory of both antiquity and Sigismund himself. The author explains that he has composed the treatise, by which he seems to have placed Sigismund "in the seats of the celestial and terrestrial councils of the immortals." Just as the fame of Hector and Achilles in the Trojan War, and of many other ancients as well, has been transmitted "by the protection of letters," so his treatise will serve the same function for the lord of Rimini. Valturio hopes that he himself might also achieve immortality through his writings.[59]

The military leader himself should be both literate and learned. Valturio stresses that he should be learned, not in "the vulgar, crass, and crude" things that they now use in camp, but in "noble and very precise subjects joined with knowledge of many things." Providing numerous examples of studious and learned ancient generals, Valturio turns from his last ancient example, Marcus Aurelius, to Sigismund himself. He describes the lord of Rimini as one who reads and hears much, who has great knowledge and the seeds of all the arts, one who takes pleasure in subtle questions, favors orators and poets, and collects books for a library that will be useful in the present as well as in the future. He praises Sigismund further for his poetry and for his buildings, which include his palace and the great fort, the

Rocca Malatestiana, which Valturio describes in detail, including its walls, towers, and munitions, as "a true marvel of the magnificence of Italy."[60]

Valturio's treatise thus depicts Sigismund as a learned soldier and a contributor to both letters and magnificent building construction. The encyclopedic work, which offers a rich display of ancient and contemporary military knowledge, including knowledge of weapons, is emblematic of Sigismund's own knowledge, power, and beneficence. It stands in stark contrast to the image of Sigismund as rapacious and cruel, which was being circulated elsewhere in Italy.[61]

Alberti and Valturio were humanists whose writings influenced learned culture for more than a century. Two points deserve emphasis. First, their treatises incorporated technical arts into learned expositions. Alberti dedicated one version of his treatise on painting to an architect-engineer, Brunelleschi, and conceived architecture as a discipline involving both design and engineering. Valturio successfully combined the technical art of weaponry and the traditional disciplines of strategy, tactics, and generalship. Second, both authors openly explicated their disciplines, including those aspects that were technical. In this they followed some of their contemporaries who undertook authorship on the technical arts, but they also followed ancient predecessors such as Philo of Byzantium and Vitruvius.

Artisan-authors in Italy

Despite their very different backgrounds, humanists and artisan practitioners shared a common cultural arena of interests and patronage. A middle ground of communication arose among skilled practitioners, learned humanists, and elite patrons. As a corollary, particularly in northern and central Italy, several men trained as artisans undertook authorship in the second half of the fifteenth century. At least one precedent had been set by Cennino d'Andrea Cennini in his early-fifteenth-century manual on painting. Although mostly a discussion of painting techniques and recipes, Cennini's work reveals pride in the craft of painting, particularly as it developed in the school of Giotto, to which Cennini belonged.[62] Most later-fifteenth-century writings by artisans are far more complex and ambitious than Cennini's manual. Yet in a broad sense they also derive from a sense of artistic pride and accomplishment. Individuals such as Lorenzo Ghiberti, Francesco di Giorgio, Antonio Averlino (called Filarete), and Leonardo da Vinci wrote one or more treatises as they also carried out numerous commissions involving goldsmithing, sculpture, painting, architecture, and engineering. Most wrote within the context of patronage; some dedicated their treatises to one or more princes.

Lorenzo Ghiberti headed a large goldsmiths' workshop that was responsible for the famous relief panels on two large sets of doors of the Florentine baptistery. About 1447, near the end of an illustrious career, he took pen in hand to write his treatise on sculpture known as the *Commentarii.* Yet there is evidence that he started thinking about such a work and collecting notes for it long before. He begins his treatise with an excerpt from Athenaeus Mechanicus, the ancient writer on siege machines (see chapter 1). The excerpt records the advice of the Delphic Apollo to be brief and relates an experience of Isocrates: the great orator was still polishing his advice to Philip when the Macedonian king ended his war without its benefit.[63]

A letter of 1430 exists in which Ghiberti attempts to borrow a manuscript of Athenaeus Mechanicus. This must have been only one of Ghiberti's many efforts to obtain and understand ancient texts. It is clear from the *Commentarii* that Ghiberti labored over and excerpted other ancient and medieval treatises, most importantly Vitruvius, Pliny, and medieval optical treatises, including those of Alhazen, Avicenna, and Roger Bacon. He believed that ancient practitioners wrote treatises to elucidate the principles of their art. Because his own age had reached new heights in matters of design, perspective, and skill, the principles of the arts would once again be laid out in writing as they had been in antiquity.[64]

Ghiberti joins skilled practice to learning, just as he connects the sculpture and painting of antiquity to the work of his own time. In the third, unfinished section of his treatise he attempts to join optical theory to artist's perspective. The difficulty and obscurity of Ghiberti's text is apparently the result of his attempt to study and use Latin texts, a project for which his Latinity was clearly inadequate. Yet Ghiberti's linguistic difficulties should not obscure his shrewd choice of excerpts or his astute use of them to develop his own notion of learned practice. Ghiberti repeatedly refers to the writings of ancient practitioners, gleaned primarily from Vitruvius and Pliny. Painting and sculpture, he insists, are "knowledge adorned with many disciplines and various teachings." It is knowledge created "with certain meditation" that is completed "through material and reasoning."[65]

Things that are fabricated "through proportions by astuteness and by reason" can be demonstrably explained, Ghiberti insists. Paraphrasing Vitruvius, he points to the weaknesses both of the practitioner who works without benefit of letters and of the learned person who works without skill. Both skill *(ingegnio)* and learning *(disciplina)* are necessary. In addition, the sculptor must develop good character traits. He must learn philosophy because it fills him "with [a] great soul," one that "is not arrogant" and is, moreover, "moderate and humble and faithful and without avarice." A work made without faith or chasteness cannot be perfect.[66]

Ghiberti repeats the Vitruvian passage that expresses gratitude to the writers of the past and suggests that such writings allow the accumulation of knowledge. He also recapitulates the Vitruvian story concerning Socrates' wish that humans had windows in their chests so that anyone could perceive their knowledge at a glance. Sculpture, he claims, is a discipline containing rational principles that advance both through fabrication and through writing. He explains in his second commentary that painting and sculpture ended with the end of the ancient world; ancient writings about these arts were destroyed. A revival began only with Giotto and continued to Ghiberti himself. He concludes his second commentary with an autobiographical catalogue of his own works.[67] Ghiberti, then, describes an ancient world in which sculpture and the other arts reached a high point of excellence and in which sculptors and painters routinely wrote about their crafts. His own work was part of the revival of the arts, and his treatise constituted an essential aspect of that revival.

The sculptor and architect Antonio Averlino, who called himself Filarete, "lover of virtue," trained in Florence, probably in Ghiberti's workshop. Filarete, wrote the first vernacular architectural treatise of the fifteenth century. After working in Rome for a number of years, he moved to Milan to work for Francesco Sforza. Sforza, one of the best-known military captains of the fifteenth century, became duke of Milan in 1450 by means of military conquest. The new prince put Filarete in charge of his major construction projects—the cathedral, the Castello Sforzesco, and the Ospedale Maggiore—despite the protests of local architects and engineers. Filarete repaid the debt with his architectural treatise, completed in the early 1460s and dedicated to his patron.[68]

The treatise describes the design and building of a new city, Sforzinda, and includes a detailed, idealized account of an architect-patron relationship represented by Filarete himself and Francesco Sforza. During the construction of the city, patron and architect enjoy a close relationship. The patron is present at the site, living in a tent. Architect and patron frequently converse. Often they dine together while they discuss issues of design, as well as the work of the following day. The project of the new city is depicted as one carried out by two near equals. The gates of the city are named after Francesco Sforza's wife, Bianca Maria Visconti, and their children. Yet one of the gates leads to the river Averlo and is named Averlina, obvious references to Filarete himself.[69]

While Filarete thus represents architecture and architects as worthy of noble company, he also endorses the value of practical skill and technology. After a dinner with the patron and other nobles, in which Filarete explains the many good omens that have occurred during the building of the city, "my Lord and the others were so taken with love for me that they gave me enough to live honorably."

Moreover, the lord's son "was so pleased by our conversation that he fell in love with this science of architecture." The son begs his father to let the architect show him everything that has been done, and the request is granted. The prince's son wants to know how metal is melted, how furnaces are made for melting bronze, and he asks for information about glass furnaces. Filarete's treatise contains one of the earliest descriptions of a modern blast furnace. It also contains a detailed description of how to make a plaster that does not stain, as well as sections on drawing, perspective, and painting. Filarete's interest in technical subjects apparently extended beyond the treatise that we have in hand. He mentions books that he is in the process of writing (none of which are extant) on "agriculture," "technical matters," and "engines."[70]

The close association of the architect and the noble patron is paralleled in the new city by institutions in which learned and craft disciplines are taught side by side. Filarete proposes a school for impoverished youth that would provide instruction not only in letters and good habits but also in "every branch of knowledge and every skill." Instructors should include doctors of law, medicine, canon law, rhetoric, and poetry. The school would be unusual, however, because "some manual arts should be taught here" by craftsmen; these would include "a master of painting a silversmith, a master of carving in marble and one for wood, a turner, an iron smith, a master of embroidery, a tailor, a pharmacist, a glassmaker and a master of clay."[71] All the children could thereby be trained under the same roof in the discipline to which their soul and intelligence were most suited.

Although the discussion of the salaries of the various instructors makes it clear that some have much higher status than others (as was of course the case in his own society), it is significant that they all work in the same school. Filarete himself recognized this significance: "This will be a thing that will last for eternity and, moreover, a thing that has never been done before." He notes that there are universities where students pay a certain amount, but this applies only to students of letters. "The other crafts," he insists, "are also necessary and noble." Filarete's city also includes a House of Virtue, in which instruction is offered in the seven liberal arts, areas are designated for military exercises, and all the crafts and trades are both practiced and taught. When the students are "judged by good masters, and if they are young and have been educated in this place," they are "given the degree like the doctors." The governors of the temple of virtue are three: the first is one of the doctors, the second is one who received honors for his feats of arms, and the third is from among the artisans.[72] Averlino created an ideal city in which learned men, military men, and artisans trained in their separate disciplines together in one place. His elaborate treatise makes manifest his own ability to traverse the boundaries between artisanal and literary practice.

Some of Filarete's contemporaries crossed the same boundary between learning and practice. For example, Piero della Francesca painted luminous, riveting paintings and also wrote treatises on mathematics and perspective. Piero's painting and his mathematical abilities were intrinsically related. Martin Kemp describes him as "a man deeply conversant with pure and applied mathematics, capable of writing treatises to match in quality, anything produced in Italy of his day." Piero wrote three mathematical treatises. The *Trattato d'abaco*, written about 1450, concerns commercial mathematics and geometry. The *De prospectiva pingendi*, written in the 1470s for Federico da Montefeltro, ruler of Urbino, treats painter's perspective; and the *De corporibus regularibus*, composed between 1482 and 1492 for Federico's son Guidobaldo da Montefeltro, concerns the five regular solids.[73]

Although he often worked in his home city of Sansepolcro, in central Italy, Piero also traveled to the courts and cities in other parts of Italy to fulfill various commissions. He worked in Arezzo, in Rome, and in Rimini, where in about 1451 he contributed to Sigismund Malatesta's Tempio. Later he went to Urbino, first as the guest of the painter Raphael's father, Giovanni Santi, and then as a painter in the court of Federico II da Montefeltro.[74]

Federico II had succeeded to the rule of Urbino after the assassination of his half brother in 1444. A brilliant soldier, he led his army in the service of the pope, Venice, Florence, and the Aragonese of Naples and against his fierce rival Sigismund Malatesta of Rimini. After he succeeded to the rule of Urbino, he initiated an extensive building program, including the rebuilding of his palace in Urbino. On the facade of the palace was constructed the "frieze of war," a series of relief panels featuring machines from Valturio's *Elenchus et index rerum militarium* and from the treatises of the Sienese engineer Francesco di Giorgio. Federico thereby represented his power and authority not only by building magnificent buildings but also by displaying images of war machines. Piero della Francesca, who painted Federico and his wife, Battista Sforza, in facing portraits, was one of the many artists who benefited from his patronage.[75]

It was under Federico's patronage that Piero wrote a treatise on painter's perspective, *De prospectiva pingendi*, about 1470. The treatise encompasses both practical and learned traditions. Perspective, Piero tells us, has five parts: first, "sight, that is, the eye"; second, the "form of the thing seen"; third, "the distance from the eye to the thing seen"; fourth, "the lines that leave from the edge of the thing and go to the eye"; and finally, "the intersection that is between the eye and the thing seen where one intends to place the thing." Piero treats these matters, not abstractly, but as they concern painting. A point is not imaginary as the geometers say, nor does the line have length without width. These things are apparent only to the intellect. However, Piero wishes "to treat perspective with demonstrations that are understood by the eye," and he provides suitable definitions. He thus defines the point, to give one example, as "a thing as small as possible for the eye to understand."[76] Piero thereby shapes abstract Euclidean geometry to the practice of painting.

Piero divides his treatise into three parts. The first treats points, lines, planes; the second, three-dimensional figures such as squares, pilasters, and round columns; and the third, more complex three-dimensional figures, such as human heads, capitals, and bases. His procedure is to begin with the statement of a general principle, the first being that "all quantity is represented under the angle in the eye," a precept derived from Euclid's *Optics*, a text with which he was thoroughly familiar. Then he shows how to put the principle into practice on a material surface. As Judith Field and others have shown, Piero used rigorously mathematical perspectival techniques in his own paintings.[77]

It seems unlikely that Piero wrote his detailed, indeed laborious text primarily for workshop use. As one scholar suggests, the techniques that he elaborates in writing could be transmitted to practitioners far more easily by spoken communication and demonstration accompanied by practice than by such a text. The evidence that we have for both the intended and the actual use of the *De prospectiva pingendi* suggests rather a learned and courtly milieu. Although the extant copies of the treatise do not contain the front matter, which would have included the dedication, evidence exists elsewhere that the treatise was dedicated to Federico. Piero dedicates his later treatise on the five regular solids to Federico's son Guidobaldo. There he expresses the hope that the treatise will find a place on the shelf next to the book on perspective that he had written for the father, Federico da Montefeltro. Both treatises were quickly translated into Latin.[78]

Another highly skilled client of the Montefeltro dukes, the Sienese Francesco di Giorgio, started out as a painter and sculptor and went on to become one of the most widely respected architects and military engineers on the Italian peninsula. Much sought after, Francesco worked frequently in Urbino, as well as in Siena and in Naples, where he was employed by Alfonso II, duke of Calabria. In addition to his extensive work as an architect and engineer, Francesco undertook numerous projects of authorship. He struggled to learn Latin, translated Vitruvius, and created pictorial notebooks and illustrated treatises on architecture, fortification, and military engineering.[79]

One example of his work, known as the *Opusculum de architectura*, is a book of drawings on vellum of machines and plans of fortification without descriptive text. The more than two hundred drawings of machines are divided into types, such as mills, pumps, siphons and water wheels, and military devices. Many of the drawings are original, yet numerous others are derived from Taccola's *De in-*

geneis. A dedication to Federico da Montefeltro, although erased in the autograph manuscript in London, appears in a sixteenth-century copy.[80] It is notable that a book filled with drawings of machines and devices, created with the expensive material of vellum, was deemed a suitable gift for a ruler.

Francesco also composed two separate but related treatises on architecture, engineering, and the military arts, *Trattato I* and *Trattato II*. The precise years in which he wrote them and his other notebooks and codices are the subject of scholarly debate. *Trattato II* clearly is the more mature work—more coherent, more knowledgeable concerning ancient writers, and containing far more accurate translations of passages from Vitruvius's *De architectura*. Although the extant manuscripts of Francesco's *Trattati* do not contain formal dedications, he wrote them while under the patronage of the Montefeltro rulers of Urbino and of Alfonso of Calabria, the lord of Naples. *Trattato II* contains a long tribute to Federico da Montefeltro.[81]

Francesco divided *Trattato I*, probably written about the late 1470s, into untitled sections according to subject matter, such as fortresses, cities, temples, theaters, geometry, water wheels and mills, ways of lifting and transporting water, and military machines. He explicates his architectural theory, which involves anthropomorphic, modular methods of design in which the proportions of a correctly designed building are considered to be in harmony with those of the human body. He also discusses and illustrates numerous machines and mechanical devices, such as artillery, pumps, and mills.[82]

Citing both Taccola and Vitruvius, Francesco insists that the architect must exhibit competence in both conceptual and material realms. If the architect does not have "shrewdness and particular ingenuity and invention," he will not be able to practice perfectly. Architecture "is only a subtle conjecture, conceived in the mind, that is manifest in the work." However, it is "not possible to assign reason to each and every thing because ingenuity consists more in the mind and in the intellect of the architect than in writing or design, and many things happen in the doing that the architect or worker never thought of." The architect must, then, be both "practiced and knowing," must have a good memory, must have read and seen many things, and must be prepared. The good architect is different from "arrogant and presumptuous" people, "who are instructed in errors" and who corrupt the world by demonstrating false things "through force of language." Francesco paraphrases Vitruvian passages on theory and practice and on the disciplines in which the architect should be competent, reinforcing a view of the discipline as one that encompasses both reason and fabrication.[83]

He stresses the impossibility of explaining all things because of the enormous variety of physical forms. In a section on pulleys and gears he suggests that one consequence is that models are necessary, but even they are no substitute for actual construction. "Many things seem easy to the mind of the architect" and appear to succeed, but when they are put into effect great deficiencies are found that are repaired only with difficulty. His own ideas, Francesco assures his readers, are based on his actual experience: "I for myself have seen a sufficient good part of experience of the inventions that here will be demonstrated, not relying on myself [i.e., my ideas]."[84] Throughout both treatises Francesco cites his own experience on a variety of issues, for example, noxious winds, types of marble, the best kind of lime, and searching for water.[85]

Following Valturio, Francesco condemns those who usurp the works of others and attribute these works to themselves. He compares such people to crows dressed in feathers of the peacock. He does not want anyone to believe that "all which is contained in this my little work" is "of my own invention." He discusses many things that have been treated "in most authentic books," especially in the *De architectura* by Vitruvius. Yet he is also careful to insist upon his own original contributions. He has taken the proportions of certain building parts, temples, and palaces from Vitruvius, yet the forms of the temples and houses, as well as his treatment of numerous other subjects, are "inventions of my own weak skill." He should be blamed for what does not please readers, just as he should be credited if there are things that offer "pleasure or truly, utility." Through himself alone many useful and delightful things "will be manifest to each that through many epochs have been secret." Francesco adds that he is well aware of modern writings, but they fail to explicate the most difficult passages of ancient books. Moreover, modern buildings are full of errors.[86] Francesco thereby credits past authorship; claims that he reveals openly things that have long been kept secret; and finally, insists that his writings are filled with his own original inventions.

Yet open authorship conflicted with the protection offered by secrecy. Francesco admits deliberating numerous times about whether "to manifest any of my machines" because he had acquired knowledge of them "with my great cost of experience and grave inconvenience," forgoing in part what was necessary to life. His reward had been ingratitude. The experiences of invention are acquired only through long work, great expense, and "impediment of other useful cares." Yet some want to have a useful design or instrument and think its invention a brief thing. They "scorn the fatigue of invention." The ignorant end up honoring themselves "by alien labors." Francesco concludes that "this vice in our time abounds," especially among ignorant architects.[87] Francesco's authorship, carried out in the context of patronage, conflicted with his work as an inventor of machines and with the artisanal ethos of secrecy; yet the rewards of open authorship seem to have prevailed.

Francesco's contemporary Leonardo da Vinci also traversed between artisanal and learned cultures. Leonardo possessed a range of skills similar to those of

Francesco di Giorgio, from painting to architecture and engineering. The two men were consulted together on at least one occasion concerning the construction of the cathedral at Pavia. Leonardo trained in Verrochio's workshop in Florence, arriving in the late 1460s. During that time Verrochio and his assistants carried out numerous commissions involving painting, sculpture, and engineering, including the task of placing the large, heavy gilded copper sphere on top of the lantern that centered Brunelleschi's cupola on Santa Maria del Fiore. Leonardo was fascinated by this project and by the machines required to carry it out. It is likely that Brunelleschi's machines for constructing the cupola itself still existed in Florence for Leonardo to see either in operation or in storage. Brunelleschian prototypes often appear among the many cranes and other lifting machines in Leonardo's notebooks.[88]

Leonardo depended on the patronage of Italian lords, who particularly valued his work as an engineer. In a 1482 letter to Ludovico Sforza of Milan he enumerated his skills, emphasizing those involving engineering. He succeeded in obtaining a position in Milan, where his new patron, Ludovico, the son of Francesco Sforza, headed the largest and most active court in northern Italy. Leonardo stayed until 1499, the year Ludovico was defeated by the French army of Louis XII. His multifarious career both in Milan and thereafter depended upon patronage.[89]

While he worked on numerous projects Leonardo filled notebooks with drawings and writing. More than six thousand autograph pages of these notebooks are extant, only a fraction of the original number. The form and content of many of the original notebooks cannot be reconstructed because of the way they were broken apart and reorganized in the two centuries following Leonardo's death in 1519. Yet his extant writings include some intact notebooks, such as the Madrid codices, two notebooks discovered in the Spanish National Library in the 1960s. Leonardo was well acquainted with the prior tradition of fifteenth-century technical authorship. He was influenced by Alberti's writings, especially *Della pittura*; he owned and annotated one of Francesco di Giorgio's treatises; and he was familiar with Piero della Francesca's treatises through his friend and tutor, the mathematician Luca Pacioli.[90]

Leonardo often carried a notebook with him to record observations and questions, using both writing and drawings. He also envisioned the creation of formal treatises, working with both codices and arrangements of loose sheets. Such planned treatises, which are distinct from his private notebooks, include remarks addressed explicitly to intended readers. He planned treatises on mechanics, painting, anatomy, and hydraulics, among other subjects, none of which were actually completed or presented either to a patron or to a printer.[91]

Madrid Codex I is an almost complete treatise on machines and mechanics that Leonardo organized into two parts, one on theory and the other on practice.

Concerning practice, Leonardo presents strikingly beautiful drawings of machines and mechanisms, as well as textual instructions for building or using them and descriptions of how they move. He also treats theoretical mechanics, which for him primarily concerned weight, force, impact, and motion. As Ladislao Reti describes it, Leonardo thought of a page as "an artistic and intellectual unit." On every page he drew the figures first and then wrote the text. Yet a disintegration occurred when he filled the page with too many notes and had to move to another page, causing the subject matter to be treated out of order.[92]

Leonardo's working methods can be glimpsed by comparing sheets in the *Codices Atlanticus*, a multivolume compilation of sheets originally from diverse notebooks that include mechanical and machine drawings, with pages of the treatise on mechanics, *Madrid Codex I*. Leonardo transferred some of the writing and drawings from *Atlanticus* to *Madrid I*. For example, at the top of one *Atlanticus* sheet appear drawings of epicycloidal gears accompanied by text that has been crossed out, indicating that it has been used elsewhere. The three drawings and a virtually exact transcription of the texts (which explain in detail how the wheels turn) have been transferred to pages in *Madrid I* (Fig. 9). The accompanying text explains the motion of the gears, while the remaining paragraphs on the Madrid pages treat the effects of motion in air and water. Leonardo asks why the motion of air through air causes a sound (as in a whistle) but the motion of water over water does not. On the opposite page, above the discussion and illustration of the gear, Leonardo asks two questions. The first concerns the action of an evenly weighted plank on water compared with that of an unevenly weighted one. The second concerns the behavior of diverse figures floating on top of a stream of water or falling through the air in a perpendicular line to earth. Leonardo clearly considered these carefully organized folios to concern various aspects of the same subject, namely, the nature of motion in air, in water, and in the mechanism of epicycloidal gears—an example of his unitary vision of the world in the 1490s.[93]

Despite the fact that he wrote backwards and from right to left, there is no evidence that Leonardo intended to deliberately conceal his work. Leonardo's mirror-writing is not encryption. Its original development can be attributed to his left-handedness and to his probably being largely self-taught. (Left-handed children often experiment with mirror-writing; Leonardo is unusual only in his fluid skill.) As Reti points out, his notebooks were known to numbers of people during his lifetime. In addition to his writings, he communicated his ideas orally both to learned friends and to artisan practitioners, including the technicians he employed, who were usually German. Leonardo's notebooks do contain examples of the withholding of information—a famous example involves the details of a submarine because, he says, disclosure would cause too much destruction.

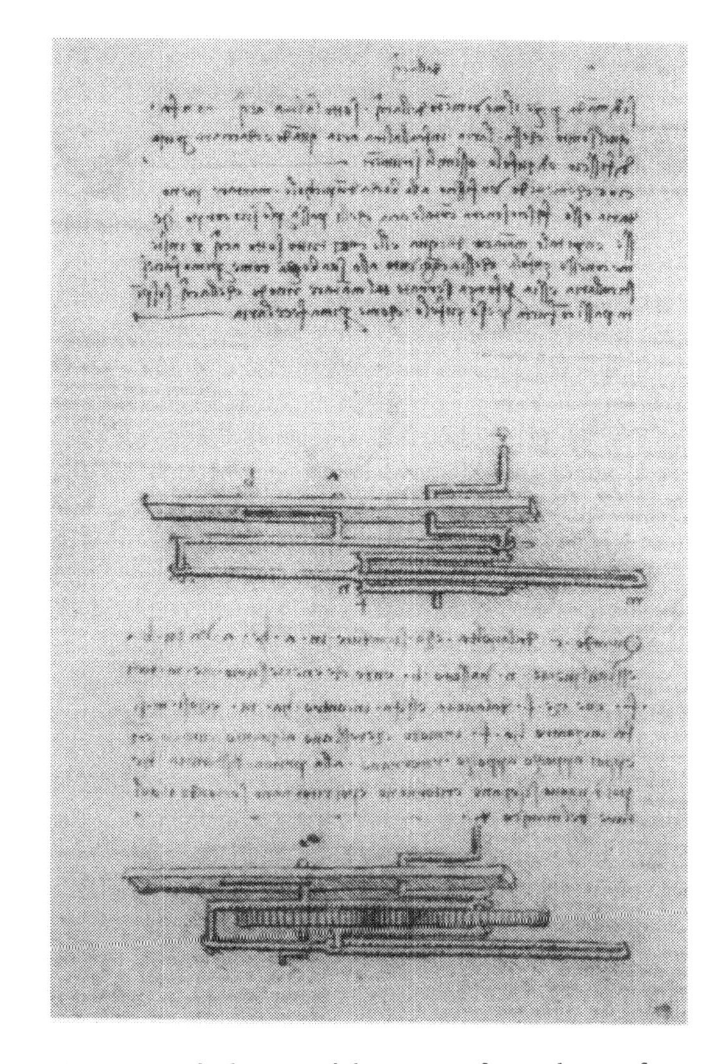

Fig. 9 Epicycloid gears and the motion of air and water, from Leonardo da Vinci's *Madrid Codex I*, fol. 111v, Biblioteca Nacional, Madrid On this sheet Leonardo asks why the motion of water forced through water (by bellows, for example) does not make a sound, whereas the motion of air forced through air (by a whistle) does. He then carefully describes the motion of the epicycloid gears that he has drawn. For Leonardo, the world of nature and the world of the machine reflect each other. Analogies between the two have explanatory force. Courtesy Biblioteca Nacional, Madrid.

Nevertheless, Leonardo produced a large number of precise drawings, including depictions of machines and machine parts and anatomical illustrations. He utilized a whole set of illustrative techniques, such as the exploded view, which demonstrates how a machine, a device, or a human body is put together.[94]

Most of Leonardo's manuscripts exist as autographs, written by himself. Works that he intended as coherent treatises are mostly unfinished. Yet, like the authorship of other practitioner authors of his day, his was influenced by patrons. We know from a sixteenth-century note written by Giovan Paolo Lomazzo that he began at least part of his book on painting (which compares the nobility of painting and sculpture) at the request of his patron Ludovico Sforza. Specialists have also recognized that Leonardo's early writings were inspired by his reading of Alberti's *Della pittura*, although he never mentions the earlier treatise by name.[95]

Yet Leonardo failed to complete any of his multifarious writing projects during his lifetime, and he seems not to have used authorship as a way of acquiring patronage. At his death in 1519 he left all his notebooks, drawings, and writings to his assistant and pupil Francesco Melzi. Melzi himself completed a treatise from these materials, creating *Trattati della pittura*, compiled from Leonardo's various notes.[96] Leonardo's failure to complete treatises can be explained in part by his complex unitary vision of nature, which inevitably produced a stream of questions rather than conclusions. In addition, his lack of a Latin education put him at a disadvantage with regard to the learned world that he attempted to encompass. Leonardo had no formal training in Latin or mathematics. His strenuous, ongoing efforts to learn Latin and to study mathematics (the latter with Luca Pacioli), beginning at midlife, in the 1490s, were never entirely successful, of which he undoubtedly was fully aware. His difficulties, despite the help of his learned friends, in being able to study fully the many facets of traditional (Latin) learning—in mathematics, mechanics, anatomy, and other areas—may have inhibited his completion of treatises.[97]

Yet it is also true that publication and the transmission of writings is only possible within the modalities available in any particular time. As Reti points out, Leonardo was well aware that his unique drawings could not be reproduced by scribes and that woodcut print technology could not reproduce their fine detail. The new method of copper engraving was adequate in theory but would have been too expensive and difficult given the large number of drawings embedded in text that he produced. Leonardo lived at a time when a fully developed scribal culture overlapped with a burgeoning print culture. He was awkwardly positioned between the cultures of scribe and print, neither of which could adequately reproduce his own books during his lifetime. Yet Leonardo owned a significant library containing both printed and manuscript books. Certainly he was aware of both modalities and of the ways that he could use them.[98]

The workshop-trained authors discussed here practiced crafts and constructive arts, from sculpting and goldsmithing to painting, architecture, and military engineering. Their writings have been studied for the most part for the particular arts that they practiced. From a more general point of view, these books mediated the world of written discourse and that of craft practice. Writing a treatise about a craft is significant in itself, and an unusual activity for craft practitioners, especially before the fifteenth century. The authors discussed here, each in his own way, rationalized craft and constructive practices. They transformed hands-on craft know-how into organized written forms of knowledge. In so doing, they created and crossed a boundary between craft practice and discursive practice involving reading and writing, contributing thereby to a transformation of the culture of knowledge. This is most apparent in the multifarious studies of Leonardo, for whom observation, experience, visual representation, and knowledge about the natural world were closely interrelated; but in diverse ways it is both implicit and explicit in the other artisan-authors treated here as well.

Fifteenth-century manuscript writings on the mechanical arts are notably diverse. They include Latin writings by physicians, German-language books by gunners, humanist treatises, and writings by workshop-trained artisans of various kinds. The topics include machines, mills, pumps, painting, architecture, perspective, and fortification. Many such books are illustrated; some are primarily or entirely pictorial The expansion of writings on the mechanical arts came out of a growing proximity of technē and praxis. Rulers used construction and ornamentation in various ways to legitimate their rule. Writing on such arts had the effect of rationalizing them and turning them into more discursive, learned subjects. Such books, whether written by university-educated men or by craftsmen, became appropriate gifts for patrons.

The expansion of authorship on the mechanical arts occurred alongside the development of craft secrecy and proprietary attitudes toward craft knowledge. In the urban centers of late medieval Europe artisanal crafts themselves expanded: palaces, churches, and other buildings were redesigned or newly built. Statues, paintings, tapestries, and objects of all kinds found an increasingly appreciative audience and market. Toward the end of the fifteenth century, especially in northern Italian and southern German cities, conspicuous consumption on the part of elites gained a foothold. In a sense the cultural value of objects themselves increased, as did the appreciation for artisanal practice. This cultural phenomenon could be exploited in various ways—by proprietary actions involving craft secrecy and patents or by open authorship in the context of patronage relationships.

Authors of books on the mechanical arts for the most part wrote openly, yet a few instances of secrecy and the meaning of that secrecy have been noted.

Fontana's code-writing encompassed an exploration of encryption itself. Taccola's references to "veiled speech" point to his efforts to obtain patronage. Those references occur in books that are full of explicit drawings and discussions of machines and apparatus. Francesco di Giorgio's ambivalence about revealing his inventions suggests his identification with traditions of craft secrecy; nevertheless, he openly displayed the arts that he described both textually and pictorially. Writing for patrons, he displayed the arts that concerned them by clear exposition or pictorial illustrations.

Authors such as Alberti repeatedly emphasized the ties binding architecture and the mechanical arts with political and moral concerns. Other authors made such ties explicit by emphasizing the moral qualities that should be possessed by gunners, sculptors, or other practitioners. The sober, god-fearing gunner, the great soul of the sculptor, the architect called "Filarete," lover of virtue, all suggest that these arts possess cultural meanings that ranged well beyond their role in material production. Such cultural meanings associated them with parallel arts in antiquity, with learning, with civic and moral concerns, and with the open, written transmission of knowledge.

Although authorship on mechanical arts brought practitioners and university-educated men closer together, it did not eliminate the differences between them. Their writings, moreover, saw very different fates. The treatises by humanists, such as Alberti and Valturio, were printed in the late fifteenth-century and widely disseminated, whereas the writings of most practitioner authors remained in manuscript form, relatively unknown or soon to be forgotten. An important exception, Piero della Francesca's treatises on the abacus and on the regular solids, were printed and widely distributed in the sixteenth century. This came about when the mathematician Luca Pacioli incorporated Piero's writings into his own treatises without specific acknowledgment, although he praised Piero highly.[99]

Whatever the individual fortunes of the writings discussed here, collectively fifteenth-century manuscript books have great cultural significance. They emerged for complex reasons involving patronage and the legitimation of political power. They combine detailed explication of machines and material practice with open authorship. Written explication in itself raised the cultural status of some of the mechanical arts. Some came to be transformed from practices involving skilled know-how and material construction to discursive bodies of knowledge, still based, nevertheless, on material practices. These books cannot be attributed solely to artisans and artisan culture or to learned humanists alone; nor can they be attributed solely to patrons, princes, and rulers. Rather, they are the products of the complex interaction of all three.

Chapter 5

Secrecy and the Esoteric Traditions of the Renaissance

As OPEN AUTHORSHIP on the mechanical arts expanded beginning in the early fifteenth century, books concerning alchemy and other traditions usually labeled "occult" also proliferated. Alchemy, Neoplatonic philosophy, Hermeticism, the cabala, and astral magic were sometimes overlapping but not identical traditions and systems of belief. Each represented a complex legacy with its own corpus of texts and its own practices. Alchemical texts from Islamic cultures reached the Latin West in the twelfth century, and humanists incorporated Platonic and Neoplatonic writings into Latin learning in the fifteenth. The assimilation of such textual traditions produced in turn a wealth of new writings and the emergence of individual practitioners and groups devoted to alchemical and other, often esoteric practices.[1]

The word *occultus* conveyed a variety of distinct meanings in the late medieval era. Scholastic writings sharply distinguished "manifest" qualities, such as the Aristotelian hot, cold, wet, and dry, and insensible qualities, such as the cause of magnetism. In contrast, students of alchemy posited the interchangeability of the occult and the manifest, in which occult qualities could become manifest and vice versa. Neoplatonic authors used *occult* in yet another way, to refer to "hidden," or unseen, noncorporeal powers of the cosmos, forces that played a part in the complex hierarchy of spiritual entities that extended through multiple cosmic spheres.[2]

Alchemical, Neoplatonic, and magical traditions are notable for their eclectic syncretism and for their diverse beliefs about openness, secrecy, and authorship. These distinct but interrelated movements gained numerous followers in the sixteenth century and beyond. In this chapter I investigate openness, secrecy, and authorship within Latin alchemy and then examine the same issues within the thought of four influential figures from the late fifteenth and sixteenth centuries: Marsilio Ficino, Cornelius Agrippa, Paracelsus, and finally Giordano Bruno, burned at the stake in 1600. Attitudes toward authorship within these sources range from the pseudonymous authorship of alchemy to accurate self-credit and claims of originality by authors such as Ficino, Agrippa, and Bruno. Far from fitting neatly under the single label "occult," these sources demonstrate an intricate mixture of openness and secrecy, exoteric and esoteric orientations. Open dissemination and articulation of the value of openness exist side by side with secrecy and the defense of esoteric knowledge.

Yet these traditions have at least one thing in common, namely, a thoroughgoing utilitarianism, the view that the knowledge purveyed can bring positive change to the material and corporeal world. Affirmation of what might be called the positive technology of alchemy, magic, and occult powers is evident. Although the uses conceived for this technology vary from one figure or text to another, the utilization of occult knowledge and cosmic powers for human and sometimes material ends is a representative feature of these texts. This utilitarian orientation is sharply at odds with much of the Neoplatonic thought of late antiquity, which conceived of higher knowledge as a movement away from earthly, corporeal concerns.

The writings treated in this chapter span the eras of manuscript and print. The earliest, Latin alchemical writings, appeared well before the mid-fifteenth-century invention of the printing press. Yet most of the authors treated here wrote after that time. Sometimes they had their books printed, sometimes not. Ficino's writings were printed in numerous editions. Agrippa circulated his writings in manuscript for many years until the early 1530s, when he turned to print. Most of Paracelsus's writings circulated in manuscript until after his death. Printing provided a choice to authors, assuming that they had access to a press. It permitted the wide distribution of texts, yet it cannot be unequivocally claimed as an instrument of greater openness. On the contrary, as the distribution of Agrippa's *De occulta philosophia* attests, it could lead to the wider dissemination of esoteric doctrines.

Alchemical Secrecy and Pseudonymous Authorship

Alchemy entered the Latin West in the mid-twelfth century with the translation of Arabic alchemical texts, which themselves derived in part from late antique textual traditions. The subsequent complex history of the assimilation and development of Latin alchemy involves philosophical doctrines combined with operational practices. Robert Halleux notes that alchemists undertook the transmutation of metals from imperfect to more noble ones, namely, silver and gold. They also undertook the artificial reproduction of natural products, thereby engaging in speculation concerning the inanimate world. Since they associated the healing of imperfect metals with the healing of the human body, they sometimes involved themselves in a universal medicine. Finally, alchemists' work in perfecting materials could move them to strive for the perfection of their own souls. Alchemical doctrine inherited some aspects of Aristotelian matter theory, but it also contradicted particular Aristotelian tenets. For example, against the traditional Aristotelian separation of nature and art, alchemists claimed to be able to perfect nature by means of art.[3]

Alchemists typically attributed their writings to "false" authors, often authoritative figures from the distant past. Chiara Crisciani and Michela Pereira emphasize the profoundly rooted pseudoepigraphical character of the alchemical tradition. They point to an early example, the treatise of Bolus of Mendus, attributed to the well-known pre-Socratic philosopher Democritus; and to the Hermetic corpus and numerous alchemical writings attributed to the semidivine Hermes Trismegistus. Numerous other examples of pseudonymous authorship extend through the Arabic tradition and to the Latin West. These include an alchemical corpus originating in part with Arabic alchemists that was attributed to Plato and numerous texts attributed to Aristotle. Alchemical texts sometimes carry multiple attributions. For example, a small treatise entitled *Spera octo figurarum de lapide philosophico* was attributed to both Aristotle and Albertus Magnus. Authors such as Arnald of Villanova wrote some authentic alchemical treatises and acquired others through false attribution. The thirteenth-century Spanish mystic Raymond Lull had fathered upon him an entire corpus of alchemical writings, all of them false attributions. Crisciani and Pereira note that texts grouped under a particular pseudoauthor often share common characteristics. Further, beyond alchemists' deliberate attribution of alchemical texts to authoritative past authors, Halleux notes the phenomenon of accidental false attributions, cases where an alchemical treatise is included within a bound manuscript of the writings of a known author and eventually comes to be attributed to that author.[4]

Writers on alchemy deliberately embraced pseudoauthorship for a variety of reasons. It helped them avoid criticism or persecution. They could also enhance the authority of their treatises by attributing them to revered authority figures from the past. Beyond these pragmatic considerations, many alchemists believed that they were participating in an ancient tradition of sacred wisdom. In both their philosophical and their operational activities they did not aim to contribute new knowledge to a growing subject. Rather, they sought to discover an already established body of ancient wisdom, to effect transformations of materials in the process, and to explicate the subject to other initiates. Assigning pseudonymous authorship to their treatises allowed them to participate more deeply in the ancient wisdom of alchemy. The subject had already been written by ancient authorities; their own role was to enter into the ancient knowledge of those discoveries and then explicate them in writings that from one point of view could be seen as having already been inscribed by the alchemists of the past.[5]

Alchemy failed to find acceptance within the curricula of the medieval universities, and it came under increasing attack with a backlash that had set in by the end of the thirteenth century. The discipline was not incorporated into university curricula in part because it included operational, workshop processes with connections to craft traditions such as dyeing and metallurgy, which were incompatible with the logical orientation of university scholasticism. Lack of an institutional base undoubtedly facilitated opposition arising from institutional religious quarters, among others. Yet alchemists mounted a vigorous defense. As William Newman has convincingly argued, one outcome was that "the alchemists and their supporters gave a conscious and articulate defence of technology, indeed, one of the earliest and most thorough to be found in Latin Christendom."[6]

Such an alchemical defense can be found in the *Summa perfectionis*, by "pseudo-Geber," written in the early fourteenth century, sometime before 1310. The *Summa* sets out the criticisms of alchemy and then mounts a systematic defense in a step-by-step refutation of the arguments of the critics. It also describes alchemical apparatus and processes. Finally, it details the nature of metals, explaining them in terms of a corpuscular theory. The whole is a well-organized compendium of alchemical theory and practice whose influence reached to the end of the seventeenth century.[7]

Combined with its forthright exposition and defense of alchemy is the *Summa*'s pseudonymous authorship. The "Geber problem," which includes the question of who actually wrote the *Summa perfectionis* has been the focus of a long scholarly debate that originated in the nineteenth century. As Newman recapitulates it, medieval witnesses were unanimous in attributing the authorship of the *Summa* to an Arabic alchemist named Geber. Scholars investigating alchemy in the nineteenth century took this Geber to be an eighth-century Arabic alchemist named Jābir ibn Hayyān. However, based on a comparison of the *Summa* with the extant texts attributed to Jābir, the great nineteenth-century scholar of alchemy Marcellin Berthelot questioned such an attribution. In the 1940s Paul Kraus argued that other alchemical texts attributed to Jābir actually had been written by a group of alchemists in the ninth and tenth centuries who used the name Jābir as a pseudonym after, as Newman puts it, "a quasi-mystical master who had died—if indeed he ever lived—perhaps a century before their composition." Newman provides detailed evidence to show that the Latin *Summa* was probably written by Paul of Taranto, a virtually unknown Franciscan from the order in Assisi. The *Summa* is clearly related in specific ways to an earlier work, *Theorica et practica*, in which Paul is named as the author. In the earlier work Paul tries to justify the power of man over nature and provides a detailed defense of applied science. In the later *Summa* he combines a rigorous defense of alchemy with a pseudonymous attribution to an Arabic alchemical authority, Geber. Thus, he defends alchemy against its critics and at the same time enhances the authority of his treatise for a readership of alchemists.[8]

Paul of Taranto was influenced by the scholastic tradition of his own day in a number of ways, including the form of the *Summa* itself, its clear expository style, and its rationalization of the discipline of alchemy. Nonetheless, he concludes his treatise by explaining how he has concealed essential points of his teachings; that is, he explains the doctrine of dispersion: "And lest we be attacked by the jealous, let us relate that we have not passed on our science in a continuity of discourse, but that we have strewn it about in diverse chapters." Had his explanations been continuous, he explains, "both the tested and the untested would have been able to take it up undeservedly." The author adds that doctrine is also hidden in places where he has spoken more openly, even though "we have addressed the artificer with a plain sequence of speech." We are, he believes, "comprehended in the mind of our God," who "extends to and withholds from whom He wishes." Thus, he who seeks will find, "not by inquiring of doctrine, but by inquiring of the motion of his own nature." He who seeks knowledge "through the goodness of his own diligence will find it." On the other hand, those who seek knowledge "by the pursuance of books" will arrive much more slowly "at this most precious art." Pseudo-Geber emphasizes his own experience as the basis for his knowledge: "But we have not written anything except that discovered by ourselves, and the method of its discovery and the techniques of those methods."[9]

Pseudo-Geber/Paul of Taranto concludes with three central notions that would remain important tenets of Western alchemy. The first is that doctrines should be hidden from the undeserving. Although there is no evidence that he actually did disperse secrets throughout his own text, both the idea and the practice would be taken up by alchemists in subsequent centuries. Second, he articulates the traditional association of alchemy with divinity. Alchemy is not a subject to be learned by reading books; rather, it is to be found by looking into one's own nature and toward God. Finally, he insists that his own experience lays at the basis of his treatise. Thereby he validates experiential knowledge in a fundamental way.

Alchemy developed as a significant and influential discipline in the fourteenth and fifteenth centuries, despite its critics and its lack of institutional protection. Numerous new treatises appeared, including a work on the "fifth essence," *De consideratione quintae essentiae* (ca. 1351–52), by John of Rupescissa; treatises written by or attributed to Arnald of Villanova; and a large group of 143 treatises falsely attributed to the Spanish mystic Raymond Lull. As Western alchemy developed, it also changed, most importantly by acquiring vitalistic tendencies emphasizing the aliveness of the cosmos and by adopting elements of prophesy.[10]

Alchemy encouraged the development of methods of secrecy and concealment, many but not all derived from ancient practices. In her classic study of alchemical visual imagery, Barbara Obrist treats the evolution of alchemical illustrations in the fourteenth and fifteenth centuries. Such illustrations came out of alchemy's traditional linguistic use of images and developed into the practice of visual representation. Alchemical illustrations were full of symbolic images representing alchemical processes that could be decoded only by initiates of alchemical learning. Other modes of concealment involved cover names, or *Decknamen*, to hide particular technical terms; *syncope*, the use of highly elliptical descriptions of alchemical processes, substances, apparatus, and theory; and syncope's opposite, *parathesis*, piling up synonyms for the same group of descriptions in order to confuse the reader.[11]

Far more work needs to be done before alchemy of the fourteenth and fifteenth centuries is well understood. Until recently there has been a notable lack of scholarly attention to the discipline in these centuries, in part a result of the view that alchemy was a kind of protochemistry that should be studied in the context of the scientific and chemical "revolutions" of the seventeenth and eighteenth centuries. More recently scholars have rejected this point of view and have attempted to understand alchemy on its own terms, and they have also begun the daunting task of untangling the complicated textual traditions of the earlier centuries. Lack of relevant sources will undoubtedly prevent contextual investigations as detailed as Bruce Moran's exemplary study for a later period of alchemy in the circle of Moritz of Hessen in the early seventeenth century.[12] Yet an understanding of who the alchemists actually were can at least partially be realized with ongoing research. It would be fascinating to learn not only their names but the social and cultural milieus in which they lived and to gain an understanding of the relationships not only of texts but of the individuals and groups who wrote, copied, and used those texts and those who undertook alchemical operations.

Marsilio Ficino and Florentine Neoplatonism

Although Neoplatonism shared some common roots with alchemy, it basically represents a separate tradition. Most of the writings and ideas of Plato and of the Neoplatonists entered the Latin medieval West indirectly. A few of Plato's dialogues, most importantly the *Timaeus*, had been available in Latin translation during the medieval period. Other avenues of Platonic transmission included the writings of the church fathers who had been influenced by Platonism, most importantly Augustine of Hippo, and Platonizing Arabic philosophical writings such as those by Al-Kindi in the ninth century and Avicenna in the eleventh. Notwithstanding these extended, mostly indirect medieval traditions, Platonism and Neoplatonism constituted a newly important intellectual movement in fifteenth-century Italy, particularly in Florence.[13]

Study of Plato's writings and translations from Greek into Latin were well under way by the time Marsilio Ficino produced his major writings, beginning in the 1470s, yet Ficino became the most important translator and commentator of Plato and of Neoplatonic texts in the early modern era. He had translated the Hermetic corpus by 1463, Plato's dialogues by 1484, the writings of Plotinus by 1492, and treatises of Porphyry, Iamblichus, and Proclus, among others, by 1497. He made these writings easily accessible to the Latin West for the first time, and he also influenced their interpretation through his many introductions and commentaries. He wrote influential original treatises as well.[14]

Ficino was a physician and a devout although unorthodox Christian who was ordained as a priest in 1473. His work was made possible by the patronage of Florentine elites, most importantly the Medici, rulers of Florence. He maintained numerous personal contacts and friendships both in Italy and throughout Europe. He wrote prolifically, and his books were published in significant numbers of editions and translations. They functioned as important catalysts for the dissemination and influence of Hermeticism, Platonism, and Neoplatonism from the late fifteenth to the seventeenth centuries.[15]

The extent to which Ficino's scholarship directly represented or reflected the power of the Medici family in Florence is a subject of debate. Eugenio Garin's view, which held sway for many years, proposed that Plato's reception in the Renaissance should be analyzed in two stages that reflect two different underlying political realities. In Garin's first stage (in the fourteenth and early fifteenth centuries) the civic, republican forms of the late medieval commune still carried some meaning. Students of Plato adopted an "academic skeptical" view of Plato, which emphasized the open-ended dialogic character of his writings. In the second stage, from 1438–39 until the end of the century, Medici rule ended free civic life and substituted for it a courtly society centered "not in the open civic space but in the suburban villa, in an ivory tower of political ineffectiveness" in which Platonists sought a new theology and a "new philosophical revelation."[16] Garin thus viewed early Platonism in Florence as an intellectual movement that encouraged and reflected an open, dialogic attitude. In contrast, the Platonism of Ficino and his circle was developed within an esoteric, elite group withdrawn from the civic polity (thus safely isolated from political activity) and supported by a patron who controlled the city by his personal rule.

More recently scholars, many of them trained or influenced by Paul Oskar Kristeller, have dissented from this view of the Florentine reception of Plato. Arthur Field insists that Ficino and his Platonic theology did not reflect the interests, narrowly conceived, of the Florentine merchant class and that the isolation of Ficino's villa at Carregi has been exaggerated. James Hankins argues that there was no Platonic Academy, that the Academy is a historiographic myth created long after Ficino's own lifetime. Hankins regards Garin's model of Plato's reception as "externalist" and therefore inaccurate to the extent that it ignores what went on in Renaissance hermeneutics.[17]

Hankins calls for a more complete view that includes study of "the history of teaching traditions, the development of interpretive principles and techniques, and the structure of interpretive communities such as schools and universities." He urges that "the habits and conditions of interpretation must themselves be considered as historical phenomena." He leaves aside the broader political context to study issues of reading, criticism, and hermeneutics involved with Renaissance attempts to introduce Plato. He believes that such interpretations of Plato are largely concerned with reconciling Platonism with contemporary Christianity or, alternatively, expelling it as anti-Christian.[18]

Yet these two points of view, although based on fundamental philosophical differences, in my view do not represent mutually exclusive viewpoints. Both "externalist" and hermeneutical interpretations of Ficino's writings are possible. Garin idealized the early Florentine commune, making it more open and less elitist than it really was, just as he exaggerated the ivory-tower quality of the Platonic Academy. Yet Field and Hankins unduly minimize the connections between knowledge and power, that is, the ways in which Ficino's great learning and huge influence also bolstered the authority and reputation of his patrons, the Medici. Taken together, the two angles of perspective can enlarge an understanding of the meaning of Platonism for Ficino and his contemporaries.

Ficino was deeply influenced by late antique Neoplatonists, especially Plotinus and Proclus. Yet his Neoplatonism differed from theirs in important ways that reflect the profound social and cultural differences between late antiquity and the fifteenth century. Ficino inherited the esoteric aspects of Neoplatonism. Yet in my view esoteric practice was not a significant aspect of his thought; he did not, for example, reserve most of his own ideas for a small group of the initiated. Moreover, he held a central place in the broadly based learned culture of Europe. His ideas and writings were widely influential because of the dissemination capabilities of print culture; because of the universalizing and proselytizing characteristics of the Christianity with which he joined his Neoplatonism; and because of the basic openness of humanist discursive practices to which he was heir.

Ficino was greatly indebted to the Medici family, and he in turn enhanced their cultural prestige and power by openly disseminating his writings—treatises, translations, commentaries, and hundreds of letters—and by extensive personal contacts that reached across Europe. His orientation was essentially exoteric, as was the humanist culture within which he lived. His open dissemination of Platonism was consonant with the interests of his patrons in that it augmented their own cultural prestige.

Yet Ficino inherited certain aspects of an esoteric doctrine. As Kristeller notes, he derived from the Byzantine humanist Pletho "the idea of an ancient tradition of pagan theology that led directly from Zoroaster, Hermes Trismegistus, Orpheus, and Pythagoras to Plato and his followers." This theology involves an esoteric doctrine indicative of divine origins. Ficino himself writes that "the ancient theologians covered all the sacred mysteries of divine things with poetic veils, that they might not be diffused among profane people." Ficino saw a history that led from ancient theology to the philosophical Christianity of his own day as represented in his own *Theologia Platonica*, as he called his major philosophical work.[19]

Yet Ficino does not sanction for his own writing the esoterism to which he was heir. In the passage cited above, for example, he discusses "poetic veils" not to endorse them but to explain that such "veils" caused the successors of the ancient theologians to develop various and mistaken interpretations of the ancient theology. After Plato, various Platonic academies misinterpreted Plato's doctrines of the soul. Only the last two academies, those of Plotinus and Proclus, interpreted Plato correctly. Discussing the doctrine of the soul as he believes Plotinus and Proclus to have understood it, Ficino concludes that Plato did not actually believe in the transmigration of souls. His discussion here, far from promoting esoterism, encompasses the humanist goal of understanding and clarifying ancient texts. His conclusions also advance his synthesis of Platonic and Christian doctrine, an effort that could not have succeeded without a serious attack on the Platonic belief in the transmigration of souls.[20]

In another departure from his Neoplatonic predecessors of late antiquity Ficino transforms the Plotinan hierarchy of being in a way that creates a more worldly doctrine. For Plotinus the hierarchy consisted of six entities—One, Mind, Soul, Sensation, Nature, and Body—with the scale of value moving in a linear direction upward from the lowest, Body, to the highest, One. In the *Theologia Platonica* Ficino changes the six entities to five: God, Angel, Soul, Quality, and Body. Thus he creates a Christianized version of the hierarchy, making God equivalent to One and, following medieval precedent, Angel equivalent to Mind. He eliminates Sensation and Nature and substitutes Quality in their stead. This allows Soul a central place of mediation between the lower and the higher realms. Ficino explains that "the soul is the intermediate degree of being. It establishes a line of unity between all the degrees inferior and superior—it rises toward the degree superior and it descends toward the step inferior." He thus elevates the importance of the human soul, and in emphasizing its mediative role he enhances the position of the lower entities, Quality and Body, as well as the dignity of human life.[21] Ficino changes the Neoplatonic hierarchy in a way that gives the corporeal world and human life on earth substantially greater value.

Ficino's wide appeal can be understood by looking at his most popular treatise, *De vita triplici (Three Books on Life)*. Published in 1489, it had appeared in almost thirty editions by 1647, including translations into German, Italian, and French. The treatise, which concerns how scholars should maintain their health, is dedicated to Lorenzo di Medici, ruler of Florence. Ficino composed the *De vita triplici* not only for his patron but also explicitly for scholars. The first book, "On a Healthy Life," treats the melancholic disposition of scholars and their propensity for black bile and provides extensive advice on how to avoid melancholy. Ficino dedicates the second book, "How to Prolong Your Life," to one Filippo Valori, a distinguished Florentine who paid for the publication of the work. The third book, "On Making Life Agree with the Heavens," is dedicated to the king of Hungary, Matthias Corvinus, a well-known patron of the arts.[22]

The third book is a commentary on a short passage from Plotinus's *Enneads* that concerns drawing divine powers to the earth (4.3.11). Plotinus suggested that men of antiquity constructed temples and statues in order to draw the gods to them and that these constructions were sympathetic to divine souls and thus able to receive them more easily. Carol Kaske observes that Platonists such as Plotinus "advocated images, but for theurgy—alluring gods into sensible forms the better to worship them." Ficino's goal is radically different: Unlike his late antique predecessors, who "scorned materialistic ends," Ficino aims at using his soul-attracting methods to affect bodily health in positive ways.[23] Rather than moving away from the material realm to higher and higher gradations of spiritual ascendancy, he attempts to harness spirits for the improvement of material and physical life.

Ficino combines a practical orientation with openness that is directed toward a particular readership, namely, scholars and intellectuals. He begins the third book with welcoming rhetoric: "Hail, intellectual guest!" (Salve hospes ingeniose). His metaphor of hospitality is inclusive rather than exclusive. "Hail to you, too, whosoever you are who approaches our threshold desiring health! See, eager guest, first of all, how hospitable I am. For certainly it was the role of the visitor, to first salute the hostelry; but I, before you could salute me, have saluted you as soon as I saw you. I have gladly welcomed you while you were entering and as yet unknown. And if you stay with me awhile, I will give you, please God, the health [*salute*] I promised you. You have gained a lodging friendly to all and now full of love towards you."[24]

The doctrine that Ficino invites his readers to learn is that the "intellect" and the "body" are connected by soul. The nature of soul is that on the one side it "conforms to the divine and on the other side to the transient." Ficino explains that the congruity between the divine and the material is mediated by the Platonic Idea. By divine power the World-soul possesses as many seminal reasons of things as there are Ideas in the Divine Mind. By these reasons, the World-soul "fashions the same number of species in matter." Ficino's therapy involves capturing the power of celestial bodies through medicine, or in some case through objects such as rings (i.e., "medicines internal or external").[25]

Explaining an aspect of his astrological doctrine, Ficino suggests that through the influence of Saturn "the substance of spirit" is recalled from the outer to the innermost, and from the lowest to the highest, faculties and subjects. For this reason, Saturn "helps one contemplate the more secret and the higher subjects." Both natural and artificial things possess occult powers from the stars. Through these powers they expose our spirit to the stars. Ficino's complex discussion of how occult powers are acquired in the material realm makes up the central thematic core of the third book. Borrowing concepts from Neoplatonic authors as well as from the magical text *Picatrix,* a compilation of Hellenistic magical texts translated from Arabic into Spanish in the thirteenth century and into Latin later, Ficino examines the ways in which celestial powers might be drawn to the earthly realm in order to influence health and life in a positive way.[26]

Ficino's own efforts to communicate the knowledge of occult things represents an attitude of openness. As noted above, he addresses scholars in particular and their melancholic tendencies resulting from too much black bile. Yet there is little evidence that he excludes others or that he is concerned to reserve his knowledge or insights for a small group of initiates. Rather, he shows how to understand and thereby harness the hidden powers of the universe for utilitarian ends.

Nevertheless, he needs to avoid the charge of being heretical or unchristian. In a final "Apology" he is compelled to write: "Marsilio is not approving magic and images but recounting them in the course of an interpretation of Plotinus." He does not, he insists, advocate "profane magic which depends on the worship of daemons" but mentions only "natural magic, which, by natural things, seeks to obtain the services of the celestials for the prosperous health of our bodies." He suggests that there are two kinds of magic. In the first kind, practitioners join themselves to demons by means of specific rites and then rely on their help to "contrive portents." Ficino rejects this practice as unchristian and advocates instead, the second kind of magic, which "is practiced by those who seasonably subject natural materials to natural causes to be formed in a wondrous way."[27]

Ficino's Neoplatonism spread across Europe. Many who absorbed his writings exerted their own broad influence. Significant early figures include Pico della Mirandola; the Italian humanist Lodovico Lazzarelli; Symphorien Champier, a physician from Lyon who was instrumental in introducing Ficinian Platonism into France; and Johannes Reuchlin. Within this complex cultural diffusion Hermeticism and Platonism were enriched by further accretions, most importantly the Jewish cabala, which was investigated by Pico and Christianized most importantly by Reuchlin; and by the doctrine of signatures and writings on the art of memory of Raymond Lull.[28]

Cornelius Agrippa: Humanist Theologian and Student of Arcana

Cornelius Agrippa, the most influential proponent of Neoplatonic magic in the early sixteenth century, incorporated many of these strands of thought into his vast compendium *De occulta philosophia.* Agrippa, who was from the region of Cologne, attended the University of Cologne and then spent seven years in Italy, where he lectured at the University of Pavia. Topics of his lectures included *Pimander,* the first dialogue of the Hermetic corpus, translated by Ficino. After his Italian sojourn, Agrippa lived an unsettled life, moving often to escape accusations of heresy and to search for patronage. He was a humanist, a Neoplatonist, a religious controversialist, a believer in the mysteries of the cabala, and a Hermeticist. A contentious, expansive, and original religious thinker, he lived in an age increasingly torn by religious conflict.[29]

Agrippa insisted on his loyalty to the Catholic Church and carefully distinguished good magic from the black magic that involved consorting with demons. In his *De incertitudine et vanitate scientiarum atque artium declamatio* (Declamation on the uncertainty and vanity of the sciences) he argues that because all arts and sciences have been the subject of extensive disagreement among scholars, it is preferable to rely upon faith in God instead of the uncertain rationality of human disciplines. Because those disciplines include magic, his remarks have been taken to contradict his great compendium of magic, *De occulta philosophia,* yet recently scholars have suggested that the two works are complementary rather than contradictory.[30] Putting one's highest faith in God does not necessarily preclude inquiring into philosophical magic.

Agrippa was a humanist theologian who was influenced by humanists such as his contemporaries Erasmus and Jacques Lefèbvre D'Étaples. He opposed scholastic theology and wanted to engage theologians in discussions concerning theological and doctrinal issues. Yet unlike Erasmus, he was unable to maintain a relatively neutral stance. He engaged in aggressively hostile arguments with conservative theologians, frequently involving himself in religious controversies. In an important revisionist study Marc van der Poel argues convincingly that Agrippa sought open disputation with theologians and deeply resented their refusal to engage his arguments directly. He also hated their habit of secretly circulating accusations and criticisms about him. Such accusations often destroyed his patronage relationships, as well as his hopes for gainful employment.[31]

Poel observes that Agrippa devoted his entire life to two things: "on the one hand the study of created things in which God reveals himself to man (i.e., the 'arcana' or 'secreta naturae'; occult philosophy) and on the other hand the study of divine things (i.e., the 'res sacrae'; theology)." The focus of my own discussion is the first—arcana, or secrets of nature. Yet it is well to keep in mind that Agrippa advocated open disputation of theological issues, while he also developed ideas about esoteric communication involving the secrets of nature. He was an advocate of openness in theological disputes and of secrecy in particular matters of arcana. Concerning the latter, he both practiced esoteric communication and articulated ideas about it. His views toward openness and secrecy were not contradictory but complementary. He merely applied them to separate areas of thought.[32]

In 1510, while he was still in Italy, Agrippa completed the first version of his great summa of natural, celestial, and ceremonial magic, the *De occulta philosophia*. After years of revision and manuscript circulation, an incomplete version of book 1 was published in 1531, followed by a full version in 1533. The *De occulta philosophia* involves, as its modern editor, Vittoria Perrone Compagni, put it, the notion of a "radical restoration of magic." It is a synthesizing encyclopedia composed of eclectic borrowings from a great variety of sources, including Pliny, Albertus Magnus, the *Picatrix*, the writings of Ficino, especially the *De vita triplici*, the Neoplatonic treatise *De harmonice mundi*, of Francesco Giorgi, and the Neoplatonic, cabalistic writings of Johannes Reuchlin. Influenced by Hermeticism, Christian-cabalist syncretism, and Neoplatonism, Agrippa provides a philosophical foundation for ancient and medieval traditions of magic and explicates a Neoplatonic, animistic cosmos in which spiritual forces can be harnessed for human spiritual and material power and well-being.[33]

In a letter to his readers Agrippa insists that he writes for those who are sympathetic to his ideas, not for those readers "weak in judgment and of hostile opinion, indeed many of evil disposition and unappreciative of our talents," who, "with their rash ignorance," may understand the name magic "in the lower form." These readers will accuse him of being a sorcerer, superstitious, demoniacal, and a magician. Agrippa provides a spirited defense of magic as a philosophically astute discipline that is praised by theology and associated with Christianity. He warns the unsympathetic not to read his book because its magical power might well drive them mad. Sympathetic readers, in contrast, will benefit from the utilitarian efficacy of magic. Magic contains many useful things "for turning away evil events, for destroying sorceries, for curing diseases, for exterminating phantasms, [and] for the felicitous preservation of life, honor, fortune." Because such things are so profitable and necessary, they can be accomplished "without offense to God or injury to religion."[34]

The *De occulta philosophia* circulated for years in manuscript form before Agrippa had it published in the early 1530s. Defending himself against charges of error, he explains that he had composed the treatise over many years and had completed the first version twenty years earlier, in 1511, when still a youth. He claims that later, as a grown man, he retracted the ideas of his childhood (in *De incertitudine et vanitate scientiarum atque artium declamatio*). Yet both the 1531 version of the first book of the *De occulta philosophia* and the complete treatise, published in 1533, represent a coherent synthesis that integrates the 1511 version into an expanded whole.[35]

Explaining why he finally decided to publish his treatise in the early 1530s after his so-called retraction, Agrippa tries to avoid the charge of heresy while defending his early writings. He had given the early manuscript to Johannes Trithemius, a Benedictine abbot, who was "a man very assiduous for arcane things." Afterwards, Agrippa reports, the work, "being intercepted before I finished it," was carried about "imperfect and unpolished." He describes a manuscript circulation in Italy, France, and Germany, claiming that some men, "whether more impatiently or imprudently I do not know," would have published it even in its imperfect state. Thus, he decided to publish the work himself, thinking that there was "less danger, perhaps, if these books were produced with my own hand with a few amendments than if published torn on account of confused fragments by the hands of others." He does not think it is a crime to save a token from his youth, and he admonishes readers to pardon the curiosities of youth if anything in the work is displeasing.[36] Thereby he attempts to protect himself from charges of heresy.

Many years before, having read Agrippa's treatise of 1511, Trithemius had responded with an appreciative letter that also contained a warning: "Yet this one precept we advise you to observe—that you communicate vulgar things to vulgar friends, but higher and arcane things only to higher and secret friends; Give hay to an ox, sugar only to a parrot." Trithemius was not concerned with the conservation of higher truth for suitable and spiritually prepared initiates but with

Agrippa's safety. He warns Agrippa that he could be trampled by oxen! The abbot's warning reflected his own troubles. He had been accused of conjuring spirits by Charles de Bovelles (the ox), who during a visit had read his *Steganographia*, a cryptographic treatise in which conjured spirits and demons carry messages back and forth. Because of Bovelles and other critics, Trithemius was removed from his position as abbot of Sponheim in 1505.[37]

Agrippa experienced similar troubles when he began to publish his works in the early 1530s. In 1529 he went to Antwerp and became archivist and imperial historian in the court of Margaret of Austria. Publishing a collection of seven short treatises in Antwerp, he also received an imperial privilege to publish his major works. In this process Margaret had the theologians of Louvain review his *De incertitudine et vanitate scientiarum*. They promptly condemned it as "scandalous, impious and heretical." Agrippa was required to answer the accusations of the theologians of Louvain at the parliament of Malines. Advised to submit to the procedure and retract offending statements, he instead refuted the criticisms point by point, attacking the theologians who had accused him.[38]

Losing Margaret's patronage and his position at court, he moved to the household and protection of his patron Hermann von Wied, the archbishop of Cologne. There, he completed the full version of the *De occulta philosophia*. A Cologne printer, Johannes Soter, had begun typesetting when the Dominican inquisitor Conrad Köllin of Ulm denounced it as heretical and nefarious. The Cologne senate ordered the printing to cease, to which Agrippa responded with a scathing attack on the senate itself. Only with Archbishop Wied's intervention was the treatise finally published in July 1533, lacking any indication of a printer's name or place of publication and including as an appendix several of the chapters of *De incertitudine et vanitate scientiarum* that criticized magic. The remainder of Agrippa's life is barely documented. His pupil Johannes Wier reports that he stayed in Bonn until 1535, then returned to France, where he was arrested on the order of Francis I for insulting the Queen Mother. Shortly after his release from prison he died in Grenoble, about 1535.[39]

In striking contrast to his lack of power and protection near the end of his life, Agrippa postulates in the *De occulta philosophia* a hierarchical cosmos in which divine power ranges through complex gradations of lesser powers for the explicit purpose of serving humans. Agrippa delineates a tripartite world, made up "of the elemental, the coelestial, and the intellectual," in which "every inferior is ruled by a superior and receives the influx of their powers." The Archetype and Highest Creator himself pours his power "through the angels, the heavens, stars, elements, animals, plants, metals, [and] stones" to us for whom he created all these things. Wise men, Agrippa assures his readers, believe it should be possible to ascend through this complex hierarchy to God himself. The result would be not only to enjoy the power already in things but to draw new potency from above.[40]

Agrippa's complex Neoplatonic explication of occult powers involves the notion that it is the world spirit that unites occult virtues to their subjects. Occult properties do not exist because of the nature of the elements; rather, they come from heaven, are hidden from the senses, and are barely known by reason. They set out from "the spirit and the life of the world through the rays themselves of the stars (which cannot be investigated by us other than by experience and conjecture)." Agrippa explains in detail which things in the elemental world are influenced by which specific celestial bodies. The sun influences solar things, the moon, lunar things, and so on through Jupiter, Mars, Venus, Mercury, and the fixed stars. Each star has a specific nature and produces a particular seal or character that impresses itself through invisible rays upon those things that it governs in the material world.[41]

Humans are uniquely positioned in this vast system of interconnections. Human nature, rather than being governed by one particular star, is "the most complete image of the whole universe, containing in itself the whole heavenly harmony." In human nature, as a result, we discover "the seals and characters of all the stars and all the celestial influxes." They include those that are "more efficacious" in that "they are less remote from celestial nature." This distinctive positioning of humans does not, however, lead to complete human knowledge of hidden things. God alone knows the number of stars and their effects on inferior things; no human intellect can ascend to this knowledge. Ancient chiromancers and philosophers can learn only a few of these things, "partly by reason, partly by experience." Many things "lie concealed in the treasury of nature."[42] Even though human nature bears the signatures of the entire harmonious cosmos, thorough knowledge of all things is nevertheless beyond its compass.

Yet Agrippa will reveal all he knows to his patron, Hermann von Wied. In a dedication Agrippa urges him to accept the second book of the *De occulta philosophia,* in which "we will make known the mysteries of celestial magic, after laying open and showing all things that the ancients skilled in these things produced"—all things, that is, about which he has knowledge. He sets forth the arcana of celestial magic for those who are studious and show natural virtue. Whoever benefits should thank Wied, for "these things, having been laid out, are released from chains." In his dedication to the third book, Agrippa emphasizes that he communicates the mystery of divine and ceremonial magic, which "I have learned without falsehood." He is, he says, "exposing to the light that which is buried in the dust of antiquity and enveloped in the fog of oblivion just as in the Cimmerian darkness to this day."[43] Thus, he reveals and explains philosophical magic that has been buried in obscurity for centuries.

The power of number suffused both the material realm and the immaterial cosmos, which the magus had to understand and harness in order to achieve his ends. Agrippa provided the means to that understanding through his extensive explanation, in which he treats each number from one ("unity") to twelve in a separate chapter. He illustrates each of the first twelve numbers with a chart in which he discusses the number as it exists in each of six different worlds For example, the worlds of the number two consist of the Archetype, the intellectual, the celestial, the elementary, the lesser, and the infernal.[44]

Influenced by the cabalistic mysticism of Reuchlin, Agrippa suggests that the power of numbers extends to letters, thereby making possible a kind of arithmetical prognostication through the study of names. If these divine numbers can be drawn out from the proper names of things, we can gain opinions "concerning things hidden and future." Through the numbers of the name much can be prognosticated. Both Pythagoreans and cabalists of Hebrew understood that "hidden and mysterious things lie concealed in those numbers, understood by few." The implicit numerology of names exists because the Creator created all number, measure, and weight, from which the truth of both letters and names originate. They were instituted "not by chance but by certain (granted unknown to us) reasons."[45]

The magician works by means of natural forces: "Celestial souls press their own powers into celestial bodies, which then transmit those to this sensible world." The magus through certain mysterious words is able to draw these powers through a natural process. Agrippa insists that every superior thing moves its proximate inferior. Thus, the universal soul moves particular souls, the rational soul governs the sensual, and the sensual rules the vegetable.[46] Noncorporeal realms intricately connect to earthly ones. The magician manipulates the powers inherent in the system for utilitarian ends.

Concerning magic, Agrippa urges secrecy. Whoever studies this discipline should "conceal in silence and hide such sacred dogma with constant taciturnity within the secret sanctuary of your religious breast." He cites Hermes to the effect that to publish something filled with the divine is a sign of "an irreligious mind." He lists ancient philosophers who advocated secrecy, including Plato, Pythagoras, Orpheus, and Porphyry. Pointing to the religious books of the Egyptians, he notes the secrecy of their ceremonies and their hieroglyphic writing. The ancients always took care "to cover the sacraments of God and nature and hide them with diverse enigmas." He cites Apuleius: "I would say, if to say were allowed, you should know it, if it were permitted to hear it, but both ears and tongue would contract the guilt of rash curiosity." He concludes by noting that even Christ spoke so that only his most intimate disciples understood the mystery of the words of God, while others heard only parables.[47]

Agrippa warns that it is improper "to commit to public letters" secret things, which should be communicated only among a few wise men by word of mouth. His own readers will indulge him if he silently passes over many things, including very powerful secrets of ceremonial magic. Yet they will not go away from the book devoid of all mysteries. In turn, such readers must not tell secrets to unworthy people but must guard them with the reverence due to them. Magical operations themselves are destroyed by publicity. "It is suitable, therefore, that the operator of magic," if he wants results, should mention his work to no one except a close companion, someone noble by nature or by education who is faithful, creditable, and taciturn. Even the loquaciousness, incredulity, and unworthiness of a companion "impedes and disturbs the effect [of magic] in all operations." An arcane and secret manner is necessary for anyone who wants to work in this art. When the mind is overwhelmed by too much commerce of flesh and occupied with the sensible soul of the body, it does not merit the command of divine substances.[48]

Elsewhere Agrippa explains that his own writing has involved strategies of concealment that will screen out the ignorant from understanding while allowing more perceptive readers to understand his meaning. Within his compendium of magic "certain things are written with order, certain things without order, certain things are transmitted through fragments, certain things indeed are hidden, having been left for the investigation of the intelligent." Agrippa explains that he has transmitted the art in such a way that it is not concealed from "the prudent and intelligent," and yet it conceals its secrets from "the truly depraved and incredulous."[49]

Agrippa tells appropriate readers how to read his book, explaining his art of dispersion: "You therefore, sons of learning and wisdom, search diligently in this book, collecting our dispersed instruction, which we have propounded in various places, and what by us is hidden in one place we make manifest in another, so that it is made visible to you wise men." He writes only for a particular kind of person, one whose soul is incorrupt, who has been taught the right order of living, who has a pure and chaste mind, who reveres God, whose hands are clean from all evil deeds and faults, and who is virtuous, sober, and modest. Only such persons will discover this doctrine preserved for them, including "secrets hidden by many enigmas." Into such persons will be insinuated the entire invincible knowledge of the discipline of magic.[50] Access to knowledge of magic and its powers is intrinsically tied to purity and other virtues of character, mind, and soul.

Agrippa himself pursued secret investigations, including those involving alchemy, and he participated in informal esoteric groups. He and a group of friends seem to have created a secret society. One piece of evidence is a letter from a friend named Landulphus to Agrippa that introduces a potential new member to their group: "And he is a curious investigator of arcane matters, and a free man, restrained by no bonds, who, impelled by I know not what reputation of yours, wishes to search through your secrets also." Make him, Landulphus advises, "if he wants to swear to our rules, an initiate of our society." Other evidence of Agrippa's personal occult interests can be found in his correspondence with his friend Jean Rogier, alias Brennonius, the pastor of a local parish in Metz. Their letters reveal an ongoing interest in occult matters and books. Such friendships and esoteric groups seem to have created lifelong bonds of secret sharing. For example, in *De incertitudine et vanitate scientiarum,* completed in 1526, he reveals that he would disclose more of the art of alchemy were it not for his lifelong oath of secrecy.[51]

Agrippa's letter of 1527 to a new friend, Aurelius ab Aquapendente, suggests that he considered a personal relationship with a master to be essential for the investigation of the secrets of nature. Aquapendente had written to Agrippa concerning books he had read, including a manuscript version of the *De occulta philosophia.* Agrippa asks, "Who are your leaders that you follow?" He warns that a student might be deceived by those who are themselves deceived. Reading books alone cannot direct you since they are "mere enigmas." He itemizes books on the magical art, astrology, and alchemy that are false but followed nevertheless. Yet since they were written by great and serious philosophers and holy men, it would be impious to think that they were lies. This shows, Agrippa concludes, that they contain concealed mysteries that have not been publicly explained by any master. This is why he doubts "if any can read the books alone without a skillful and faithful master, unless illuminated by the divine, which is given to very few." Concerning the version of the *De occulta philosophia* that Aquapendente has read, Agrippa regrets the imperfections of the present version and reveals his plan to complete and revise it, "the key of the work nevertheless having been reserved only for the most intimate friends," including his new friend Aquapendente.[52]

Agrippa saw occult philosophy as a way to harness the hidden powers of the universe to effect change in the world. Such an orientation is consonant with other of his technical interests and activities (beyond his intermittent practice of medicine) to which he intriguingly refers in his writings. He claims that he undertook not only diplomatic but also military service for the emperor Maximilian. In 1508 he went to Spain with some friends for some unknown military adventure (perhaps to put down a peasants' revolt on a friend's estate in northern Spain). Whatever the reason for the excursion, he used some kind of military contrivance that he had invented, probably pyrotechnic. At another point he attempted to obtain patronage from Francis I and the Queen Mother by making reference to a treatise he planned to write on engines of war. He also claims to have begun to write a treatise on mining and to have supervised some imperial mines. Although none of these activities are otherwise documented, they point to ongoing practical and technical interests along with the practices of medicine and law.[53]

This same interest in practical and technical efficacy undergirds Agrippa's occult philosophy. In a fine article on Agrippa's 1533 struggles in Cologne, Charles Zika puts it succinctly: "The most critical and distinctive characteristic of Agrippa's occult philosophy or magic, . . . was its concern with 'operation.'" For Agrippa and other sixteenth-century scholars magic did not involve "a mystical escape from the society and the world." Rather, it "involved a knowledge which could 'operate' in the world, which could influence individual consciousness, provide models for group behaviour, help ensure future success, mitigate the influence of evil and misfortune. And in this sense it shared common aims with the humanists, who . . . were concerned to influence behaviour through eloquence and persuasion."[54] Agrippa's utilitarian orientation had been shared by an influential predecessor, Marsilio Ficino, and would be characteristic as well of the thought of his younger contemporary Paracelsus.

Openness and Secrecy in the Philosophical Medicine of Paracelsus

The iconoclastic physician Theophrastus Philippus Aureolus Bombastus von Hohenheim, called Paracelsus, led a life of wandering, involved himself in numerous vitriolic conflicts, and suffered intermittent persecution. Paracelsus was both a physician and a religious thinker who opposed the traditional book learning promulgated by the universities, including traditional medicine based upon the Galenic notion of the balance of the four humors. His early career as an unconventional medical practitioner initially showed great promise. He won the support of patrons and gained renown by his relatively noninterventionist approach to medical treatment, as well as by a few spectacular cures. His early education and experience included his father's tutelage in natural philosophy, botany, and metallurgy, experience in mining, possibly a medical degree or at least some medical training at the University of Ferrara, and a stint as a military surgeon in the service of Venice and other powers, as well as instruction by one of Cornelius Agrippa's early mentors, Johannes Trithemius. Paracelsus practiced alchemy and was indebted to medieval alchemical traditions, which he combined in important ways with his medical theory and practice. In the practice of medicine he advocated experience over books. Yet for him experience did not represent the equivalent of a modern notion of empiricism; rather it combined practical experience, the experience of wandering through diverse places, and the religious experience of a pious nonconformist of the early Reformation who had adapted a Neoplatonic cosmos to his own particular system of beliefs.[55]

Paracelsus failed to establish himself in any given locality despite his good early prospects. He was forced to leave Salzburg, barely escaping prosecution for sympathizing with the peasants in the Peasants' War of 1525. After several other stops, he set up a medical practice in Strasbourg but then accepted an invitation to Basel, where he successfully treated an illness of Johannes Froben, the well-known humanist publisher. With the help of Froben and Erasmus, who was staying in Froben's house at the time, he was appointed municipal physician and professor of medicine at the University of Basel. The other professors opposed him, especially after he announced his disagreement with Galenism and promised a new course based not on traditional texts but on his own experience as a naturalist and practicing physician. He then publicly burned the *Canon* of Avicenna, and he irritated his critics further by giving lectures in German rather than Latin. The untimely death of Froben, his patron, gave the upper hand to opposing academic physicians and hostile apothecaries (whose profits he had decried). His 1528 denouement in Basel came in a dispute over fees, a result of his habit of charging exorbitant fees to the rich and insisting upon their collection, while treating the poor for little or nothing.[56]

Thereafter he became a kind of itinerant lay preacher and medical practitioner. His volatile personality led to frequent harangues, arguments, and hasty departures, but he also taught disciples, led followers, and treated patients on a regular basis. In one of his seven "Defensiones," written in 1538, he defends his peripatetic lifestyle. His defense of wandering centers on its role in the acquisition of knowledge. "The arts are not all confined within one's fatherland, but they are distributed over the whole world." They can be found neither in one man alone nor in one place. Rather, "they must be gathered together, sought out and captured, where they are." The heavens influence all parts of the world. Therefore, just as one who seeks God must go after him, one who seeks experience should go after that, "competently enquire," and then "move on to further experiences." A man should also travel if he wishes to recognize many diseases. If he travels far, he will have great experience and learn to recognize many things. Paracelsus concludes that it is praiseworthy to have traveled. "For this I would prove through nature: He who would explore her, must tread her books with his feet. Scripture is explored through its letters; but nature from land to land. Every land is a leaf. Such is the *Codex naturae*; thus must her leaves be turned."[57]

During his lifetime Paracelsus's influence seems to have been primarily personal, a result of oral communication and individual religio-medical practice. Most of his voluminous writings circulated only in manuscript during his lifetime, unpublished in his view because of the obstructions of his enemies. He died in poverty in Salzburg. He was buried at Salzburg Cathedral, and his grave became a pilgrimage site for the sick and afflicted.[58]

In a recent study, Andrew Weeks rightly insists that Paracelsus's "science" and his religious philosophy should not be sharply distinguished. Weeks argues that Paracelsus's essential context includes the religious upheavals of early Reformation Germany and suggests that the separation of his writings in the modern edition of the Paracelsian corpus into medical, scientific, and philosophical writings, on the one hand, and religious writings, on the other, is misleading and anachronistic. Yet Weeks unnecessarily minimizes the equally important Neoplatonic context of Paracelsus's thought.[59] Paracelsus's theoretical system as a whole is predicated in essential ways on the Neoplatonic cosmos promulgated by the writings of Ficino, Agrippa, and others in the early sixteenth century. His system also involves the application of alchemy, both philosophical and material, to the practice of medicine.

Paracelsus rejects traditional Galenic medicine based on the balance of the four humors within the whole organism. His intricate system takes two traditional alchemical principles of activity, mercury and sulfur, and adds a third, salt. He combines these principles with the four traditional elements—earth, air, fire, and water—elements conceived, however, in very non-Aristotelian ways. He individualizes these principles, utilizing a complex doctrine of signatures in which particular things exhibit the signs of related things in the natural world and in the heavens. He expounds the Platonic view of the human being as a microcosm, or small world, that reflects the great cosmos. The complex intricacies of this cosmos are indebted to medieval alchemy and to the writings of Ficino, Agrippa, and Johannes Reuchlin, and they are also informed by biblical and Reformation theology. Infused with spiritual and noncorporeal entities, characterized by transformative powers, Paracelsus's thought is a syncretic synthesis of many parts that is similar to and also different from the thought of Ficino and Agrippa.[60]

Paracelsus's enormous posthumous influence can be attributed in part to the rich nominalism of his system and to the many ways in which physical explanations, remedies, corporeal subsystems, diseases, and alchemical processes are linked to a multifaceted spiritual universe. Paracelsus's system validates individual everyday experience and offers innovative therapeutics and hope for medical reform. While he lived, I would argue, his reputation was based not on his theoretical system as a whole but on his activities and reputation as a medical lay preacher and on his vivid rhetoric against academic physicians and money-mongering apothecaries.

Paracelsus's mode of communication of various aspects of his complex interrelated system of the body and the world is at times highly obscure and at other times reasonably clear. It is not for the most part deliberate obscurantism. In a discussion of the historiography of openness and secrecy in chemistry, Jan Golinski has pointed to scholarship on Paracelsianism after 1550 that demonstrates that it was at the center above all of a medical-reform movement. Concerning Paracelsus, Golinski points to his insistence on speaking and writing in the vernacular and to the scholarship that shows that ideal of openness began not with opponents of Paracelsianism but with Paracelsus himself.[61] In addition to openness, Paracelsus articulates a doctrine of restriction and esoterism. His esoteric doctrines, however, should be seen in the broad context of his life's work, which as a whole combines proselytization and openness with tenets of secrecy.

Paracelsus uses authorship to carry on his lifelong polemic against traditional academic medicine based on the books of the ancients. He also writes to expound his views, dedicating many of his books to potential patrons and addressing followers, whom he often calls his "disciples." In the *Archidoxis*, of the mid-1520s, a work influenced by the writings of John of Rupescissa and by pseudo-Lullian alchemical treatises, Paracelsus admonishes his "dearest sons" to consider "our wretchedness and destitute condition." Given the misery of human life, as long as we follow ancient medicine and the books of the ancients, we will be imprisoned in destitution, weighted down and bound in chains. There are many doctors who through the old medicine have arrived "at great riches," but they have done so "with many lies." Paracelsus, in contrast, wishes to come "to more certain ends and practices." He does this by approaching the "mystery of nature" and freeing it or separating it from its impediments. Compared with the art of the apothecaries *(aromatariorum)*, his own art of separation is "as light in a darkness."[62] He is bringing into the light a new art of medicine based on knowledge and investigation into the mysteries of nature, which are both theological and medical. Despite the difficulties of those mysteries, he portrays his work as an enlightened medicine, unlike the obscure and profit-motivated practices of traditional physicians and apothecaries.

Yet Paracelsus here addresses disciples. He wants to restrict his readership to a small group of followers and keep his teachings from the "common people." He has written the *Archidoxis* "as an aid to the memory for us." He specifies that "we wish to include ourselves and only with ourselves, to speak only to our [followers], to write to the same with sufficient understanding, and not to write for the commonality of the people." He explains that "we wish not to show and give our sense and thought, our heart and mind to the deaf; and thus we lock [them] up with a good wall and with a key." In case his work should become known, however, and not be kept "from such idiots who are enemies to all the arts," Paracelsus will not publish the final (tenth) book of the *Archidoxis*, which concerns "the uses of all the others [i.e., the other books]." Thus, we "will not push a happy ape to the idolatrous gods but will not be less sufficiently comprehensible to our own [disciples]."[63]

The subject of the *Archidoxis* is the microcosm, or body, and the entities by which the body is related to the cosmos. Those entities include "quintessence," the four arcana—*prima materia*, philosopher's stone, *mercurius vitae*, and tincture, and the "magistries" and their extraction from various substances (e.g., pearls, growing things, and blood). The work also includes sections on "specifics," as well as elixirs, remedies, and internal and external diseases. Medicine and its practice are profoundly dependent on a knowledge of the cosmos and its interrelationships with bodies and things in the world.[64]

Knowledge of medical matters comes not just from writings, in Paracelsus's view, but from practice and experience. He gives an example concerning the "quintessence"—a "nature, a force, a virtue, and a medicine"—which is shut up within things but when extracted is free of any corporeality. He writes only briefly concerning the separation of the quintessence and insists that "no one should wonder about the reasons for our shortness of hand and pen." His discussion is sufficiently long that "it demonstrates fundamentally and clearly the work thereupon that should and will happen." Too much writing on the subject will "drag [the reader] into peevishness" and does not take into consideration "that the work and practice of such demonstrates everything sufficiently."[65] These remarks, which occur within a discussion of the separations of various substances, underscore Paracelsus's belief that the experiential knowledge from practice, what we might call tacit craft knowledge, is an essential complement to any written account.

In a treatise titled *Liber de longa vita* (On long life), composed about 1526 or 1527, Paracelsus again emphasizes that he writes for his own disciples, who have practical experience in the relevant arts, and not for academic physicians, who do not. He wishes to speak only with his disciples, who through experience know about the properties of things, discovered through art and practice. Such things, he continues, "are hidden and unknown to common physicians" and to these he does not write. If his treatise is difficult to understand, it is only so "to those who understand neither us nor nature." He is not concerned with these people, nor is he interested in making his treatise comprehensible to them. On the other hand, to those who have a foundation "we wish to have described our processes and to have disclosed the same sufficiently."[66]

Paracelsus again stresses the necessity of experience for knowledge in his treatise on miners' diseases, written in the 1530s. He notes that only those who are actually experienced with mining and metallurgy will be able to fully understand his written discussion. This is because knowledge of the subject is based upon interaction with the natural world rather than exclusively on books. The nature of things should be regarded highly. Since things must occur in nature, one should not only study books: nature should be further investigated. From nature "comes the correct teaching and the correct instruction." Although Paracelsus does not deliberately conceal or obfuscate his discussion of miners' diseases, he does insist that the tacit knowledge gained from mining experience is necessary for full and clear comprehension. If the introductory sections of each book "will be difficult and strange for the ordinary physicians, the cause is that the mines and that which belongs to them are also foreign to them, therefore it is reasonable that I proceed with experience in the light of nature." Yet despite limitations posed by the necessity of tacit experiential knowledge, Paracelsus here advocates openness: "Since every doctrine comes from God and physic is created for the patient, it should not be concealed nor remain hidden."[67]

Paracelsus warns that the subject will be difficult for the inexperienced: "the prescriptions will be difficult and very difficult for some to understand." They should be satisfied that for those experienced in mining, "especially for the masters of the mines and for those who are experienced with metals, enough has been said and they understand it sufficiently." He continues, "For how can a silk-embroiderer turn a rope maker with his cords into a silk-embroiderer?" Paracelsus's theory and remedies are specific to place (the mines) as well as to diseases (in this case those engendered in the mines). The knowledge that he imparts involves, he suggests by his analogy of silk embroidery and ropemaking, a kind of specific craft knowledge that cannot be generalized or totally explicated in writing. "Thus the disease," he reiterates, "remains for the mine, and the book too for the mine, therefore the understanding also is to be acquired in the mine."[68]

A vivid portrait of Paracelsus can be found in a letter written about 1555 or 1565 by Johannes Oporinus, who had served as Paracelsus's secretary and assistant for two years in the late 1520s. Oporinus's letter, whose accuracy was disputed by earlier scholars, is hostile in that it disparages Paracelsus's medical abilities and describes his drunkenness in graphic detail. Yet I concur with Weeks's recent assessment that there are many reasons for accepting it as an authentic portrait.

Oporinus lived with Paracelsus during and after the Basel sojourn. Assuming the accuracy of Oporinus's description—"he was in all the days and nights that I lived with him . . . given to inebriation and intoxication so that one could discover him sober barely one or two hours"—perhaps the master's drunkenness was related to the disruption of those years. Yet inebriation apparently did nothing to hinder his stellar reputation: "Among the rustic nobles [of Alsatia]" he was considered "just as another Aesculapius." Nor did drink appear to affect his lucidity. When he was very drunk, Oporinus reports, "having returned to the house, he was accustomed to dictate something of philosophy to me." The things that he dictated "seemed to one to cohere beautifully, so that they seemed incapable of being made better in great sobriety." Oporinus also reports that Paracelsus would arrive home drunk and fall onto the bed fully clothed, without unsheathing the sword (a gift from an executioner) that he always carried with him. At times Oporinus feared for his life, such as when Paracelsus would suddenly jump from his bed in a rage and start hitting the floor and walls with his sword. He spent money freely, bought a new set of clothes every month, and tried to give away his old ones, which no one wanted because they were so dirty.[69]

There is no way to confirm most of these reports, although authenticity is suggested by their bizarrely idiosyncratic character. Other aspects of Oporinus's portrait suggest values and beliefs that are amply reflected in Paracelsus's writings. Oporinus notes that Paracelsus continually worked at alchemical operations. He "always had his workshop furnace prepared with perpetual fire" and was busy preparing various concoctions. In addition, "he represented himself as a certain one of the prophets, and he carried knowledge of certain arcane things, as something secret," which, Oporinus confesses, he himself feared. Paracelsus created many medicinal substances and bragged that he alone had been able to "restore the living from death." He cultivated evangelical doctrine but often condemned both Luther and the pope along with Galen and Hippocrates. He complained that no writer, ancient or contemporary, "had brought out every kernel of scripture correctly, but had clung to the outer part or skin."[70]

The vast posthumous influence of Paracelsus reveals the great appeal of his thought. He brought together alchemical operations and theory, Neoplatonic notions of cosmic powers and forces, a fundamental appreciation of experience and craft practice, and an idiosyncratic, highly personal religiosity. All of these elements were combined with his reputation as a medical healer. The appeal of his thought for the sixteenth century had to do with its utilitarian efficacy, its validation of experiential knowledge, and its intense spirituality, or more precisely, the combination of these attributes. His writings are not wholly consistent and offer options to various kinds of readers, including the value of openness, which often depended on the reader's tacit craft or experiential knowledge, and the allure of secrecy.

The Esoterism of Girodano Bruno

Like Cornelius Agrippa and Paracelsus, Giordano Bruno lived a life of wandering and upheaval, driven in part by threatened and actual religious persecution. A prolific and eclectic author who wrote in both Italian and Latin, he utilized the Hermetic corpus, ancient Neoplatonists, and the writings of Ficino and Cornelius Agrippa, among others. Entering a Dominican monastery at age seventeen, he left eleven years later, in 1576. He traveled to France, where he earned a doctorate in theology at the University of Toulouse. His many other travels included a stay in England in 1583–85 as a guest of the French ambassador, Michel de Mauvissière. There he wrote and published *Ash Wednesday Supper*, discussed in detail below. Bruno eventually returned to Italy by invitation of the Venetian nobleman Zuan Mocenigo, who in May 1592 turned him over to the Venetian Inquisition. Although he recanted at his trial, he was nonetheless delivered to the Roman Inquisition. Imprisoned in a Roman dungeon for seven years, he was convicted of heresy; he refused to recant and was burned at the stake in 1600.[71]

Bruno held a view of the cosmos that was influenced by the heliocentric system of Copernicus. He believed in an infinite cosmos with an infinite number of worlds within it; and he thought that the ability to understand and harness the harmonic sympathies of this infinite cosmos could bring about world peace. In her classic study *Giordano Bruno and the Hermetic Tradition* Frances Yates argues that Hermeticism was the underlying synthesizing ingredient in Bruno's thought. Yates's interpretation recently has been challenged by Karen Silvia de León-Jones, who argues instead for the overriding significance in his thought of the cabala.[72] Yet Bruno's thought is a highly eclectic synthesis that cannot be attributed to only a single influence.

Bruno's fusion of theology and cosmology is most evident in *La Cena de la Ceneri*, or *Ash Wednesday Supper*, a dialogue published in London in 1584 that explicates a form of Copernicanism and also concerns the Ash Wednesday Supper, or the Eucharist on the eve of Lent. For Bruno, an understanding of the harmonious, infinite cosmos, infinitely full of heliocentric worlds, is emblematic of the transformative redemption that results from the sacrament of the Eucharist. He adopts the Neoplatonic Ficinan view of ensouled, animated celestial bodies and adds Copernican heliocentrism, thereby creating a harmonious Neoplatonic cosmos. Properly understood, this divine and harmonious universe can be manipulated to bring about harmony and peace between political entities (England and France) and between religions (Protestantism and Catholicism) for which the nature of the Eucharist had long been a major bone of contention.[73] Bruno's manipulation of occult powers had the utilitarian goal of bringing peace and harmony to the mundane world.

Ash Wednesday Supper is structured around a banquet in which the interlocutor speaking for Bruno, Theophilus Philosophus (philosopher lover of god), argues for Bruno's heliocentric cosmos against the Aristotelians. Yet before the discussion gets underway, there is an introductory dialogue and then a long description of how the party of interlocutors struggles from the French ambassador's residence through the mud and mire of London, fighting off the vulgar and insulting London rabble, to the door of the House of Glanville, where the supper debate takes place. Since England is a Protestant country and the House of Glanville is a Protestant household, the Ash Wednesday ceremony of the Eucharist is to be replaced by the Protestant Lord's Supper, which, however, does not take place.

Bruno's views are expressed by Theophilus: "Then, thank God, the ceremony of the cup did not take place." The reasons for his relief emphasize the vulgarity and barbarism of the Protestant substitute for the Eucharist. He describes the Protestant passing of "the goblet or chalice" in physically revolting terms: "After the leader of this dance has detached his lips, leaving a layer of grease which could easily be used as glue, another drinks and leaves you a crumb of bread, an other drinks and leaves a bit of meat on the rim, still another drinks and deposits a hair of his beard." Instead of this repulsive ceremony, a discussion takes place that concerns the structure and rationale for the infinite heliocentric universe.[74]

Bruno's depiction of the vulgarity of the London rabble and of the coarseness of the Lord's Supper are not mere diversions but one aspect of the self-conscious elitism that is an inherent aspect of his philosophy. He believes that only a few can gain true understanding and also emphasizes the originality of his own thought. In the first dialogue, for example, the interlocutor Theophilus relates that sometime before, two men had come to Bruno on behalf of a royal retainer who "longed for his conversation on [and thus his exposition of] Copernicus and other paradoxes in his new philosophy." Bruno replies "that in judging and determining he saw through neither the eyes of Copernicus nor those of Ptolemy, but through his own eyes." He owed much to the observations of these and other mathematicians. From time to time they added "light to light," and they established principles sufficient to lead up to wisdom. Yet wisdom could be brought forth only after many stages. Bruno suggests that these men were only interpreters. There are others "who penetrate into the sense, and they are not the same ones." Ptolemy and Copernicus are "like the country folk who report the circumstances and shape of a battle to a captain who was absent." It is not they who understand "the proceedings, the reasons, and the art by which victory had been gained"; rather it is the captain who has "experience and better judgment of the military art."[75]

Bruno fully praises Copernicus but insists that he did not get to the crux of the matter because he was "more a student of mathematics than of nature." Yet Copernicus "was ordained by the gods to be the dawn which must precede the rising of the sun of the ancient and true philosophy." Bruno makes clear that it is he himself who brings about the rising sun of true philosophy. The interlocutor Theophilus, praising the ancient Tiphys, who invented the first ship and crossed the sea with the Argonauts, and Columbus, who likewise set out on unknown waters, asks whether Bruno should not also be honored, for he "has found the way to ascend to the sky, compass the circumference of the stars, and leave at his back the convex surface of the firmament."[76]

Other explorers often bring violence and destruction, whereas Bruno "has freed the human mind and the knowledge which were shut up in the strait prison of the turbulent air." This freedom has come about because he has eliminated the darkness caused by "the sophists and blockheads" who "extinguished the light which made the minds of our ancient fathers divine and heroic." Bruno here refers to the illumination of the *prisca theologia* and insists upon credit for his own role in allowing the light to shine again. He "has surmounted the air, penetrated the sky, wandered among the stars, passed beyond the borders of the world." By means of the light of his senses and reason, "he laid bare covered and veiled nature, gave eyes to the moles and light to the blind."[77]

Bruno's discoveries are restricted to a limited readership. His is an esoteric learning, for "this burden is not for the shoulders of everyone"; it is for people like Bruno himself, "who can bear it, or at least can move it toward his ends without experiencing perilous difficulty, as Copernicus was able to do." Those who do possess Bruno's truth "should not communicate it to every sort of person." Bruno reiterates the point in a concluding remark of the first dialogue: "In the end it is safer to seek the true and the proper outside the mob, because it [the mob] never contributes anything valuable and worthy. Things of perfection and worth are always found among the few."[78]

Those who understand Bruno's Platonic philosophy reveal its efficacious good through their character and ability to govern. Such people are "moderate in life, expert in medicine, judicious in contemplation, unique in divination, miraculous in magic, wary of superstition, law-abiding, irreproachable in morality, godlike in theology, and heroic in every way." The manifest effects of good character extend to both the corporeal and social realms. Such people live long lives, are healthy, and make "lofty inventions." Their prophesies are fulfilled, their people are peaceful, their sacraments are inviolable, their actions just.[79]

Bruno's catalog of benefits that accrue to those who understand his philosophy underscores his profoundly utilitarian view of the ability to harness cosmic powers. Benefits range from individual health and longevity to peace, harmony, and justice. Those who understand Bruno's philosophy are, it seems to go without saying, princes and rulers. In *Ash Wednesday Supper*, at least, Bruno reserves true understanding only to those capable of the deepest insight. That excludes the vulgar rabble, which Bruno has described in highly uncomplimentary terms, but it also excludes contributors to his cosmological system, such as Copernicus.

Bruno's defense of the Copernican heliocentric cosmos is not only profoundly un-Copernican in his vision of infinite worlds but also in its fundamentally religious and spiritual grounding. Bruno is not an astronomer but a religious philosopher, deeply indebted to Ficino, Agrippa, the Hermetic corpus, and the cabala. His syncretism allowed him to develop a rich, multifarious set of writings and a fruitfully imaginative philosophical outlook that are distinctly his own. His insistent claim to his own originality is notable and is, I suggest, a reflection of his immediate cultural context, in the late sixteenth century, which was characterized by a growing appreciation for the value of originality.

The alchemical and Neoplatonic writings that are the focus of this chapter are eclectic, complex, and diverse, yet they share a number of common values. First, each in its own way embraces the positive value of technical efficacy. The alchemists believed that their art equaled or even improved upon nature. Ficino in *De vita triplici* shows how to manipulate cosmic powers for long life and good health. Agrippa's entire complex synthesis of magic integrates the celestial and mundane worlds rather than separating them. The astute magician can harness the huge potential powers of that system for use in the world. Paracelsus creates an amalgam of alchemy, Neoplatonism, and religion to reform medicine and combat diseases. Bruno suggests that his own knowledge of the cosmos allows him to bring about peace and harmony. For these authors and many of their contemporaries utilitarian values were paramount. They lived at a time when crafts and the mechanical arts flourished, elite classes engaged in conspicuous consumption, and power, including military power, was a central concern. Their own utilitarian spirit very much reflects that of the wider culture.

Renaissance Neoplatonists repaired the Apuleian split between efficacious magic in the world and the noncorporeal life of the spirit. They brought together utilitarian operations, astral magic, the cabala, and various forms of philosophical Neoplatonism. Unlike most of their late antique predecessors, they tried to harness occult powers for utilitarian ends in the material and corporeal world. Bringing together centuries-old traditions of practical magic and Neoplatonic philosophical traditions, they provided legitimation for the former and brought efficacious use to the latter. Utilitarian values undoubtedly attracted some to an interest in alchemy, which had always embodied the potential for simultaneous spiritual and material manipulation and change.

In addition, both alchemy and Neoplatonism seemed to many to offer the possibility of more highly intense and more personal spirituality than did institutional Catholicism or even some of the newer forms of Protestantism. From this point of view, the development of Neoplatonic, alchemical, and related traditions can be seen as part of the religious upheaval of the sixteenth century. This is true whether or not Neoplatonists or alchemists attempted to integrate their philosophy with orthodox Christianity, as many of them did. Yet they risked and sometimes incurred charges of heresy, especially as religious unorthodoxy became increasingly dangerous during the sixteenth century. A view of the religious crisis as a conflict between institutional Catholicism and Lutheranism, Calvinism, and other forms of "Protestantism" is too narrow. Alchemy and various forms of Platonism and magic were also central to the history of spiritual crisis and religious change.

Both alchemical and Neoplatonist writers, relying on ancient traditions, participated in the renewed development of esoteric doctrines and techniques of concealment. Authors utilized a variety of techniques, such as the dispersion of secret doctrine throughout a text. Secrecy was based for the most part on the belief that the moral integrity and purity of the knower or magical operator were crucial. For alchemists, as well as Agrippa and Bruno, knowledge of the world and of the cosmos intersected with the knower's purity of soul. Numerous printed texts publicly disseminated such esoteric doctrines. Doctrines inculcating secrecy and esoterism were disseminated widely through the medium of print.

The attitudes toward authorship expressed in the writings treated in this chapter are notably diverse. The pseudoepigraphical values of alchemy are nowhere apparent in the writings of Ficino, Agrippa, Paracelsus, or Bruno. Ficino published his writings under his own name. They contributed not only to the fame of his patrons but to his own vast reputation. Agrippa circulated his magical treatise in manuscript for many years. When he finally undertook publication in the early 1530s, he preserved intact the writings of his youth, which allowed him to respond to criticism by attributing any problems to youthful writing. Yet his publication of works in their entirety also suggests his interest in preserving and receiving proper credit for all of his writings. Paracelsus's many writings and his boasts concerning his medical prowess suggest a well-developed sense of his own originality and entitlement (to credit if to nothing else). Bruno's view of the superiority and originality of his own thought is explicit. These figures viewed their writings as original and desired credit for their own authorship. In part this can explain why they often explain things openly, while their books were widely distributed in print. For them, the values of openness and secrecy often existed side by side.

Chapter 6

Openness and Authorship I

Mining, Metallurgy, and the Military Arts

SIXTEENTH-CENTURY WRITINGS on mining, metallurgy, artillery, and fortification emerged as a result of specific economic, social, and technological developments. Many of these books contained detailed technical information, which their authors for the most part purveyed openly. Print made possible the rapid reproduction of multiple copies, facilitating wide dissemination. Yet printing by itself should not be credited with the promulgation of the value of openness; it could also promote, as previously noted, the greater circulation of esoteric doctrines. The press was not the unambiguous herald of openness that some have claimed it to be.[1] Although the greater availability of books at lower cost facilitated the development of a larger and more diverse readership, printing cannot be isolated from other aspects of the social, cultural, and economic context in which it operated.

Mining, metallurgy, and the military arts were closely tied to the power and wealth of princes and rulers. From antiquity the military arts had been a focus of authorship. Fifteenth-century manuscript books, as we have seen, tied the praxis of tactics, strategy, and military leadership more closely to the technical arts of weaponry. In the sixteenth century technological developments, including the great expansion of gunpowder artillery and a new form of fortification, the bastion fort, further motivated military authorship. Mining and metallurgy were tied to the development of artillery by virtue of the manufacture of guns. Mining and practices of metallurgy such as ore processing and assaying became a new focus of authorship in the sixteenth century. Authors on both mining and the military arts treated their topics openly; indeed, some explicitly advocated open communication.

The display of courtly magnificence and learning was never the only issue. Aggressive and efficacious mining, ore processing, and assaying had a very real impact on the actual wealth of the princes of central Germany and elsewhere, a fact that they themselves undoubtedly never forgot. Assaying was essential to the integrity and value of the coinage of the realm, a constant concern in the face of the chaos of early modern specie. Military effectiveness, with its complex requirements of armaments and other technologies, organization, leadership, and supplies might have everything to do with whether a particular ruler ruled or not, and *where* he or she ruled. Military success depended upon complex factors, including technical competence in many areas. Cultural studies concerned with status and representation in the courts and elsewhere that neglect such realities of material wealth and military power present an incomplete picture.

Mining, metallurgy, and the military arts provided a focus of intense interest on the part of a range of individuals, from noble patrons to middle-level practitioners. Authorship devoted to such practices expanded. Given the ease of creating multiple copies, what might be called the performative range of books expanded as well. Dedicating a book to a patron was a time-honored method for author-clients to create a gift that could result in patronage. Yet by producing books for market, the printing press facilitated a new mode of authorship, that undertaken for the purpose of selling books. The book as a commodity encouraged the growth of a middle-level readership. A book could function simultaneously as a gift within the system of court patronage and as a commodity in the book market. A book could explain to investors the nature of the practices in which they had invested, such as mining. It could also instruct current and potential practitioners and help to shape the self-image and group identity of particular kinds of practitioners, such as military captains and engineers. Finally, the printing press provided a new choice, between producing a manuscript book and producing a printed edition.

Ultimately, I suggest, such authorship had broad epistemological significance. Mining, metallurgy, the mathematics of aiming cannon, and the design and construction of forts continued to function as practices carried out in numerous specific locations. Authorship on such topics meant that these practices also took the form of discursive disciplines structured by principles, whether mathematical or otherwise, presented in writing. Authorship created discursive forms out of skill-based practices, and it created physical books suitable for libraries, books that treated topics such as ore processing or the quality of soils needed for earthworks in fortification. The expansion of authorship in general on such technical arts helped to connect the world of empirical practice to the world of learning. Authorship created disciplines of knowledge out of practices formerly based primarily on craft skill.

Writings on Mining and Metallurgy

In the first half of the fifteenth century a gradual rise in population as well as other factors led to a shortage of metals, which were needed especially for specie and for guns. Existing mine operations, which were primarily small, local undertakings and shallow mines, could not meet demand. New expectations of profits, a result of the metal shortage, provided motivation to dig deeper mines and to solve the technical problems that accompanied greater depth, such as water removal. Deep mines brought about a change in the organizational structure of mining across Europe. More costly to construct and operate, they necessitated greater outlays of capital. Small, local cooperative groups of miners were replaced by wage earners who worked within larger-scale operations and were paid increasingly by absentee shareholders, who provided needed capital and also reaped profits. Sharing these profits were princes and other rulers who held regalian rights over the land.[2]

Wealthy investors and holders of regalian rights profited immensely from this mining boom. They became ready patrons and consumers of treatises on mining and metallurgy. Authors of such books wrote for rulers and other wealthy investors who wanted to maximize the productivity of their mines. They wrote as well for the expanding number of new practitioners, whose skill in prospecting, mining, and processing metals provided the key to profits for their employers. Treatises on mining and metallurgy not only had practical utility; these books transformed mining from a relatively low-status occupation into a learned subject with ancient precedents, a contribution to humanist learning. Mining treatises often included striking illustrations of mining sites, operations, tools and machines, ore processing, and metallurgical procedures.[3]

Two early pamphlets on mining and ore processing emerged in the context of central European mining. The author of the *Bergbüchlein* is accepted on sixteenth-century evidence as Calbus, the town physician of Freiberg, an important mining town in Saxony. The *Probierbüchlein* is an anonymously authored work on assaying whose full title asserts that it was "compiled with great care for the benefit of all mintmasters, assay masters, goldsmiths, miners and dealers in metals."[4]

Calbus of Freiberg served on the town council and helped to establish a humanist Latin school. He invested in mines at various locations and enjoyed great prosperity as a result. He wrote the *Bergbüchlein* in the form of a dialogue between Daniel, "the mining expert" (Saint Daniel was the patron saint of miners), and Knappius, a young miner. It discusses the birth and growth of ores, how to discover them, and how the shares of the mine are divided. Knappius is pleased to gain this knowledge because "I shall be given a reasonable understanding which mines can be worked gainfully so that my investment will not be wasted but will show a profit." Daniel hastens to add that "as a mere side issue," profits should not be spurned, but if profits became more important than knowledge of the generation of metals, it would "cheapen and condemn this little book and the art." Daniel emphasizes the importance of knowledge and its close relationship to practice.[5] He stresses both knowledge and practice over profits, which nonetheless are not to be spurned. A written dialogue on the subject itself reinforces the notion that mining is a discipline that encompasses practice and profits but also learning.

The second pamphlet, the *Probierbüchlein*, consists of a group of recipes for testing metals that suggests a preexisting collection. Evidence from various editions suggests a readership of both practitioners, including those concerned with minting and coinage, and individuals interested in mine operations. Some editions contain an anonymous dedication to one Hans Knoblach, an administrator of the Harz Mountain mining operations of Elizabeth, duchess of Braunschweig and Lüneburg. Duchess Elizabeth was a key figure in the renewal of iron mining and the introduction of steelmaking in the upper Harz. Her efforts, which brought economic prosperity to the entire region, led to her being eulogized as, among other things, *inventrix metallorum*. The dedication to the booklet on assaying informs us that Elizabeth's mine administrator, Knoblach, had encouraged the unknown author to publish his collection of information on the assaying of ore, which he had gathered "from writings and from his own experiments."[6]

Far more ambitious than these German pamphlets was the Italian treatise *Pirotechnia*, by the Sienese Vannoccio Biringuccio, published posthumously in 1540. Biringuccio wrote with remarkable freshness and self-confidence, largely from his own practical experience. His expertise is evident in the technical descriptions and explanations of the treatise, which contains a wealth of information on ores, assaying and smelting, the separation of gold and silver, alloys, bronze casting (including the first detailed description of the bronze casting of guns), metal melting, furnaces, fireworks for warfare and festivals, and numerous related topics.[7]

Biringuccio's expertise in mining, metallurgy, and gun founding led to his varied and successful career, which was supported by the patronage of the nobility. He was a loyal client of the ruling Petrucci family of Siena, and his good and bad fortunes, including wide travels but also exiles and confiscations of property, were closely tied to theirs. It is significant that one of Biringuccio's earliest patrons, Pandolfo Petrucci, aggressively exploited mining wealth by constructing many iron plants in the Boccheggiano Valley, near Siena. The tumultuous Petrucci rule of Siena ended with their final expulsion in 1524.[8]

Biringuccio traveled widely in the German states and in Italy, gaining first-hand knowledge of mining and metalworking operations. His patrons included princes such as the Farnese of Parma and Alfonso I d'Este, lord of Ferrara. He worked as well for the Florentine and Venetian republics. At one point he was given a monopoly for saltpeter, a key ingredient of gunpowder, in the territory of Siena. His various positions included those of overseer of a silver mine in Carnia, in northern Italy, supervisor of the iron mines in the Boccheggiano Valley, head of the Sienese armory and of the Sienese mint, director and architect of the Opera del Duomo in Siena, and head of the papal foundry and munitions in Rome, where he died about 1538, before his treatise was published.[9]

The Venetian printer of the *Pirotechnia*, Curtio Navo, dedicated the work to the "Magnificent" Bernadino di Moncelesi of Salo, who is also mentioned in the text itself. This individual, who was probably a patron of the enterprise, as well as other prospective readers, clearly belonged to an elite class that was unskilled in metallurgical practice. Biringuccio's wellborn prospective readership is made especially evident by his statement that he has written extensively and in detail "because I have thought that you had not hitherto had the slightest shadow of knowledge of what I have described in this treatise of mine." Elsewhere, discussing precious stones, he remarks that he includes the subject because "it is a fine accomplishment for a gentleman to have some knowledge concerning gems."[10]

Biringuccio urges aggressive exploitation of mineral resources, another indication of his orientation toward a powerful, elite readership. Italy is rich in copper, he contends, but very little is mined there. Launching into a polemic about why these copper ores had not been sufficiently mined, he suggests that the problem is cowardly Italian avarice, laziness, or indolence. Perhaps seafaring commerce seems easier. He offers a harrowing account of the perils of such commerce, which he sees as a far from adequate substitute for mining. Worse are those who give themselves over to theft and usury. Just as bad are merchants: every adversity that comes to them "is fitting punishment"; they forsake "the natural good and just way" of extracting metals from the earth, produced liberally for us by nature. Minerals and metals are "copious blessings conceded by heaven." Biringuccio believes that men "wrong themselves, their fatherland, and the province where they were born" by failing to mine them.[11]

On the positive side, he praises the courage and persistence needed for successful mining operations. Turning to the empire, he describes a copper, lead, and silver mine in Austria where the owners persisted despite a layer of very hard limestone. He is amazed by their habit of "working in both night and day shifts," which seems marvelous to him. If these owners had begrudged expense or time, or if they had despaired of finding ore and abandoned the undertaking in a cowardly way, their expenditure of money and effort would have been in vain, and they would have failed to profit their superiors, their relatives, their native country, or their poor or rich neighbors. But they did profit them "through their strength and goodness of soul" and their "hope and tenacity."[12] Mining had become not just a practice that could bring profits but a moral good that rewarded the persistent miner and benefited society as a whole, rich and poor.

It is a corollary to this ethical dimension of mining that Biringuccio openly communicates its technical details, tying his writing to the progression of knowledge as a whole. Discussing gold ore, he gives his reasons for writing: "I have done this [writing] willingly in order that you may acquire more learning and because I am certain that new information always gives birth in men's mind to new discoveries and so to further information. Indeed, I am certain that it is the key that arouses intelligent men and makes them, if they wish, arrive at certain conclusions that they could not have reached without such a foundation, or even nearly approached."[13] By putting his writing within a framework of knowledge and further discoveries, Biringuccio makes clear that his authorship has import well beyond merely communicating known technical details of practice.

Condemning secrecy, particularly the secret operations of alchemy, Biringuccio explains that he derides the alchemists so that the inexperienced might be prevented from throwing away their talents by following the same path. Indeed, alchemists themselves might be encouraged to share their knowledge openly: "I am also content because, in order to show my ignorance to the world, the desire may come to some worthy philosopher and alchemist to bring to light at least the open arguments for their art, if not the completed work." If this happened, Biringuccio jests, great utility would result because the art would be made clear and "all good men of ability" would begin to make gold in great quantities and thus "make men rich, secure, and happy."[14]

An important aspect of openness is the accurate crediting of authorship. Biringuccio expresses incredulity at the alchemical custom of disguising the true authorship of a work. The hopes of the alchemist's "fantastic writings are but masked shadows," and "in order to lend authority to their recipe books they head them with the name of an author who not only did not write them but perhaps never even thought about the subject."[15]

Just as Biringuccio scorns the pseudonymous authorship of alchemy, he also derides craft secrecy. Noting the differences of opinion on how to make the chamber of a gun, he suggests that secrecy was used fraudulently to suggest expertise and special technique that did not exist: "Under this veil these men pretend to have a great secret and puff up their reputations by telling lies which deer could not leap over, promising that from their guns not only balls but lightning flashes will issue." In the end, they make only what others have made, and when asked what theory is behind their work, they give "only a surly answer."[16] The *Pirotechnia* itself stands in stark contrast to this practice of secrecy, revealing in detail how to make gun chambers.

Biringuccio provides abundant technical detail throughout his treatise, and he frequently emphasizes that he is revealing craft secrets. Referring to metal melting, he promises to tell "some methods that are held as secret by the masters." Concerning techniques of the goldsmith, he does not wish "to fail to tell you of some things concerning their operations which they withhold from most people almost like secrets, so that you may know these as well." In a section on ironwork, he lists what he calls "secrets." While perhaps, as the editors suggest, he took these "secrets" from editions of the *Kunstbüchlein* (small books of craft recipes), which may have been known to him, he was undoubtedly thoroughly familiar with the practice of ironworking. Elsewhere he discusses intarsia work and admits his own difficulty. It is "a very great secret, and one that is still not well known to me although I have practiced it diligently in order to learn it." Finally, in what is the earliest clear discussion of an amalgamation process, he describes how he paid to learn the secret of using mercury to extract gold and silver from sweepings: "Wishing to know this secret, I gave to the one who taught it to me a ring with a diamond worth twenty-five ducats, and I also pledged myself to give him the eighth part of whatever profit I should gain from this operation." In turn, Biringuccio wants to reveal the secret to the reader, "not in order that you would repay me for teaching it to you, but in order that you should esteem and value it so much more."[17]

Biringuccio was a practitioner and overseer from the middle level of Sienese society who gained unusual access to the rich and powerful, such as the Petrucci family of Siena and other patrician families on the Italian peninsula. His position at his death as head of the papal foundry and munitions suggests his authority as a practitioner and also, perhaps, his skill as a client within systems of patronage. His lengthy, detailed treatise, far more extensive than previous writings on the subject, suggests an effort to rationalize the practice of mining and metallurgy, primarily for the benefit of patricians, princes, and investors such as his patrons. By means of patronage Biringuccio was able to remove himself from the local artisanal culture, whose secrecy he decries. In contrast, he openly presents a rationalized body of knowledge about mining and metallurgy, a suitably learned subject for an elite readership.

The *Pirotechnia* is a printed book, yet its essential context includes not only the culture of the printing press but also the mining boom of the first half of the sixteenth century and, further, Biringuccio's own practice as an overseer and the patronage system within which he worked. Sixteenth-century technical authorship developed in part within the context of print, but not entirely. Manuscript books, including those pertaining to mining and metallurgy, continued to be written throughout the century. While occasionally a manuscript copy indicates merely an accident of fate that prevented a work from reaching the stage of print, in other cases manuscript production bears a special meaning, usually suggestive of unusual value.

An important case in point is the beautifully hand-copied *Schwazer Bergbuch*, created in the mid-1550s, which exists in at least seven manuscript copies. The treatise comprises an extensive compilation of mining law, custom, and regulations with more than a hundred hand-painted miniature illustrations of various mine activities, probably by the painter Jörg Kolber (see Fig. 10). It is the most important sixteenth-century source for Tyrolian mining law and custom, mine technology, and the conditions and responsibilities of mine officials and workers. The author was almost certainly Ludwig Lässl (d. 1561), an official in a mine court in Schwaz, in the Tyrol, between 1543 and 1555.[18]

Erich Egg has reconstructed some aspects of the life of Lässl. He was born into a peasant family, and his career exemplifies the upward mobility that the sixteenth-century mining industry could sometimes provide. Lässl obtained his post as clerk of the mining court through his father-in-law, Hans Möltl the younger, who occupied the position before him, from 1530 to 1543. Lässl's appointment as a mine clerk and his later retirement (with pension) because of ill health are recorded in the papers of the archduke Ferdinand, ruler of Austria and one of Lässl's patrons. Lässl is also known as the founder of the first paper mill in the Tyrol.[19]

Egg suggests that the *Schwazer Bergbuch*'s emphasis on the localities of particular mines (which is irrelevant to mining law) strengthens the presumption that the work was not written primarily for mineworkers. He proposes that the prospective audience was much further afield and was conceived in the context of a financial crisis in the early 1550s. Capital investments for Tyrolian mining came primarily from commercial firms in Augsburg, most importantly the Fuggers but also many others. In 1552 two Tyrolian mining firms, plagued by the overextension of credit and the high costs of deeper mines, went bankrupt. Creditors from Augsburg were pulling back. In 1553 the Augsburg firm Baumgartner, the most important investor next to the Fuggers, gave up its Schwaz mining interests Egg suggests that the *Schwazer Bergbuch* was intended to rouse both Augsburg investors and rulers to provide financial help in the form of mining investments.[20]

There is good evidence for Egg's hypothesis in Lässl's text. Lässl argues that the wealth produced by mines is a gift of God and points to the great riches and improvements brought about by mining. Many dukes and others had risked great sums and goods to build more extensive mines. Not only workers and miners but also all other persons of high and low station, as well as towns and businesses,

Fig. 10. Silver refiners, from the *Schwazer Bergbuch*, Kod. Dip. 856, fol 97r, Tiroler Landesmuseum Ferdinandeum, Innsbruck. A manuscript book created in the mid-1550s, the *Schwazer Bergbuch* included more than a hundred hand-painted miniature illustrations of mine activities, probably by the painter Jörg Kolber. Courtesy Tiroler Landesmuseum Ferdinandeum, Innsbruck.

had benefited. Many had gathered in lightly populated areas, property values had increased fivefold, land had been developed, and what once had been worth little or nothing was bought and sold for much money. All this showed that mining was a divine gift, created for the sustenance and benefit of man. Lässl insists that because of mining's great benefits, the welfare and rights of mineworkers should always be considered. Lassl's concern for mineworkers is a remarkable sentiment for the time period, shared by Paracelsus, perhaps, but not by well-known authors of mining and metallurgy such as Georgius Agricola. Lässl claims that he is writing his treatise because over the years mine laws and decisions have become confused. Often two or more regulations refer to the same topic. He correctly lays out the old regulations in new form.[21] As he brought order to mine regulations, Lässl also created an emblem for the riches that mining might bring in a book beautifully copied and illustrated by hand, a book that shows miners diligently creating wealth for princes and investors, a book fit for the libraries of wealthy burghers, nobles, and princes.

In the same years that Ludwig Lässl wrote the *Schwazer Bergbuch*, his contemporary Georgius Agricola was at work on the most famous mining treatise of the sixteenth century, the *De re metallica*, published in 1556. Agricola is rightly known as a learned humanist, but it is also relevant that he was from a family of skilled artisans and maintained lifelong connections to practitioners. He was born in Glauchau, Saxony, at a time when the region was experiencing an expansion of metal mining, particularly of silver, which greatly enriched the Saxon princes as well as other residents. Although he came from an artisanal family, Agricola, along with two brothers, was university-trained. His family gave him a close and lifelong association with artisans, a social circumstance that may have informed his appreciation for empirical knowledge and practical techniques. His father (probably Gregor Bauer) was a dyer and woolen draper, a profession followed by his younger brother Cristoph. Two of his sisters were married to dyers. His first wife, Anna (née Arnold) was the widow of Thomas Meiner, director of the Schneeberg mining district. His second wife, Anna Schütz, was the daughter of a guild master and smelter owner, Ulrich Schütz.[22]

Agricola's matriculation at Leipzig University at the age of twenty was uncommonly late for the time but consonant with his social background and upwardly mobile status. He received a bachelor's degree in 1515, remaining to lecture on elementary Greek. His first work was a booklet on Greek grammar. He later traveled to Italy, studied medicine in Bologna, Padua, and possibly Ferrara, and then remained three years in Bologna and Venice to help edit the Aldine editions of Galen and Hippocrates. Thus steeped in humanist culture and editorial practice, he returned to the empire. He first went to Saint Joachimsthal (now Jáchymov, Czechoslovakia), a mining town on the eastern slope of the Erz Mountains in Bohemia, close to the Saxon border, one of the most productive mining areas of central Europe. As the town physician and apothecary, Agricola tended the sick but also visited mines and smelters day and night, learning as much about mining and metallurgy as about the diseases of miners. In 1533 he moved to the quieter town of Chemnitz, in Saxony, to become the town physician. There he continued his medical work, wrote treatises, and invested in mining. His knowledge allowed him to profit—by 1542 he was one of the twelve richest inhabitants of Chemnitz. He was given a house and plot by the Saxon prince Maurice in 1543, and he was made burgomaster by command of the same prince in 1546. He was also appointed a councilor in the court of Saxony and was sent on various diplomatic missions on behalf of Charles V.[23]

Agricola wrote his first metallurgical book while working as a physician in Joachimsthal. The *Bermannus sive de re metallica*, published initially in 1530, is a dialogue among the interlocutor physicians Johannes Naevius and Nicolaus Ancon and a mine overseer, Bermannus, as they stroll through the mountains near the town. The book consists mainly of a discussion of regional ores and those mentioned in ancient writings. The introductory letter, by Erasmus, was obtained by Petrus Plateanus, a teacher who at that time was rector of the local Latin school. In addition to his Latin-German glossary, Plateanus contributed his own letter of introduction, dedicated to Heinrich von Könneritz, the region's mine superintendent.[24]

The introductory letters of the *Bermannus* emphasize the ideal of openness. Erasmus praises the work for its vivid descriptions of "those valleys and hills and mines and machines," almost as if he had seen them rather than read about them. Plateanus also praises open writing. None are more deserving, he says, "than those who transmit to posterity through writings the secrets of either the arts or of nature invested by oneself or by others." Men are endowed with powers of reason, understanding, and knowledge, making them superior to the mute beasts. They are capable of virtue and of various skills and disciplines; they are even able to be inventors and therefore can "penetrate into every very concealed thing of nature." Yet knowledge would be completely narrow if it were limited to one person's experience. Plateanus points to the very learned men of former ages, who made discoveries after much work and committed them to writing. He condemns those predecessors who lost these writings or allowed them to be destroyed, admonishing that we should take care that the same fate does not overcome our writings or those of our successors.[25]

Agricola himself also stresses the value of openness. He has written the *Bermannus* to give the studious a taste of work to come. He wishes to motivate his contemporaries to more diligent investigations. Finally, he wants to bring to light useful things to be found in German mines that had been unknown in antiquity. As for the ancients, they provide a model not only by their learning but particularly also because they transmitted their own knowledge and that of others to their successors in writing. Agricola emphasizes that the Greeks, "the most learned people of all," transmitted their own written accounts as well as those of foreigners. Yet his own countrymen have failed to do the same. "It is shameful for us," he laments, "that our things through our own negligence and idleness indeed now are almost concealed by darkness and lack their own light."[26]

Agricola portrays the interlocutor Bermannus as a model who combines direct observation and experience with knowledge of ancient texts. Only near the end of the dialogue do we learn that Bermannus is the overseer of a particular mine. And when he leaves his new friends briefly to talk to the mine captain, the other two praise him for, among other things, his openness in sharing his knowledge: "That which he discovers with great labor, he explains very easily and very diligently to others, and by no means is one who, with a certain envy, conceals, as in mystery and arcana, a very bad habit of not a few."[27]

Along with openness, Agricola advocates capital investment in mining. One of the interlocutors, Ancon (the physician trained by scholastic methods), suggests that miners lose money. Bermannus scoffs and points to miners in the area who had begun to excavate with little means and had become wealthy as a result. Ancon later insists that he would not pay money just for hope. He says that mining involves great expense for such hopes, and he would not spend "what was certain for uncertain things" and thus rashly give up his fortune. Bermannus rejoins that Ancon is too cautious and that such extreme caution will always be in his way. Ancon's attitude reveals a good Aristotelian, but he will never be a good miner or a rich man, says Bermannus. If a farmer had such a view, fearing catastrophe, he could never sow; if a merchant had it, fearing a shipwreck, he could never trade; nor could anyone go to war, because of the uncertainty of the outcome. On the other hand, "all hope for good and often it turns out well. No one truly with an abject and timid soul ever did anything or indeed ever will do anything."[28] Thus Agricola persuasively argues for the risktaking involved in mine investments.

During his years in Italy, especially while working at the Aldine press, Agricola had been steeped in the values and practices of Italian humanism. His advocacy of openness was influenced by Roman writings of Vitruvius, Pliny, and Columella. In his treatise on mineralogy, *De natura fossilium*, for example, he lists his authorities, just as Pliny did in the *Naturalis historia*. As Agricola notes, "Pliny gives credit openly and frankly to those whose writings he uses and likewise I shall give credit by name to those whom I quote."[29]

Agricola's masterpiece, *De re metallica*, published posthumously in 1556, is a comprehensive treatise on mining and metallurgy that contains spectacular technical illustrations. Agricola's topics range from finding ores and veins, surveying, digging shafts and tunnels, tools and implements used in mining and ore processing, and the assaying and processing of various kinds of ores. The woodcut illustrations depict various aspects of mining and ore processing. Illustrations depict diverse kinds of pumps and water wheels for draining mines, shafts, and particular phases of ore processing, including sorting, crushing, and smelting. Various kinds of furnaces, bellows, and other machinery and numerous tools are shown, along with numerous mineworkers and metallurgists, all soberly and industriously absorbed in their work (see Fig. 11).

In an introduction modeled on the defense of agriculture by the Roman author Columella, Agricola provides a masterful defense of mining. Following Vitruvius, he lists the disciplines necessary to the miner: philosophy, medicine, astronomy, surveying, arithmetic, architecture, drawing, and law. He defends mining against every critic, dismissing those who point to the dangers and unhealthiness of mining. A physician of miners and thus in a position to know better, Agricola

Fig. 11 Miners smelting ore, from Georgius Agricola's *De re metallica libri XII* (Basel: H. Frobenius & N. Episcopius, 1556), 319. The illustration depicts two miners smelting ore. There are two furnaces. One smelter breaks the material frozen around the tap hole of the furnace. The other carries a basket of charcoal. Tools and materials are labeled, including the wicker wheelbarrow in which coals are measured. Courtesy of the Smithsonian Institution Libraries, Smithsonian Institution®.

suggests that accidents were rare and were caused by the carelessness of workmen. Mining is profitable to the competent and useful to the rest of mankind, he says. The wealth it generates has many good uses (against those who point to the evils of riches). The dignity of mining and of investment in mining is greater than that of commerce and equal to, although more profitable than, that of agriculture.[30]

Agricola advocates openness and credit for authorship. Past writers should be properly credited. "No one should escape just condemnation who fails to award due recognition to persons whose writings he uses, even very slightly." He discusses openness in terms of the clarity of technical language, criticizing alchemists in particular because all of their writings are "difficult to follow, because the writers upon these things use strange names, which do not properly belong to the metals, and because some of them employ now one name and now another, invented by themselves, though the thing itself changes not." Agricola also complains about the inefficacy of alchemy, which he says consistently fails to produce riches, and about alchemical frauds. Finally, he condemns the alchemical practice of pseudonymous authorship.[31]

Agricola was a physician and humanist scholar close to the rulers of the empire. Other men who wrote books on metallurgy worked as assayers, overseers, or minters and in other areas of practice. One such practitioner was Lazarus Ercker, an assayer and miner who promoted his career by means of authorship. Ercker was born in Saint Annaberg, Saxony, a boom town that Calbus of Freiberg had helped to lay out. His marriage in 1554 to Anna Canitz led to his appointment in 1555 as assayer at Dresden. The elector Augustus chose him after the intervention of his wife's relative Johann Neef, whom we can recognize as the interlocutor Naevius in Agricola's *Bermannus*. Neef worked as town physician of Annaberg and as personal physician to the electors Maurice and Augustus. Augustus in particular was an enthusiast of mining, metallurgical, and alchemical experiments.[32]

Ercker's practice of authorship illustrates how both manuscript and print books could be utilized in the quest for patronage and advancement. Less than a year after his appointment as assayer at Dresden he completed a small book on assaying, *Das kleine Probierbuch*. Hand-copied by a scribe and dedicated to Augustus, it is a practical handbook that treats the construction of an assay oven, assaying, weights and measures, and cementation. It also provides assorted metallurgical recipes. Shortly after Ercker presented it to the elector, he was appointed general assay master for all matters relating to the mineral arts and minting for Freiberg, Annaberg, and Schneeberg.[33] We can surmise that Ercker's *Probierbuch* served two functions—first it provided instructions for use in the assaying operations of the prince; second, as a gift to the prince, it led to the reward of a promotion.

Although he was demoted (for unknown reasons) to warden of the Annaberg mint, Ercker found a new patron in Prince Henry of Braunschweig, who appointed him assay warden at the mint at Goslar, in the Harz Mountains. Prince Henry, the grandson of Duchess Elizabeth, had continued her work of expanding the productivity of the region's mines. Much of the reign of this Catholic prince was spent in armed conflict in an effort to gain or regain and consolidate territory under his own power. For Henry, the consolidation of political and territorial power and the development of his most important economic base—mining—were prime motivations and went hand in hand. Encouraged by his friend Duke George of Saxony, the father of Augustus and Maurice, he had revived the ancient silver mines of the upper Harz, investing his own income and encouraging other investors. In 1552, after years of struggle, he conquered the imperial (but Protestant) city of Goslar, and from that time on he controlled the mines in the Rammelsberg, an ore-rich mountain to the south.[34]

Ercker found himself, therefore, in a familiar environment—working in a mint, the appointee of a prince who was deeply interested in the productive exploitation of mining. Once again he turned to authorship as a way of achieving advancement. He wrote a *Münzbuch*, a treatise on minting, which he presented in 1563 to Henry's son Julius, duke of Braunschweig-Wolfenbüttel. Julius also pursued the aggressive exploitation of mining in his territories. Most significant economically by this time were the iron mines and the accompanying manufacturing industries, particularly of artillery, to which Julius contributed numerous of his own inventions and experiments. He also opened many new mines, expanded old ones, and made administrative reforms to prevent corruption. He himself undertook authorship in the mechanical arts. His *Instrumentenbuch*, an illustrated manuscript treatise that according to the subtitle was "in part conceived by Julius and drawn and painted by his own hand," exists in a single manuscript copy. It depicts machines for removing ores from mines and transporting them. A second section includes material on ships. Lazarus Ercker apparently understood Julius's interests well. Shortly after he dedicated his *Münzbuch* to him in 1563, he was promoted to master of the Goslar mint.[35]

In the *Münzbuch* Ercker specifies why he is presenting a practitioner's knowledge of mining to a ruler. If nobles who control mines and mints are not well-informed of such practical operations, unfaithful servants will take advantage of them and they will be unable to distinguish between true and untrue employees. Conversely, if they understand metallurgical practice, they can cast off false subordinates, appreciate true service, and not be subject to overreaching from unfounded hope. Ercker insists that his information, based on the efficacy of experience, will often be useful to nobles and dukes in relation to new mines.[36]

The *Münzbuch*, a manuscript full of information about techniques, was not meant for widespread distribution. Ercker criticizes alchemy, but for its lack of practical results, not for its traditions of secrecy. He admits that many of the practices of assaying, silver and gold refining, and similar arts had their origins in alchemy. Yet few alchemists of his own time had maintained assaying as a useful art by practicing it correctly and becoming experienced in it. Concerning the mint, Ercker supports its traditional secrecy. He cautioned Prince Julius "not to let this work come before everyone, so that it remains a beautiful art as up to now it has been."[37]

In the mid-1560s Ercker was once again seeking employment. After the death of his first wife, Anna Canitz, he married Susanne, daughter of a Dresden official. His new brother-in-law, Caspar Richter, was a minter in Prague. Through him, Ercker was appointed control assayer *(Gegenprobierer)* in Kutná Hora (Kuttenberg), Bohemia. Susanne herself also served for many years as the manager of the mint in the same place, with the title "manager-mistress." They had two sons, Joachim and Hans, both of whom became assayers.[38]

Ercker's writings until this time were small manuscript works concerning workshop practices whose purpose was possibly to instruct assayers and minters and certainly to obtain patronage for himself. His last work of authorship, *Treatise on Ores and Assaying*, was something quite different, a major printed treatise containing woodcut illustrations that was meant for wide public distribution. First published in 1574, it was dedicated to Emperor Maximilian II. The treatise was written for the benefit of the emperor's vast mineral resources and for those who made their living from them. He expresses the hope that these resources would be further developed and long maintained "through serious effort stimulated by complete information." He provides detailed discussions on the ores and assaying of silver, gold, copper, lead, tin, and saltpeter.[39]

Clearly, Ercker's treatise was inspired by Georgius Agricola's *De re metallica*, as its ambitious scope and striking illustrations make evident. He claims that his experience is greater than that of his predecessors, an unmistakable reference to Agricola. A treatise that explains ore processing and assaying in rich technical detail, Ercker's treatise represents not a practitioner's handbook but rather an organized body of knowledge on ore processing and metallurgy. The treatise seems to have furthered Ercker's career. Shortly after its publication the emperor named him courier for mining affairs and a clerk in the supreme office of the Bohemian crown. Maximilian's successor, Rudolf II, appointed Ercker chief inspector of mines and then knighted him in 1586.[40]

By the mid-1550s the mining boom in the German states had come to an end. Rich veins became less productive, and removal of ores from poorer veins proved more costly, while the growing influx of precious metals from the Americas lowered the value of the metals that were extracted. The oversupply of precious metals both from the German states and from across the Atlantic contributed to the inflationary trend known as the price revolution. Exacerbating the declining value of money was the chaos of specie that had long prevailed in the German states and had encouraged widespread fraud in minting. The mint became a particular focus of attention, with accurate assaying an emerging priority. Efficient methods of assaying, extracting, and refining metals came to be increasingly crucial to overall productivity. Not surprisingly, writings on metallurgy in the last third of the century focused on assaying. Examples include the small treatises of the assayers Ciriacus Schreittmann, Modestin Fachs, and Samuel Zimmermann, each of whom in different ways advocated openness, clear explanation, and precision.[41]

To summarize, authors of books on mining and metallurgy in the sixteenth century wrote within the context of the great expansion of mining in the first half of the century. Except for Biringuccio, they were Germans who lived close to the areas of central Europe that experienced a mining boom. Most were clients or employees of German rulers who held regalian rights and who filled their coffers with profits from mining. Yet mining was sustained by a broad range of wealthy investors beyond rulers and princes. Similarly, authors on mining and metallurgy came from a wide spectrum of backgrounds. Calbus of Freiberg and Georgius Agricola were physicians and humanists who also invested in mining. Biringuccio and Ercker were practitioners who functioned as high-level overseers. Authors included assayers such as Ciriacus Schreittmann and nobles such as Julius, duke of Braunschweig-Wolfenbüttel. Most authors wrote in the context of patronage relationships. Yet these writings did more than advance the careers and reputations of individual authors. Collectively they created out of mining and metallurgy a learned discipline worthy of a wellborn readership.

The most important uses for the products of mining and skilled metallurgy, particularly for copper and iron, were in cannon and other artillery, armor, and military equipment. During the fifteenth and sixteenth centuries numerous technological experiments and developments involving gunpowder artillery led to an increased demand for metals essential to the fabrication of many varieties of cannon, bombards, and smaller guns, as well as cannonballs and protective armor.[42]

Mining and metallurgy provided the materials for the technology of armaments. Beyond material and technological considerations, political, territorial, and religious strife fueled the endemic warfare of the sixteenth century, providing a fertile arena for military authorship. Throughout the century, literally dozens of books were published on armaments and artillery, fortification, the organization of armies, the ordering of battle, and military strategy.

Military Authorship in a Century of Conflict

Warfare in the sixteenth century brought about complex changes in technologies, tactics, and organization. Most basic was the rapid development of gunpowder artillery, which in turn induced changes in the design of fortification. The new bastion forts, built lower than their predecessors, featured angled bastions, accompanied by ditches, detached forts (ravelins), artificial slopes to keep attacking artillery away from the central walls, and earthworks behind the walls. New tactics also developed in conjunction with artillery warfare. The size of armies increased significantly, as did the complexity of their organization. This century of warfare produced tens of thousands of soldiers, large armies composed of both mercenaries and citizens, a wide variety of career opportunities for military captains, huge supply trains, and whole industries devoted to producing military supplies, including arms and armaments.[43]

Endemic warfare is the essential context for the proliferation of sixteenth-century military books. As John R. Hale points out, Venetian printers alone published 145 books devoted to military matters between 1492 and 1570. Books on military matters published in Venice and elsewhere treated artillery and gunpowder, fortification, tactics, strategy, and instruction for captains. Military culture involved guns, fighting, fortification, armies, supplies, death, and destruction but also an expanding arena of authorship and readership on the art of war.[44]

Niccolò Machiavelli's *Art of War* (1521) became the best-known military treatise of the 1520s. Machiavelli wrote his treatise in the light of the dire experiences of his native city. Medici rule of Florence collapsed in 1494, when Piero di Medici failed to mount an effective defense against the invading army of Charles VIII and too quickly handed Pisa and other Florentine possessions over to the French king. When the Florentines instituted a new republic led by a Great Council, Machiavelli served in the chancery. Eventually he came to be closely associated with the republic's *gonfaloniere*, or chief magistrate, Piero Soderini, appointed in 1502. In 1507 he was named secretary of the Nine of Militia, taking responsibility for the enlistment of ten thousand infantry. In the siege of Pisa in 1509 he supervised the three camps of the Florentine army, directed their supplies, kept the soldiers' behavior under surveillance, and informed the government of all developments.[45]

The fledgling Florentine republic collapsed in 1512, when Spanish and papal troops reinstalled a branch of the Medici family as rulers beholden to the Spaniards. Machiavelli was forced into exile to his small country estate nearby. There he wrote all of his well-known books, including *The Prince*. He dedicated his writings to various individuals whom he hoped might get him back into Florentine politics, his singular and unsuccessful ambition. His writings explore the crucial role of force in political life. Among other things, he advocated the creation of a citizens' army, which he regarded as superior to mercenary troops as a defensive force.[46]

The Art of War was the only one of Machiavelli's writings to be published in his lifetime. It takes the form of a dialogue in which an experienced military captain, Fabrizio Colonna, converses with a group of friends in the garden of Cosimo Rucellai, a leading citizen of Florence. The primary subject of the conversation is the organization, training, and leadership of an army. Machiavelli uses numerous examples from ancient military writings, yet his aim is practical; he believes that the military practices of the ancients, particularly the Romans, offer insight and advice for the creation of a citizens' army in his own day.[47]

In a recent article Marcia L. Colish asks why Machiavelli chose Fabrizio Colonna as the interlocutor who defends the superiority of the citizens' militia over mercenary armies. The real Fabrizio was a mercenary captain employed by Ferdinand of Aragon. His military leadership contributed significantly to the success of the Spanish conquest that led to the fall of the Florentine republic and the return of Medici rule in 1512. Colish's detailed argument suggests that the use of Colonna in subtle ways indicates Machiavelli's opposition to the Medici family. Machiavelli makes numerous political allusions based on the actual alliances and hostilities of his own day, yet he also follows humanist literary practice by creating an interlocutor who in life represents the opposite of the values that he expresses in the dialogue. In this way Machiavelli fully exploits the ironic potential of the dialogue form.[48]

Irrespective of Machiavelli's opinions, paid mercenaries fought most of the wars of the sixteenth century. The occupations of soldier and military captain be came attractive options for employment among young men and even women, who carried out many of the cooking and provisioning tasks of early modern armies. Military captains new to their occupation apparently welcomed instruction. For instance, Battista della Valle's small handbook for captains seems to have enjoyed great popularity. Della Valle first published his manual in Naples in 1521. It is a small book, of a size that might fit into a knapsack or the pocket of a soldier's shirt or uniform. This little treatise in four books went through eleven editions in thirty-seven years.[49]

Battista della Valle was a military captain from Venafra (present-day Isernia, in southern Italy) who served under Francesco Maria della Rovere, duke of Urbino. He dedicated his manual to Henrico Pandone, lord of Venafra. In abject deference he insists that his own mind is weak but that "the opinions, reasons, and military precepts" that he lays out are taken not from his own "rude ingenuity" but from Henrico's "excellent . . . concepts," which the lord of Venafra has "reported and exquisitely taught" to his faithful servants. Della Valle emphasizes how much he has learned from the prince, "hearing these precepts [concerning military plans and councils] with curiosity and attentiveness in my weak mind and memory as imprinted on a hard marble." Since hearing these precepts he has discovered and proven their truth "with experience and long practice."[50]

Despite his deference to the lord, Della Valle frequently emphasizes his own experience and skill. Many military authors, he writes, have composed polished works based "solely on authority and imitation of other authors and not through true experience." In contrast, he writes from his own personal experience. "But I, who from my tender and young years have exercised myself in the practice of arms," have written only about those things "that I have carried out through long experience and proved with continuous labors, sweat, and dangers." His book will reach a diverse group of readers, not only the learned. His wish was not to write elegantly only for the learned and intelligent but "to speak with the low, rude, and common, and to all understanding men," explaining what is appropriate to strong, brave soldiers.[51]

Della Valle provides detailed advice on many aspects of military leadership, advice that seems especially appropriate for new and inexperienced captains. The first chapter concerns the color of the captain's uniform and its correlation with the character traits he might want to project. Dark blue signifies "fulminating jealousy," which delights in vigilance, perseverance, penetrating ingenuity, and attractive judgment; white, purity and modesty; black, firmness and stability; and red suggests cruelty. These colors influence the effect of the captain's personality on the troops, perhaps enhancing his ability to control them. What is most important, Della Valle admonishes, is that the soldiers obey the captain.[52]

The book contains a multitude of instructions, both practical and technological. It treats appropriate punishments for the disobedient and traitorous; instructs how the camp should be organized; and discusses the surveillance and fortification of the camp. An early Italian proponent of earth fortifications, Della Valle provides information on bastions and other aspects of fortification. He discusses the preparation of mortars and other weaponry to defend the walls, including incendiary devices. He teaches how to make torches that resist wind and water to guide an army through storms and how to mix gunpowder for various kinds of artillery. One chapter provides a moving exhortation to battle for the captain's use, outlining in graphic detail the consequences of defeat for men, women, and children. It then describes the contented soul of the soldier who dies in battle. Della Valle also explains how to organize and change the guard, how to construct clocks to time the changing of the guard, and how to send messages by fire signal.[53]

He instructs how to capture a territory, where to engage in battle, how to place artillery, and how to dig trenches. He explains how to build various kinds of gabions (earth-filled wickerwork structures that protect infantry and artillery) for defensive and offensive operations and provides detailed illustrations. In explaining how to attack a walled fortification he discusses siege ladders, portable bridges, and pumps for extracting water from trenches. He describes how to make tunnels and mines. He details various orders of battle suitable for organizing battalions of various sizes and combinations of guns and pikes, information that is useful for actual battle plans as well as for the organization of military parades. Finally, he praises the nobility of the profession of arms and provides extensive advice concerning how the captain might defend his honor through the duel.[54]

Coincidentally, Machiavelli's and Della Valle's tracts were both published in 1521. Machiavelli's was a humanist dialogue that treated numerous aspects of the organization and leadership of the army, using the ancient Roman army and the ancient virtuous soldier as models to be emulated. Whereas Machiavelli directed his learned treatise to princes and rulers, Della Valle's was more of a practical handbook. A captain himself, he offered day-to-day advice to captains in the field. The range of Della Valle's instruction is notable. He treats, as we have seen, the color of uniforms, rhetorical aids to persuading an army, and innumerable technical matters relevant to both attack and defense. In addition, he considers the honor of the soldier and captain, integrating the technical aspects of warfare with issues of leadership and honor. While Machiavelli's and Della Valle's treatises are quite different one from the other, together they presage the significant proliferation of military writings during the sixteenth century. This tradition of authorship came to provide a common ground of communication for princely rulers, military captains, and engineers.

A theme of subsequent writings, particularly evident in the books of Niccolò Tartaglia, was the practical mathematics of gunnery—questions such as the angle and the distance from the target at which the cannon should be set to obtain the most desirable trajectory of the cannonball. Tartaglia was a mostly self-taught mathematician and mathematics teacher from Brescia who spent much of his working life in Venice. His authorship was closely tied to the printing press, yet he clearly used it to seek patronage as well. He produced the first Italian translation of Euclid and editions and translations of some of the writings of Archimedes, as well as small, original treatises. These last—the *Nova scientia* (New science), published in 1537, and *Quesiti et inventioni diverse* (Diverse questions and inventions), appearing in 1546—take the form of question and answers or reports of conversations between himself and numerous other named individuals, many of them nobles and wellborn men, others gunners and practitioners of various kinds.[55]

Although Tartaglia describes the *Nova scientia* as a treatise of five books, only the first three books were actually published. The first concerns "the nature and effects of uniformly heavy bodies in the two contrary motions [Aristotelian natural and violent motion] that may occur in them, and their contrary effects." The second demonstrates geometrically the proportionality of trajectories of bodies ejected or thrown through the air and the proportionality of their distances, and the third concerns the determination of distances by sighting and calculations. Recent studies by Henninger-Voss and Cuomo emphasize the representational significance of the treatise in terms of the relationship between, in Cuomo's words, "the key concepts of power/control and knowledge/experience/learning." Tartaglia brought together issues of practice and issues of knowledge involving mechanics and mathematics. The numerous interlocutors and questioners who appear in his texts suggest a common ground of discussion and interest among practitioners, mathematicians, and noblemen. Problems such as where to place cannon and how to aim them were of intense interest to nobleman and practitioner alike. Proposing numerous solutions, Tartaglia presents himself everywhere in his writings, emphasizing his own novelty and originality and the correctness of his solutions.[56]

Tartaglia dedicated the *Nova scientia* to Francesco Maria della Rovere of Urbino, the same nobleman under whom Battista della Valle had worked. Francesco had been engaged by the Venetians to organize an anti-Turkish defense because Turkish control of the eastern Mediterranean had seriously damaged Venetian trade. Thus signaling the larger military significance of his writing, Tartaglia mentions that a close friend and expert bombardier in Verona had inquired about how to aim a cannon to achieve the farthest shot. He discusses the mathematical reasoning that he used to prove that the correct angle is forty-five degrees. In a similar discussion the following year, Tartaglia and a chief of bombardiers made a wager concerning the range of the cannon shot. Describing the trial that proved him correct, Tartaglia explains the mathematics of projectiles discharged at various angles.[57]

He begins the treatise with a series of definitions, suppositions, and axioms modeled on Euclid's *Elements* that concern falling bodies and trajectories. Thus, he subjects the practice of gunnery to mathematical analysis while at the same time discussing it in terms of Aristotelian physics. Because he analyzed the motion of projectiles and falling bodies mathematically, Tartaglia's writings have been viewed as contributing significantly to the sixteenth-century mechanics that led directly to the mechanics of Galileo. The way Tartaglia brings together mathematics, Aristotelian physics, and the concerns of practical gunnery is notable.

Similarly, in the social sphere he worked within an arena of communication involving noble patrons concerned with real-life military problems, issues of mechanical knowledge relevant to learned culture, and skilled gunners from a lower social stratum, and he depicts such communication in his treatises.[58]

Tartaglia wrote his treatise of the 1540s, *Quesiti et inventioni diverse*, in a way that dramatically displays its social and political framework. His portrait on the frontispiece points to the importance of his own authorship. Dedicating his new work to the English king Henry VIII, he explains that he decided to write the treatise when his previous work *(Nova scientia)* "provoked many people (and for the most part not ordinary men, but men of high intellect) to seek me out anew with various other questions or interrogations, and not merely about matters of artillery, ammunition, saltpeter, and powder." Such questions, Tartaglia reports, led to "knowledge and discovery" of many other particulars that he would never have discovered or considered. Tartaglia explains his open publication of his new discoveries as follows: "I reflected that no small blame attaches to that man who, either through science or industry or through luck, discovers some noteworthy things and wants to be their sole possessors." If all the ancients had done the same, Tartaglia notes, we would be like irrational animals. For this reason he decided to publish his questions and inventions.[59]

In his dedication to King Henry VIII, Tartaglia notes that his English friend and student Richard Wentworth had spoken to him of the magnanimous generosity of the king and of how "your majesty took great delight in all matters pertaining to war." He admits that what he presents are first fruits only, not elegant and polished. Although they "are mechanical things, plebian, and written as spoken in rough and low style," he offers them "only as new things."[60] Here Tartaglia combines the rhetoric of novelty with the claim of originality.

Quesiti et inventioni diverse is divided into nine books, each comprising a certain number of questions. The persons asking the questions are identified; some are nobles, such as Francesco Maria, duke of Urbino and condottiero of Venice, whereas others are ordinary gunners and other practitioners. The nine books concern, respectively, shooting artillery pieces; cannonballs; gunpowder; ordering infantry ranks; mapping terrain and the use of the compass; fortification; the principles of the balance (in preparation for a discussion of statics); the "science of weights," derived primarily from the medieval treatise of Jordanus of Nemore on the same subject; and problems of mathematics, including the solution to cubic equations. This last book contains Tartaglia's accusations against the mathematician and physician Geronimo Cardano that led to their well-known priority dispute over the solution to cubic equations. In the mathematical sections

Tartaglia often gives the exact date on which a particular conversation took place, seemingly to establish his own priority for the solutions to the problems under discussion.[61]

The Priority Dispute between Tartaglia and Cardano

Openness, secrecy, and priority were central issues in the famous dispute between Tartaglia and Cardano, who conducted his side of the dispute through his student Ludovico Ferrari. The contest concerned a general rule for the solution to algebraic equations to the third degree, or cubic equations. Early in the century the well-known mathematician Luca Pacioli had declared that no general method was possible for solving equations higher than those of the second degree (i.e., quadratic equations). Pacioli was proved wrong when the general solution to cubic equations was discovered. That discovery brought on the acrimonious dispute between Tartaglia and Cardano in the years 1547–48.[62]

The single area of agreement among all the parties was that original authorship and invention, in this case involving the solution to a mathematical problem, had great value and should be accurately credited. The origins of the dispute in terms of its mathematics can be dated to 1505 or 1515 (reported thus diversely by Tartaglia and Cardano, respectively). At that time an obscure lecturer in mathematics at the University of Bologna, Scipione del Ferro, discovered, but did not publish or widely communicate, a general rule for solving the third-degree equation. By 1530 Tartaglia was working on cubic equations after receiving some problems from Zuanne de Tonni da Coi, a Brescian mathematics teacher. In 1535 he became involved in a mathematical contest with Antonio Maria Fiore, a former student of Scipione del Ferro's. The contestants followed standard procedures, in which each gave to the other thirty problems to be solved within a certain time period. The loser was to pay for thirty banquets for the winner and his friends. All of Antonio's questions to Tartaglia involved cubic equations. After frantic work, Tartaglia discovered a general solution to the cubic equation and won the contest. (For unknown reasons, he renounced his banquet prize.)[63]

News travels fast. The Brescian mathematics instructor Zuanne de Tonni da Coi visited Cardano in Milan and related the story of the contest, including Tartaglia's discovery of a general rule. Cardano, who was writing a book on mathematics at the time, wanted to know the solution. He sent a Milanese bookseller, Zuan Antonio da Bassano, to Tartaglia in Venice. As Tartaglia relates it, Zuan Antonio mentioned that Cardano's book *The Practice of Arithmetic, Geometry, and Algebra* was ready to go to press and relayed Cardano's request: "His Excellency begs you to be good enough to send him this rule that you have found, and if it suits you, he proposes to publish it under your name in his present work, but if you do not see fit that it should be published, he will keep it secret." Tartaglia reports his reply: "Tell his excellency that he must pardon me, that when I publish my invention it will be in my own work and not in the work of others, so that his Excellency must hold me excused." Zuan Antonio then asked for the thirty problems and Tartaglia's solutions, a request that Tartaglia also refused (to acquiesce would in essence reveal the rule). Cardano's messenger returned with only a copy of the thirty questions Tartaglia had received from Antonio Maria Fiore.[64]

Cardano responded to Tartaglia's refusal with an insulting letter in which he lamented the difficulties of being a mathematician, noting that it was little wonder that laymen considered them to be almost among the insane Yet along with insults, Cardano offered a carrot—the possibility of patronage. He advised Tartaglia that both he and his patron, the Spanish governor of Lombardy and commander of the imperial army stationed in Milan, Signor Marchese Alfonso D'Avalos, had enjoyed Tartaglia's book on artillery *(Nuova scientia)*. Cardano criticized Tartaglia's descriptions of motion in the treatise but excuses him because he had said that gunnery was not his own subject. "On this point I reply," retorts Tartaglia, "that I take delight in new inventions and treating new things that others have not discussed, and it does not please me to proceed as certain others who fill up their volumes with material stolen from this or that author." He admits that "talking about artillery and how to aim it at a target may not in itself be a very honorable matter" but says that since it is a new topic and includes some speculation, it seems worth talking about.[65]

Along with his reply, Tartaglia sent two books, presumably copies of the *Nuova scientia*, and two instruments, one "a square to aim the said artillery to place it at level or to examine any elevation" (the *squadra*, or gunner's square) and the other an instrument to determine distances on the plane. Cardano in turn implored Tartaglia to come to Milan and stay in his house, emphasizing that he was writing at the urging of the marchese. Undoubtedly enticed by the possibility of patronage, Tartaglia traveled to Milan During his visit Cardano again urged Tartaglia to allow him to publish the solution to the cubic equation in a separate chapter that would give him full credit. Again Tartaglia refused. He finally agreed, however, to divulge the solution if Cardano swore a solemn oath not to reveal it.[66]

Cardano reneged on his oath ten years later, in 1545, when he published the solution in his *Ars magna*, giving Tartaglia due credit. In the meantime he had discussed the solution and its ramifications with his student Ludovico Ferrari, who in turn found the solution to quartic equations. Together Cardano and Ludovico had traveled to Bologna, where the family of the deceased Scipione del Ferro showed them the professor's notebook Clearly the Bolognese master had discovered the solution before Tartaglia, although he had not communicated the solution widely and Tartaglia had discovered it independently. Tartaglia's response to Cardano's publication was harsh criticism in the ninth book of his *Quesiti*, published in 1546, where Tartaglia quotes Cardano's oath of secrecy in full and their correspondence.[67]

The response to Tartaglia's published attack came in the form of the first *cartello*, signed by Cardano's student Ferrari. Cardano would not lower himself to dispute with a low-status personage such as Tartaglia and never responded directly despite Tartaglia's many attempts to get him to do so. The cartelli consisted of broadsheets that were widely distributed to eminent persons over Italy—a list of 55 recipients was placed at the end. The first cartello listed many errors and objections to Tartaglia's work and challenged him to reply in thirty days. In all, a total of six cartelli and six replies from Tartaglia were exchanged.[68]

In one of the cartelli Ferrari provides a rationale for publishing the solution to the cubic equation despite the oath. "Cardano obtained from you this bit of a discovery of yours . . . and this languishing little plant he recalled to life from near death by transplanting it in his book, explaining it clearly and learnedly, producing for it the greatest, the most fertile, and most suitable place for growth. And he proclaimed you the inventor and recalled that it was you who communicated it when requested." Ferrari accuses Tartaglia of keeping his invention secret because he wants no one else to enjoy it. For this Ferrari calls him "un-Christian and malicious, almost worthy of being banned from human society. Really since we are not born for ourselves only but for the benefit of our native land and the whole human race, and when you possess within yourself something good, why do you not want to let others share it?" He continues by showing that originally "this was not your invention" but rather the invention of Scipione del Ferro.[69]

Tartaglia finally agreed to travel to Milan for a public dispute with Ferrari (Cardano himself left the city for the occasion). It has been conjectured on the basis of some evidence that Tartaglia went because he had been promised a lucrative lectureship in mathematics at Brescia if he disputed successfully in Milan. Tartaglia traveled to Milan with his brother. The public dispute began on 10 August 1548 in the Church of Santa Maria del Giardino. Numerous nobles and distinguished guests were present, including the arbiter Don Ferrante di Gonzaga, governor of Milan. The topics were the problems posed in the cartelli and the suggested solutions. Yet the proceeding did not go well for Tartaglia. He left Milan before the dispute was completed, claiming later that he was not allowed to proceed in proper form.[70]

Tartaglia's writings and the cartelli of the dispute provide rich evidence for the interrelationships of practical and technical concerns with mathematical and learned culture in mid-sixteenth-century Italy and for the increasing value of original authorship. Tartaglia clearly believed that his feat in solving the cubic-equation problem gave him the right to claim ownership as well as the right to decide whether to publish it. The value of original authorship was attached not to writing per se but to the solution of a mathematical problem, or, as they called it, an "invention."

The dispute between Cardano and Tartaglia was a public one in which numerous individuals took part, including witnesses, judges, the audience at the public disputation in Milan, and the many recipients of the cartelli. As Mario Biagioli has pointed out, issues of social status are central to understanding the conflict.[71] Status issues are clearly evident in the physician Cardano's refusal to communicate directly with Tartaglia, a mathematical practitioner who occupied a lower social position than his own. Despite these status considerations, the cartelli and Tartaglia's other writings reveal a high degree of interaction between individuals of varying social status who were interested in gunnery and other military practices, in practical and theoretical mathematics, and in mechanics. This interaction involved issues of both practice and knowledge.

The Technical Arts and Mathematics in Writings on Fortification

Tartaglia was a mathematician who involved himself significantly in the practice of artillery. Other authors on various military arts were practitioners or military captains, many of whom saw mathematics as intrinsically important to their topics Much more study is needed of the literally dozens of military treatises published in the second half of the sixteenth century. The writings of two authors on fortification, Battista Zanchi and Giacomo Lanteri, exemplify the complex social context of this authorship.[72] These treatises treat both mathematics and issues of practical construction. Just as important, they depict and report conversations and the exchange of substantive information among wellborn men and skilled practitioners. Such representations of the interchange of knowledge and practice and of conversations between noble princes and skilled practitioners, whether or not they reflect actual conversations, are culturally significant.

Giovan Battista de' Zanchi, a military captain and engineer, served in the papal army in Germany and as an engineer for Venice in Cyprus and in Ragussa. Zanchi's *Del modo di fortificar le città* (On the method of fortifying cities), first published in 1554, is a short treatise dedicated to Maximilian of Austria, who in 1564 would become the emperor Maximilian II. Zanchi explains that fortification was understood by "many brave soldiers and experienced and very judicious captains," none of whom left an account to posterity. They were "envious that others, learning it in leisure and for pleasure, might be able to procure the honor that they had striven to acquire with very great labor and long experience."

Zanchi particularly emphasizes the importance of experience. Reading thoroughly and judiciously in the military art was not enough since the development of artillery had brought about the need for new experience in fortification. Yet many had practiced in the militia for a long time, and through "diverse trials" they had been able "to establish a number of conclusions as upon very firm foundations." Yet they lacked confidence to express themselves, or "they attend only to the practice of the army and not [to] learning." He says that he does not claim to be an expert in war but desires to accomplish a noble thing, one that is useful and serves Maximilian. If his treatise does not perfect the subject, at least "more noble and understanding spirits" in Maximilian's court may bring perfection and better form and increase the number of inventions.[73] Yet Zanchi effectively suggested that the experiments and experience of captains and engineers in the field was absolutely necessary for understanding and writing on the topic.

Zanchi lays out the general principles of fortification. He explains that he strives, not to teach how to plan particular buildings, but to show "a certain way and universal rule of such to build and fortify the city." He discusses the number of ways that a city can be besieged, balancing considerations of people and of material construction—cities can be defeated because of traitors or because of the negligence of those who take care of the walls, or because of defects in the walls themselves. Zanchi treats offensive weaponry and fortification of both ancients and moderns, the effects of gunpowder artillery, and how to fortify based on the site, the size of the fort, and the placement of artillery.[74]

Publication of Zanchi's treatise was arranged by Girolamo Ruscelli, himself a prolific writer, who was employed by the Valgesi publishing firm in Venice. Ruscelli inserted a letter at the end of the treatise addressed to one Dottor Nicola Manuali. In it, he describes conversations among several learned doctors and their patients about fortification, the importance of the subject, and their desire to learn it. He says that before departing from Venice, Zanchi left him the treatise, which he studied after dinner, and that upon his return Zanchi gave him permission to publish it. Concerning the debate about the relative value of letters and arms, Ruscelli emphasizes that the two go hand in hand.[75] Ruscelli's fascinating postscript describes discussions about fortification among both practitioners and learned men. His respect for the author-practitioner Battista Zanchi is indicated by the inclusion of Zanchi's portrait as the frontispiece of the treatise.

Similarly, Giacomo Lanteri's two treatises on fortification explicate technical details but also portray dialogues and discussions, indicating that it was a subject worthy of civilized and genteel conversation. Lanteri was a military engineer from Brescia who served the Spanish king in Naples and northern Africa, as well as the Venetians, the papacy, and other princes. He composed his treatise *Due dialoghi . . . del modo di disegnare le piante delle fortezze secondo Euclide* (Two dialogues . . . on the way to design the plans of fortresses according to Euclid) as a dialogue between the military engineer Girolamo Cataneo of Piedmont, a Veronese engineer named Francesco Trevesi, and a "young Brescian," who possibly represents himself. He dedicates the first dialogue of his treatise to Marc Antonio Moro, count D'Arco, a nobleman from Brescia, emphasizing the greatness, nobility, and generosity of the count and "my low and small gift" produced by "my weak intellect." Lanteri emphasizes "the affection that I carry and will always carry to your lord to whom I am a suppliant, and I kiss your virtuous hands."[76]

Lanteri's letter "to kind readers," however, suggests a broad prospective audience. He praises men of the past who were devoted to honest studies, stressing how obligated he and his contemporaries are to those forebears. Inspired by them, he developed a great desire "to create some work from which the world might take some use." He began the study of mathematics, which was more certain than all other disciplines except sacred letters, and realized that "one of the most necessary things of the world" was the order that must prevail in fortifying the city, which saves the city itself and its inhabitants "from the furor of enemies." He decided to write on the plans of fortified cities. He would also discuss how to use the compass and how to make models. Finally, "because all those who write must try to write certain and not false things," he uses the propositions of Euclid.[77]

Yet before plunging into the substance of the treatise, namely, the design of polygonal fortresses according to Euclidean principles, Lanteri establishes that the subject is an extremely pleasant one that enjoys high social status. The three interlocutors—Giulio, the young Brescian, Francesco Trevesi, the engineer, and Girolamo Cataneo, the military engineer—are friends. It is a feast day. The subject is suggested as "some light chat in order to pass leisure time and the heat," especially since, as Girolamo says, "the friendship of us three being perhaps (as I judge) inferior to none others." Giulio mentions "the very beautiful design of the city" that Francesco had discussed the previous Thursday, and Girolamo mentions "the high delight" it gives him "to hear you propose things that, disputed, can render honor and utility together."[78]

Cataneo underscores the nobility of the subject of fortification by recounting that thirteen years before he had been a guest of the count D'Arco. The count Felix, the count D'Arco's cousin, who had just returned from war to visit his family and friends, highly recommended the study of the mathematical science of fortification, of which, Felix informed his cousins, Girolamo Cataneo was a master. The cousin Francesco believed that "the perfect understanding of this beautiful art can sooner be perfect and clear with the study that you have mentioned [i.e., mathematics] than with the experience of war." All three lords, Cataneo relates, came to him, the guest, to get instruction on the mathematical design of forts. He emphasized how greatly honored he was and how great his obligation was to the lords to instruct them well. He insisted that it was necessary to know the first six books of Euclid in order to do this. The art of fortification is both noble and mathematical. As Cataneo told his noble hosts, "This virtuous desire [to learn the art of fortification] is very worthy of your very noble souls."[79]

Lanteri dedicated the second dialogue of the treatise, concerned primarily with how to make models, to the Brescian nobleman Giovanbattista Gavardo. He praised Gavardo as one "who knows how equally to render a very good account of letters and of things of war." Now Francesco Trevesi, the engineer, takes over the role of instructor. He begins with a lengthy speech reminiscent of Alberti in praise of architecture. Does not architecture maintain humans "in a tranquil and quiet state," render them secure from fire, from their enemies, and from heat and cold? Does not architecture provide access "to all the other arts, liberal as well as mechanical"? The negligence of writers who ignore architecture is unfortunate, as is the view of the vulgar ignorant that "it seems a mechanical thing to practice it."[80]

Giulio challenges Francesco's interest in the mechanical arts. He asks, "You would like perhaps that nobles practice these arts manually?" to which Francesco replies in the negative but also allows that he would like the youths of the city who waste their time ambling up and down the streets to apply themselves to architecture instead, and he advises "all virtuous men to have some knowledge of this (by way of science, not of practice)." He recalls the admonition of Vitruvius that the architect know many disciplines, that he "not only be a good humanist, but philosopher, physician, astrologer." He concludes, "It is enough for me to have demonstrated to you and proved also in the end that architecture can after agriculture obtain to first place."[81]

The interlocutors turn to a discussion of models that includes both practical and theoretical considerations. They discuss the appropriate size and proportions of the fort in relationship to the size of the artillery, the characteristics of the particular site, and the nature of various materials to be used. "One always has respect thus to the circumstances of the sites as to the quality of the materials that one has to adopt in the construction," Lanteri writes. Very large forts require larger guns, more costly and more difficult to move than medium-sized cannon. These difficulties provide reason for building smaller fortifications. Other practical considerations include the size of the bastions relative to the walls of the fort, the size of the parapets, and the size of the embrasures (portholes for firing artillery). Materials are also crucial for the construction of the model, including the type of wood, stucco, and wax.[82]

Lanteri instructs how to design forts and models in the real world, but he also presents fortification as a subject worthy of the attention of nobles. He underscores the nobility of the subject again near the end of the dialogue, when the interlocutors mention numerous nobles from Brescia and elsewhere in laudatory terms. The dialogue ends with a third dedication, this one to Oliviero D'Arco, a relative of Marc Antonio D'Arco.[83]

Lanteri dedicates a second treatise, *Duo libri . . . del modo di fare le fortificationi* (Two books . . . on the method of constructing fortifications) (1559), to Alfonso d'Este of Ferrara. Lanteri emphasizes the "great utility" that he has seen drawn from ancient and modern writings, such that the skill of writing makes an author more similar to God than man in his power to help others. Having measured "my ingenuity and small powers," he fears that he is not up to the task of producing this written work. Yet he realizes that "he who fears the high deed too much is as much to blame as he who fears nothing." He decides to dispel fear, therefore, realizing that he who fails to begin also fails to advance and achieves nothing. He claims the novelty of "the beautiful and very useful subject of modern [i.e., earthwork] fortifications," which he claims have not been discussed previously in writing. He knows that he is certain to be lacerated by those who "either by envy or by evil intent blame the things of others." Yet he knows that good men will accept the products of his labors, having regard for his good intentions and appreciation when someone who has practiced an art adds "form and perfection by writing."[84] Lanteri thereby validates the authorship of practitioners such as himself.

In a letter to readers, Lanteri emphasizes the special importance of the labor of authors and of the success of their writings. He hopes that his own writings "are useful to you and equally enjoyable." Nothing is more necessary to the soldier "than understanding of the order of fortifications." Lanteri praises the Romans because they, including even the ordinary Roman soldier, understood fortification. He has long considered how important the topic is in his own time for soldiers. Again turning to his own authorship, he notes that he was taken with a great desire to be useful "to the public with some record and some rule." He explains that his own understanding of the subject increased greatly "when . . . four or five sheets of paper came into my hands with a summary of this subject written down by a very valiant man of this profession." This summary, although "written as badly as possible, so that one can scarcely understand the meaning of it, nonetheless had satisfied me of many doubts" and made him, he adds, more secure in his own writing.[85]

Thus noting the essential contribution of a virtually unlettered soldier, Lanteri stresses the importance of the knowledge of form, "which one is not able to possess perfectly without geometry." Above all, the soldier must be practiced in geometry. Citing Vitruvius, he emphasizes that "all architecture is born of construction and discourse." After geometry, knowledge of sites is necessary, and it should either be measured with a compass or, if knowledge of the compass is lacking, drawn on paper approximately, or the soldier can measure the site by paces. The soldier also needs to know the quantity and quality of artillery that must be defended against, and he must judge the appropriate size of the fort on the basis of this estimation of artillery power.[86]

Lanteri clearly writes not just to provide general information; he writes to instruct those who actually need to supervise the construction of fortifications. He provides a lengthy discussion of the appropriate size of the fort and the proportions of each of its elements, from bastions to curtains to ditches, giving a labeled illustration as an aid to readers (Fig. 12). He discusses the proportions and construction of each element. He does not focus solely on design but takes up many practical issues of construction, including the amount of time available to build the fort, the nature of the ground on which it should be built, which soils are suitable to build upon, and how many workers are required, including bricklayers, carpenters, manual laborers, and contractors. How should these workers be obtained—should they be paid or commandeered? Lanteri urges the payment of wages as a way of obtaining the best workforce. Other considerations are the quantity and quality of available materials, the weather, the season, the site, and the availability of provisions such as food (how many mills in the area?).[87]

After pointing out such general considerations, Lanteri tells the reader how to build the fort from the foundations up. He begins with laying it out, discusses the nature of the soil (clay or sandy?), putting in the stakes, and excavation. He treats how to handle the earth in detail—digging, throwing, carrying, leveling, and putting it into the work. He describes various types of spades and hoes suitable to diverse types of soil, providing illustrations of several. He itemizes the numerous different ways of hauling earth, including various kinds of wicker baskets, carts, wheelbarrows, and other devices, and assures the reader that both men and women can haul. He discusses some of the tacit craft knowledge that is needed. For example, on the important subject of making bricks, how should the soil suitable for brickmaking feel? If after rubbing soil in the hands when it is wet, it feels rough and "like pasta doesn't handle well," it is not good for bricks. Moreover, it should be "not too sandy, rocky, or crumbly." He treats the quality of the mud and how to put it into the works, how to make bundles of wood *(fascia)* to put into the construction, materials for embrasures including kinds of wood to use—oak, elm, lotus, cornel wood, olive.[88]

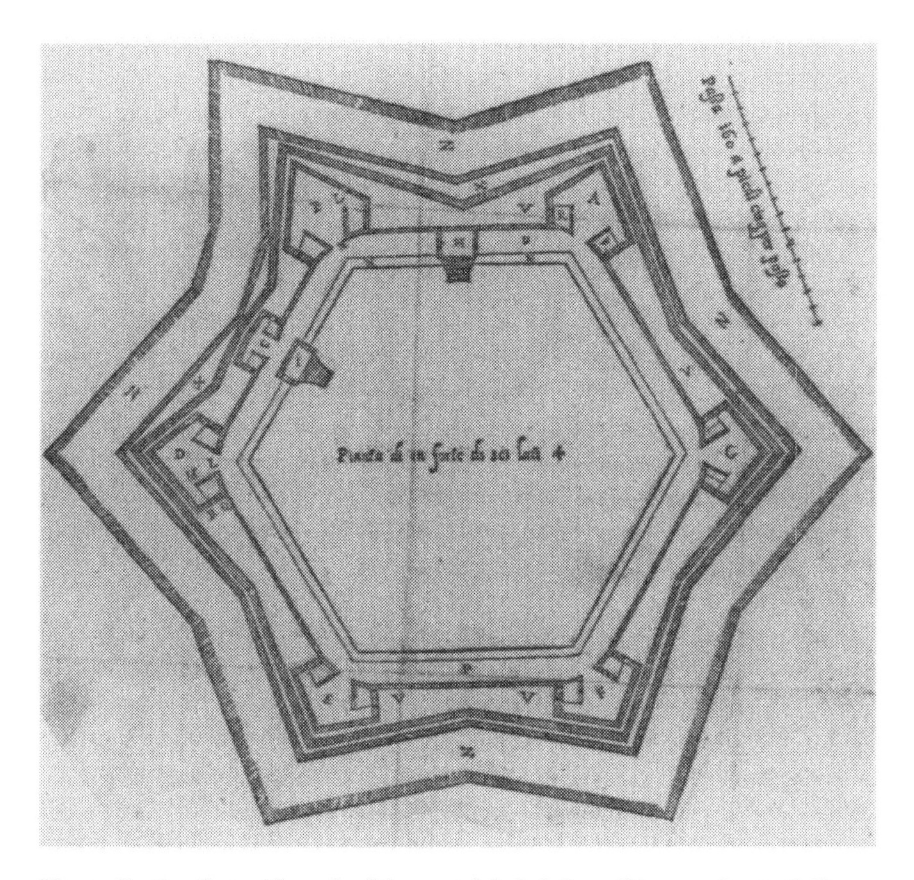

Fig. 12. Bastion fort, with each of the parts labeled, from Giacomo Lanteri's *Duo libri . . . del modo di fare le fortificationi* (Venice: Bolognino Zaltieri, 1559), fold-out between pp. 100 and 101. Each labeled part is described in the text By permission of the Folger Shakespeare Library.

Lanteri's concern for the actual construction of forts is also evident in his instructions concerning workers and masters. He urges that every effort be made to get good masters. "Above all, one must have the best and most efficient masons that it is possible to have, striving above all to have them." There should be one master for every two hundred workers. In the second book—on how to make earth fortifications in the countryside—he again insists that it is better to pay workers than to conscribe them.[89]

Finally, Lanteri advises concerning the cost of the fortification, warning that "one will always have regard for the power and will of the Prince," that above all one should not make expenditures that might be regarded as excessive. There are princes who condemn expenses and then blame the construction itself that brought them about. Lanteri explains how "many times the souls of the great remain clouded." He suggests that they sometimes "do not recognize true usefulness, but led on by the bad-mouthing of false courtier flatterers, and many times by the avaricious, they deprive those who serve them with faith and loyalty not only of their thanks but also very often of the rewards for much labor that they do in their service."[90]

In these exhortations Lanteri addresses overseers and engineers, not the princes and nobles themselves. Yet his two treatises on fortification address both practitioner and noble readers. He presents his subject as suitable for civil discourse but also fundamentally pertinent to those supervising the actual design and construction of real forts. Such actual construction may be a low mechanical art (as one interlocutor's comments suggest), but it is nevertheless necessary for the wellborn and ordinary soldier alike and thus worthy of written treatments. Indeed, such writings contribute to the public good.

Sixteenth-century writings on mining, metallurgy, and the military arts addressed topics of interest to elite groups whose wealth was based on the real products of mining and ore processing and whose power required a grasp of the realities of warfare. Yet these books do not simply purvey technical information; they also present technological subjects as worthy of written explication. Many were dedicated to wealthy and powerful patrons. Their subject matter implicitly represents the interests of such patrons, but they include much useful material for practitioners as well. Military authors especially construed their topic as mathematical, thereby enhancing its epistemological value. Writings on mining, metallurgy, and military matters point to a community of interest among elite rulers and practitioners. In some of the books, the dialogue form presents practical and technological topics as the focus of intense discussion between nobles and practitioners.

Printing allowed the production of multiple copies with relatively greater ease and thereby increased the performative range of books. They could function within client patronage relationships and could also provide instruction to practitioners. Writings on topics involving practical know-how and technological expertise transformed certain disciplines into subjects worthy of a learned readership and of discussion between wellborn individuals and technically competent practitioners. Technological subjects often were presented openly in writing; sometimes authors specifically advocated openness, condemning the secrecy of craftsmen and alchemists alike. Open writings on technical and practical arts closely tied to the praxis of rulership created disciplines of learning suitable for a readership of both the wellborn and the technically skilled.

Such writings represent an expansion of authorship by practitioners and a validation of practice. Many of the books discussed in this chapter contain pictorial depictions of workers performing technological tasks. At the same time, particularly in writings on gunnery and fortification, mathematics is invoked as essential to the topic. Authorship itself is emphasized. Authors such as the mathematical practitioner Tartaglia and the military engineer Zanchi displayed their portraits on the frontispiece of their books. Open display of technological practices and of practitioner-authors developed in tandem with the growing value of novelty and priority. At times, as in the case of Tartaglia's solution to the cubic equation, the values of openness and priority came into conflict. Yet, open authorship often could be used to establish priority. Pride in authorship on technological and mathematical topics and the glory of first discovery went hand in hand.

Chapter 7

Openness and Authorship II

Painting, Architecture, and Other Arts

DURING THE SIXTEENTH CENTURY, artisan practitioners and their patrons increasingly construed arts such as painting, sculpture, and architecture as liberal arts, characterized by both learning and skill. This rising cultural status came about for complex reasons, including the appearance of books devoted to such arts; the development of theories of art and architecture that stressed their foundation in mathematics; a growing emphasis on the ingenuity and originality of the individual artist; and the founding of academies of art as alternatives to the workshop, which included in their membership both practitioners and learned men.[1]

Painting, architecture, and other decorative arts flourished as families of oligarchs, bankers, and patricians increasingly engaged in conspicuous consumption. The display of power and wealth required the construction, furnishing, and ornamentation of palaces; it necessitated elaborate dress based upon the manufacture of luxury goods such as silks, brocades, and precious jewelry; and it required decorative items and fine wares—majolica pottery, panel paintings, tapestries, inlaid wood cabinets, and other luxuries of all kinds. In newly extravagant ways rulers and patricians signaled their wealth and power by large-scale construction projects, including enormous and enormously expensive fortifications that served both practical and symbolic functions, and by the conspicuous consumption of luxury goods.[2]

As worldly goods and ornamentation gained symbolic value, both practitioners and learned men wrote books that described particularly those arts that provided the material foundations for conspicuous luxury. As we have seen, this activity of authorship developed significantly within the manuscript culture of the fifteenth century. Aided by the printing press, authors continued to write books on painting, architecture, and sculpture as they also treated topics such as pottery and goldsmithing. Commentaries on the *De architectura* of Vitruvius and independent treatises on architecture aroused particular interest. Authors continued to dedicate such books to wealthy and powerful patrons. Practitioners and patrons at times became collaborators. Practitioners became authors, sometimes struggling to learn Latin and to acquire knowledge of ancient texts and modern learning. Their more learned contemporaries, including patrons, sometimes aided them. Meanwhile, certain princes and wealthy patricians schooled themselves in the constructive arts with the aid of skilled clients and other practitioners.[3]

Authorship on architecture and other arts furthered communication and collaboration among the wellborn, the learned, and the skilled. Numerous interactions developed in the form of conversations and in the actual planning and execution of projects such as the construction of villas, loggias, and gardens. As important as actual collaboration are the representations of collaborative discussions within dialogues and other kinds of writings, similar to those we have seen in military writings. Communication across social boundaries concerning architecture and other arts of design and construction became commonplace in mid-sixteenth-century circles of painters, architects, and humanists and could blur (although not extinguish) hierarchical social distinctions. Elite individuals interested in the constructive arts communicated with trained artisans who wrote treatises and who understood their work within a cultural sphere far more extensive than fabrication pure and simple. A broad middle ground of communication, a trading zone of knowledge, developed between the wellborn who were seriously interested in the constructive arts and the skilled individuals who became their employees, clients, teachers, and sometimes friends.

Authorship and the Arts in the Southern German Cities

Southern Germany, as we have seen, in the fifteenth century produced a significant number of manuscript writings primarily on gunpowder artillery but also on machines and other mechanical arts. In contrast to this manuscript tradition are printed German books on Gothic design techniques, published in the 1480s. One of the authors, Mathes Roriczer, a mason, wrote booklets on designing pinnacles and on practical geometry. He was a member of a well-known family of masons in Regensburg. His father, Conrad Roriczer, supervised the construction of the Regensburg cathedral and intermittently served as overseer and adviser, often at a distance, for cathedral construction in nearby towns, including Nuremberg. Conrad trained his son Mathes as a mason and subsequently had him placed as his undermaster *(Parlier)* in the church of St. Lorenz in Nuremberg. The hands-on supervision of St. Lorenz fell to Mathes, who eventually worked his way up to the post of *Werkmeister*, a position of direct authority made possible by his elevation from journeyman to master mason. Mathes remained in this position until, for unknown reasons, he was fired in 1466. Thereafter he worked on cathedrals in nearby towns until he assumed the position of master mason of the Regensburg cathedral after the death of his father in the 1480s.[4]

While he was supervising the Regensburg construction, Mathes acquired a print shop. Whether he set his own type and whether he designed, carved, and cast is unknown. In any case, he printed on his own press the two pamphlets that he wrote—one on the correct design of pinnacles, *Büchlein von der Fialen Gerechtigkeit*, published in 1486; the other, known as *Geometria deutsch*, published in the late 1480s. In addition, Mathes wrote a four-page leaflet on the gablet, a decorative element of a building in the shape of a small gable.[5]

Roriczer dedicated his pamphlet on designing pinnacles to the bishop of Eichstätt, Wilhelm of Reichenau, who had been "a lover and patron of the liberal art of geometry" and had also very much wanted those who made a living by this art (i.e., masons) to come "to a deep understanding and comprehension of it beforehand." In this way the deficiencies of those who did not know what they were doing could be rooted out, while an art of great usefulness would "be beneficially spread around and openly brought to light." Roriczer reports that the bishop held many conversations with him, asking him to write the pamphlet in order "to serve the common good."[6] Roriczer's pamphlet, then, involved an attempt to reform the practice of the master masons by making them aware of a geometric method of designing pinnacles. He, along with the bishop of Eichstätt, wanted to openly disseminate such methods for the common good. Roriczer reports that the pamphlet itself came out of frequent conversations between himself and the bishop and that he wrote it at the bishop's request.

He reiterates that designing pinnacles involves "the fundamentals of geometry through the manipulation of dividers." Sensitive to issues of authorship and credit, he notes that the explanation of how to do this comes "not from me only, but also from the old timers who knew this art, namely the junkers of Prague." He assures the bishop and those who understand the art that he has not undertaken the publication for his own fame, only for the common good. Whatever can be improved upon, he will improve it, "for whatever brings forth fruit will truly purify and clarify this art."[7]

The treatise itself is a how-to manual of constructive geometry that explains the technique of laying out the ground plan and elevation of pinnacles step by step. Roriczer's pamphlet on the gablet and his booklet on geometry share this how-to orientation. The latter instructs, for example, how to make certain geometric figures, such as a right angle, a pentagon, and an octagon. Roriczer's small treatises do not stand alone. A goldsmith from Nuremberg, Hanns Schmuttermayer, wrote another book on pinnacles, the *Fialenbüchlein*, published about

1489. Another master mason, Lorenz Lechler, wrote an *Unterweisung*, or instruction book, for his son, Moritz in 1516. The latter exists in a single hand-copied version. Lon R. Shelby rightly emphasizes the "how-to-do" quality of these manuals as opposed to the elaborate theoretical orientations of architectural treatises such as those of Alberti and Filarete.[8] Yet the emergence of written accounts of design procedures for certain structures within Gothic cathedrals is notable. Addressed to other masons, at least in part with reforms in mind, the writings point to a shift from exclusively oral communication within apprenticeship systems to written and illustrated accounts. This suggests a certain pride in the techniques and a manifest interest in explicating their details in written form.

Another indication of a shift in habitual practices among masons is the remarkable portrait of Mathes Roriczer by the Augsburg painter Hans Holbein the Elder (Fig. 13). The inscription on the portrait identifying Mathes is a later addition. Yet the mason's mark to the left is a slightly modified version of Conrad Roriczer's mason's mark and thus identifies his son, Mathes. The mason's mark on Mathes's portrait represents a new use for a traditional mark. Masons normally placed such marks on the building stones that they cut, thereby identifying their work and guaranteeing its quality. Goldsmiths placed marks on their crafted objects for similar reasons. Holbein uses the mark in a very different way—to identify the mason himself in a remarkably individualized portrait.[9] Mathes's authorship on design techniques is consonant with the individuality that his portrait suggests. His description of procedures in writing with illustrations is highly innovative from the point of view of the longstanding tradition of oral transmission within the practices of Gothic cathedral construction.

Both Roriczer and Schmuttermayer appreciated that they came from a common tradition of master masons who developed the geometric techniques they were describing, and both wanted to openly disseminate these techniques to other masons. Such openness appears to conflict with the evidence from masons' organizations, called "lodges," in which secrecy was explicitly dictated, just as in many other craft guilds. For example, in Regensburg in 1459 a congress of master masons and journeymen from many areas of the empire gathered in order to unify practices and rules governing building construction. The final declaration of the congress, which includes requirements of secrecy concerning building methods, was signed by seventy-two masters and thirty-four journeymen, but it was not signed by Conrad Roriczer, who as the master mason of the Regensburg cathedral would have hosted the conference, nor by his son. (A later endorsement of the declaration includes the name of Mathes.)[10]

The reason for the Roriczers' failure to sign the Regensburg declaration is unclear. One view is that by failing to sign, they were exempt, or exempted

Fig. 13. Portrait of Mathes Roriczer by Hans Holbein the Elder, KdZ 5008, from Staatliche Museen Preussischer Kulturbesitz Kupferstichkabinett, Berlin. Although the writing added later identifies the portrait as that of Mathes, the mason's mark on the left also identifies him. The mark is a slightly altered version of the mason's mark of Mathes's father, Conrad, and here it identifies the individual in the portrait. This represents a new use of the mark, which traditionally was placed on stones to identify the mason who worked the stone.

themselves, from the obligation to secrecy. Joseph Rykwert dissents from this view, suggesting that there was a rich oral tradition transmitting masons' knowledge, not only rules of thumb but also architectural theory. He contrasts a secret, private, orally communicated Euclidean tradition (that of the master masons) with an open, public Vitruvian tradition transmitted in writing and suggests that masons could easily participate in both. The pamphlets, he suggests, may not have been publications in the true sense of the word; that is, they may not have been made available for wide distribution. As the tiny number of copies suggests, although the pamphlets were printed, they may have been intended primarily for other masons and not for general distribution.[11] Rykwert's point is well taken. *Publication*, meaning general distribution, refers to an activity that took place from ancient times. *Printing*, defining a method of reproduction, is not a synonym for *publication* (i.e., distribution). On some occasions, as, perhaps, in the case of the Roriczer and Schmuttermayer pamphlets, printing could occur without publication.

It is also true that Roriczer and Schmuttermayer lived in a transitional culture in which written exposition had become a well-established method for raising the status and visibility of particular crafts and for turning them into more learned disciplines. It seems likely that the authors of the pamphlets were aware of the implications of authorship. It is notable that their famous younger contemporary Albrecht Dürer emerged from the same milieu and exploited the medium of print in new ways as well.

Dürer was the son and namesake of a prominent Nuremberg goldsmith who initially apprenticed to his father. When the younger Durer was fifteen years old, his father granted his wish to train as a painter rather than a goldsmith. Thereupon he moved to the tutelage of the Nuremberg painter Michael Wolgemut, head of the largest painter's workshop in the city. Wolgemut was the first major German painter to contract directly with a printer. He hired his own woodblock cutters to execute designs created by draftsmen in his workshop for book illustrations. Clearly, Dürer learned from Wolgemut's woodblock activities. As his recent biographer Jane Campbell Hutchison notes, it was in woodcut design that Dürer "was to make the most revolutionary of his contributions to European art."[12]

Dürer's artistic output was prodigious. Still extant are about 72 paintings, more than 100 engravings, about 250 woodcuts, more than 1,000 drawings, 3 published treatises, and a substantial number of manuscript writings. Influenced by Italian methods of perspective and by the higher social status of Italian painters, he created a new position for himself—and for subsequent graphic artists and painters—within the context of southern German culture. He achieved this position by the use of the printing press, by his association with learned humanists, by his own self-portrayal as an artist, and by writing treatises.[13]

Printing gave Dürer a certain economic autonomy and control over his artistic production that he otherwise could not have achieved. By family circumstance he had close ties to the burgeoning print trade. His godfather, Anton Koberger, was a practicing goldsmith who gave up that craft to become a printer. Koberger succeeded spectacularly well, becoming one of the most important publishers in Germany. Dürer himself used the medium of print in original ways. Rather than following the customary practice of waiting for orders, he created a stock of woodcut prints and copperplate engravings. He kept his worked plates and blocks in his possession and used them whenever he needed a new supply of prints. He hired salesmen to go from city to city to sell his prints. He also sold them himself in Nuremberg, which, as the trade center of the empire, attracted numerous markets, fairs, and traders. His wife, Agnes Freye, worked as his business manager and handled many of his print sales, both in Nuremberg and at markets in nearby cities such as Frankfurt. Yet Dürer's success in this enterprise was based not only on his entrepreneurial skills, which were considerable, but on his artistic ingenuity. He cut much larger woodblocks than were usual and "raised this crudest and most old-fashioned graphic medium to the status of fine art."[14]

In the early 1490s Dürer created the famous monogram that appears on numerous of his panel paintings, drawings, engravings, and woodcuts. In an essentially new use of traditional German markings, such as the stonecutter's and the goldsmith's mark, Dürer uses his monogram to signal his authorship. Having purchased his own press around 1497, he controlled every aspect of the production of his prints. In at least two instances he took legal action against individuals who made unauthorized copies. He brought one lawsuit in Venice, against an Italian engraver, Marcantonio Raimondi, who published pirated editions of the *Life of the Virgin* and the *Great Passion,* two large woodcut series. Vasari tells us that the Venetian senate ruled that Raimondi could no longer use Dürer's monogram. It is important to note that the senate protected the monogram but not the original images. In another legal action, which Dürer brought in his own city, the Nuremberg council ruled that a foreign copyist should refrain from using the monogram. Whereas Dürer clearly used his monogram to declare his original authorship and ownership of his images, judicial bodies in both Venice and Nuremberg saw it as a kind of commercial mark that signaled that he had actually made the copies being sold; they notably failed to protect the images themselves. They protected Dürer as a printer and commercial entrepreneur, not as a creator of original images.[15]

Printing enabled Dürer to acquire far greater wealth and autonomy than would otherwise have been possible. Yet printing was not the single or even the predominate cause of his sense of his own originality. Other circumstances also contributed, including the influence of Italian, particularly Venetian, culture. Dürer undertook two journeys to Italy, in 1494–95 and 1505–7. Many of the details of the second trip are known from surviving letters to his friend Willibald Pirckheimer, the first known personal letters of an artist to a friend. It is clear from Dürer's extant letters and diaries that he loved to travel. He was much appreciated in Venice, where his work was greatly admired and where artists, or at least Dürer himself, enjoyed high status. As Dürer wrote to Pirckheimer, "Here I am a gentleman; at home only a parasite."[16]

His lifelong friendships with Pirckheimer and other Nuremberg humanists, such as Conrad Celtis, allowed Dürer to cross the boundaries of social class. Pirckheimer, a humanist from one of the leading families of Nuremberg, opened many doors to Dürer, both social and cultural. The learned Latin inscriptions on some of Dürer's paintings can be attributed to his influence, as can the complex mythological and philosophical references in some of his paintings and prints. Pirckheimer and Celtis were engaged in the rehabilitation of the reputation of Germany and German culture, which was frequently maligned by Italian humanists, and in the explication of the ancient roots of German culture. They pointed to Dürer's well-known accomplishments in painting and the graphic arts as examples of high German culture. That Dürer routinely participated in social circles above his own was due not only to his wit and artistic ingenuity but also to the helping hand of Pirckheimer and other humanist friends.[17]

Dürer and Pirckheimer seem to have mediated their differences in social class and educational background through appreciation of each other's very different talents and through humor. In the letter cited above, for example, Dürer expresses his delight in Pirckheimer's growing fame as a man of letters, reporting "the great pleasure it gives me to hear of the high honour and fame which your manly wisdom and learned skill have brought you." Emphasizing the unusual youth of Pirckheimer considering his accomplishments, Dürer identifies his friend's fame with his own: "It comes to you however as to me, by a special grace of God. How pleased we both are when we fancy ourselves worth somewhat—I with my painting and you with your wisdom." He continues in a bantering tone, joking about their mutual love of flattery. And he concludes, "Now however that you are thought so much of at home, you won't dare to talk to a poor painter in the street any more; to be seen with the painter varlet would be a great disgrace to you."[18] The jocular, ribbing tone that characterizes many of Dürer's letters to his friend suggests that this remark must be read as part of an ongoing, humorous repartee, one that mediated their very real differences in social status.

Upward social mobility and a heightened sense of the value of artistic creativity are evident in the way Dürer presents himself in his numerous self-portraits. For example, in the famous self-portrait of 1500 he is dressed in elegant furs, his famous monogram on the left marking the ultimate transformation of the goldsmith and mason's marks into a sign of the artist himself (Fig. 14). Yet Dürer presents himself in this Christ-like image not simply as an individual but, as Joseph Koerner especially has explicated, as a divinely inspired artistic genius. Such a view of artistic creativity is derived in part from Ficino and the Italian Neoplatonists. Dürer's friend Pirckheimer had been deeply influenced by this tradition through his personal friendship with Pico della Mirandola before the latter's untimely death in 1494. Painters and other artists also contributed to this view. Especially important is Leonardo da Vinci's emphasis upon *fantasia*, or imagination, as intrinsic to artistic creativity and closely associated with the intellectual faculty, a view that may well have influenced Dürer.[19]

Dürer's view of the painter's creativity coincided with his rising social status and his desire to partake in material well-being and dress similar to that of the Nuremberg patriciate. He immensely enjoyed fine possessions and clothing and was able to acquire them to an extent unusual for an individual of his social background. In letters from Venice to his friend Pirckheimer, he humorously celebrates his recent purchases: "My French mantle greets you and my Italian overcoat too!"; and a few weeks later, "My French mantle, my Hungarian Husseck and my brown coat sends you greeting." Dürer made his most spectacular purchase, however, some years later, in 1509, when he bought a large house in Nuremberg near the city wall and the imperial castle. It had been the home of the famous astronomer and mathematician Regiomontanus (Johannes Müller of Königsberg) and had subsequently been owned by Regiomontanus's pupil Bernhard Walther, the wealthy merchant and astronomer. The house contained Walther's library, mechanical workshop, printing equipment, and observatory, for all of which Pirckheimer served as the custodian. Dürer's status as a property owner was soon recognized: the same year he was made a member of Nuremberg's Great Council.[20]

Dürer increased his reputation further by publishing three treatises in the 1520s; two appeared while he lived, and one appeared shortly after his death in 1528. Yet long before these publications, he labored on writings, only some of which became part of published treatises. Considering his precocious use of the printing press for woodcuts and engravings, the publication of his treatises in the 1520s is remarkably late. Much earlier he was at work on a treatise of instruction for young painters that was never published; the earliest dated manuscript fragment for this early work is from 1512.[21]

In a draft of a preface dated 1512 Dürer suggests his reasons for writing. Hundreds of years earlier some famous painters, including Phidias, Apelles, and Polyclitus, "wrote about their art and very artfully described it and gave it plainly to the light." Yet these writings had been lost, and he knew of no later writer that he might read for his own improvement. "For some hide their art in great secrecy

Fig. 14. Albrecht Dürer, self-portrait, 1500 Oil on lindenwood panel, Alte Pinakothek, Munich. Durer here portrays himself as Christ and as a member of the patrician class. Individuality and divinely inspired universal genius are invoked at the same time. The famous Durer mark on the left represents the ultimate transformation in the use of the goldsmith's and mason's marks to represent the signature of a highly individualistic artist
Courtesy of Foto Marburg/Art Resource, New York.

and others write about things whereof they know nothing, so that their words are nowise better than mere noise." Dürer therefore "will write down with God's help the little that I know." He says that although many will scorn what he writes, he will not be troubled, for it is easier to blame than to improve something. Moreover, he continues, "I will expound my meaning as clearly and plainly as I can." He further urges "all who have any knowledge in these matters that they write it down."[22] Dürer advocates open, clear, written communication of craft knowledge.

Yet he did not actually publish a treatise until more than a decade later, in my view because he had far less confidence in his abilities as an author of treatises than as an author of prints and paintings. By 1523 his treatise on human proportions was ready for the press. But in that year he acquired ten books relevant to painting that his friend Pirckheimer had selected from the library of the late Bernhard Walther, whose house Dürer now owned. The exact titles of Dürer's acquisitions are unknown, but they must have included Alberti's *Della pittura*, which was not published until 1540 but which existed in a manuscript copy in Walther's library. Shortly after his book purchase, in 1525, Dürer published a treatise, not on human proportion, but on various kinds of geometric constructions and measurement. Perhaps, as Walter Strauss suggests, Dürer's study of his new books guided his decision to publish a manual on measurement before publishing his more ambitious work on human proportions.[23]

Dürer's unpublished draft dedication to Pirckheimer for his planned treatises on measurement and on human proportion is dated 1523. The dedication indicates numerous discussions concerning these subjects between himself and his friend: "It has often come to pass that the speech between us has turned upon the different arts." Dürer had asked Pirckheimer whether books on human proportion existed from earlier times. Hearing the negative reply, he began to investigate the matter himself and brought his friend "what of such matters I then found out and invented." He feared censure or the discovery of books by the ancients that would discredit his own theories. Yet at Pirckheimer's urging, he decided to "give it to the light."[24]

In another unpublished letter, Dürer suggests points for a preface to the treatise on human proportion. He says that nothing should appear boastful or envious. The preface should deal with nothing that is not in the book and should introduce nothing that is stolen from other books. The preface should emphasize that Dürer writes "only for our German youth" and that he admires the Italians for their naked figures and their perspective. Finally, Dürer admonishes, "I pray all such as know anything instructive for art to publish it."[25] Modesty, openness, and abstinence from plagiarism emerge as predominant values.

Dürer's 1525 treatise on measurement is a practical manual of instruction that treats a variety of topics, such as lines, including spirals; plane surfaces, including polygons; solids such as columns and a tower; the construction of alphabet letters in the Roman style; and perspective. Dürer explains how to fabricate sundials, several measuring instruments, and perspective devices. In his dedication to Pirckheimer he laments that talented young painters in Germany fail to acquire "real foundation." Some do achieve skill through continuous practice, "but their works are made intuitively and solely according to their tastes." Therefore, they have "grown up in ignorance like an unpruned tree." The sole reason such painters have "never learned the art of measurement without which no one can become a true artisan," is that they have derived pleasure from their errors. Dürer declares that his manual "is well meant and intended for everyone desirous of learning about art—not only for painters, but also for goldsmiths, sculptors, stonemasons, and carpenters."[26]

Although Dürer emphasized a readership of artisan practitioners, it is clear that in the process of writing he collaborated with his learned humanist friend Pirckheimer. Pirckheimer aided him by providing books for consultation (the volumes of 1523) and by translating relevant passages from ancient texts. For example, Dürer treats the problem that involves the doubling of the size of a cube while retaining its shape by providing a solution derived from Eutokus of Ascolon, an ancient commentator of Archimedes. He obtained the solution from a manuscript translated by Pirckheimer, who also translated for him relevant passages from Euclid. Dürer's sources include ancient writers such as Euclid and Vitruvius. He was also significantly indebted to such treatises as the *Divina proportione*, by Luca Pacioli, and the pamphlets of Schmuttermayer and Roriczer. From the latter's *Geometria deutsch* he directly borrowed several constructions. It is relevant that Roriczer's dedicatee, Wilhelm von Reichenau, bishop of Eichstätt, was in charge of the education of Dürer's friend Pirckheimer for a time in his youth. In those same years (1486–88) Schmuttermayer and Roriczer wrote their manuals. It is possible that Pirckheimer was personally acquainted with Roriczer, just as it is likely that Dürer knew Schmuttermayer, who was a Nuremberg goldsmith like his father.[27] Yet Dürer's own treatise is far more ambitious than those of his immediate predecessors.

Dürer's broadly cosmopolitan point of view is most evident in his treatise on fortification, *Etliche underricht zu befestigung der Stett / Schloss / und Flecken* (Some instruction on the fortification of cities, castles, and places) (1527). In this, his second published treatise, he addresses a readership of princes, kings, and rulers of cities. In his dedication to Ferdinand I, king of Hungary and Bohemia, he suggests that the work would be useful "not to Your Majesty only, but to all other Princes, Lords, and Towns that would gladly protect themselves against violence and unjust oppression." Within the body of the treatise Dürer immediately takes up the ever-present issue of cost. Think of the kings of Egypt, he suggests, who spent huge sums on the pyramids, "which indeed had no use." Further, the prince ruled over many poor people; it would be better to give them daily wages to work on fortifications than to give them alms. It would also be better for a prince to spend a great sum on fortifications than to be driven from his land by an enemy.[28]

The treatise provides a comprehensive design of a fortified city. Dürer describes methods of building bastions and of constructing a blockhouse. The king's castle is the central focus of the city. The plan details the location and design of the castle, including its internal palace and the dwellings of the king's advisers, the locations of provisions, artillery, horse stalls, huts for the foot soldiers, and the church. Within the city walls Dürer prescribes the location of inns, foundries, the market, the *Rathaus*, or town hall, houses of the lords, dwellings of military personnel, including captains, dwellings of merchants, and houses of foundry workers, carpenters, and many other artisans. He also specifies the location of the slaughterhouse, the brewery, and bakeries. He treats the problem of strengthening the fortifications of an already existing city, and he briefly discusses the mounting of artillery. The woodcut illustrations include a panoramic bird's-eye view of a city under siege, a view that displays the same combination of specific detail and comprehensive overview that characterizes the treatise as a whole.[29]

In sum, Dürer's view of his visual and textual production involves a developed sense of his own originality and ownership. His writings explicate in detail that painting involves knowledge of mathematics; his treatise on fortification envisions an entire fortified city based on the needs of its ruler. In his lifetime, Dürer transformed himself from artisan practitioner to author of original prints and paintings and author of books. His friendships and numerous associations with social elites who were university-educated men points to a development in which artisanal production and discursive knowledge came to be closely associated. Yet this process of joining artistic production to rational discourse was not the work of a single individual but reflected a widespread cultural phenomenon. Many sixteenth-century practitioners associated themselves and their crafts with learning and with open written communication.

Practitioners and Patricians in the Vitruvian Renaissance

An important arena of collaboration among skilled artisans, patrons, and university-educated men was architecture, including the design and construction of buildings, the study of ancient ruins, and Vitruvius's *De architectura*. Vitruvius's treatise became the focus of intense interest in the late fifteenth century. After the publication of Giovanni Sulpicius's first printed edition of 1486, numerous editions, translations, and commentaries appeared as authors also continued to write independent treatises. Editors of Vitruvius, following the Roman architect's view that architecture required both theory and practice—*ratiocinatio* and *fabrica*—promulgated the notion that handwork and construction should be combined with reason and mathematics. Vitruvius's attempt to join theory and practice remained a rather isolated effort in the ancient world but found fertile ground in the sixteenth century.[30]

Cesare Cesariano's Italian translation and commentary on the *De architectura* were published in 1521. As we know from recently discovered documentation, Cesariano was born into a family of notaries. He attributes his own interest in learning to his father, who was in the service of Duke Galeazzo Maria Sforza of Milan. His father died when he was a young child, leaving him in the care of a woman whom he calls his "stepmother" (but who was actually his mother), Elizabeth. In his youth, he spent a brief time in the workshop of the architect Donato Bramante before the latter left for Rome, in 1499. He reports that he was forced to flee from the care of his stepmother because of the "innate violence" of the man she had married. Thereafter he worked as an itinerant artisan in "various towns and regions." He notes that he worked as a painter and architect, that he studied diligently, and that he was highly self-sufficient. "I employed myself with painting and architecture. With my daily gain, I conversed and studied much in order to observe and understand diverse talents and practices of men. I was helped by God and myself." Cesariano's presence is documented in several towns of northern Italy, including Ferrara, Reggio Emilia (from which he fled in 1507 after murdering a man), and Milan.[31]

Providing a dramatic full-page allegorical portrait of his own life, Cesariano describes his troubles and his ambitions for his commentary (Fig. 15). He calls the allegory "The Chosen of the World Configured by Cesare Cesariano." The "chosen," namely, the great and wealthy, can be found in their seats of immortality on the wheel of fortune, on the right. The company of the impoverished appears on the left. Cesariano portrays himself with his back turned to the viewer and his hand extended toward Fortune. A sign on his back reads, "Exigitur Tandem A Paupertate Ductus" (The learned man at length is cast forth from poverty). He is dressed in the clothes of a patrician, a fat purse clasped to his waist. His head is encircled with a wreath of laurel leaves, symbol of learning. In his hands he carries a compass and rule, tools of the architect.[32]

He explains that he has boldly risked himself to "touch the long-haired brow of Fortune" so that he could reach the seats of immortality occupied by "great wealthy men, kings and princes " He has done this with the advice of Patience and Prudence; they in turn handed him "to Audacity with these my commentaries with the compass and the rule. Thus I might seek to flee the dark sphere as a rat from the hands of stepmother and poverty." Cesariano emphasizes the monumental labors that his commentary has entailed, and he says that since "God and nature do nothing in vain," he believes he was especially created and educated so that he could explain this divine work "for the great advantage and necessity of the world." He finally ends his account of the "unspeakable disasters" he has suffered so that the reader will not be forced to "burst . . . into tears" over his misfortunes.[33]

Cesariano worked as a painter, architect, and a military engineer while he labored on his Vitruvian commentary. He places special stress upon the Vitruvian ideal of *ratiocinatio* and *fabrica*. "Not only architecture," he says, "but every other art" is made up "of the work or fabrication and reasoning." The reasoning concerns the "well-calculated and -considered" rational aspects of each art and involves general rules. The work itself constitutes the particular application. It is necessary "to know how to say and how to do." The work, the doing, is "almost of greater necessity" than the rational part, the saying. The rational part itself involves "the speaking with reason about the handmade thing" and a demonstration of the object from section to section.[34]

Handwork is closely associated with reason. The treatment of materials is the "drawing out of the sense of the thing through explanation, as does the skilled teacher of some technical skill, who demonstrates not only with words but with actions in order to teach the uneducated workers." The ability to understand an object is associated with the ability to handle it skillfully. Nothing arises "in this life except as a result of handling." Those people "who know how to work through handling things themselves give shape to elegance . . . in order to be recognized for their knowledge."[35] In sum, knowledge and handwork are closely interrelated.

Cesariano brings together the rational and the mechanical in his discussion of machines. He defines "machination" as "the contriving, effecting, and inventing of manual operations." He insists, moreover, that "this machination [is] intellective, since it is the cause of the formation of crafted instruments or artists adept at explaining the effect of whatever we want to complete." Manual operations and causal explanations are closely interrelated. "This ingenious mechanical knowledge is necessary not only in the military arts but in all liberal demonstrations and operations." Without the mechanical arts almost no convenience would be available for the use of ordinary life, neither clothing nor any number of other artificial things necessary for human use. Elsewhere Cesariano praises the "noble philosophers" who invented machines. They are to be admired for their "under-

Fig. 15. Cesare Cesariano's allegory of his own life, from his Vitruvian commentary of 1521. Cesariano represents himself in patrician clothing with architectural instruments in his hand. He reaches up to the wheel of fortune on the right. The sign on his back translates, "The learned man at length is cast forth from poverty." He stands among a crowd of the impoverished, including his wicked "stepmother." The impoverished are governed by the three figures on the platform—Invidiousness, Ignorance, and Persuasion.
Courtesy of the Smithsonian Institution Libraries, Smithsonian Institution®.

stood contemplation," which preceded their "great knowledge." In addition, they are to be praised for their "burning desire to produce in sensible works with their own hands that which they have reasoned with their mind."[36]

Cesariano advocates the open, written transmission of knowledge, and he notes the benevolence of writers who had left their treatises to posterity: "O great goodness of wise men to bequeath the most precious things." No worldly treasure could reward sufficiently works of "such great distinction" that they seem "divine and not human." Writings from the past allow humans to consider carefully and with reason "the effects and obvious examples" of the ancients, thereby suggesting to them the most suitable methods of proceeding with their own work. They tell people "in what way they must use human practices and technical skills."[37]

Cesariano's own traumatic effort to produce his commentary stands in striking contrast to his beneficent portrayal of writings handed down through successive generations. About 1520, he gained the support of two fellow citizens, a Milanese mathematician, Aluisio Pirovano, and one Agostino Gallo, of Como. Together they provided financial support and made arrangements with the Milanese printer Gotardo da Ponte to print the work. While Gotardo was printing the completed parts of the book, Cesariano boarded with Agostino's brother, Sebastiano Gallo, in nearby Como and worked on the commentary to books 9 and 10, the last of which concerned machines. A dispute arose over mistakes in the proofs. Cesariano left the Gallo house with his belongings, including his unfinished commentary. His sponsors subsequently forced the manuscript materials from him, and had the commentary completed, in a highly perfunctory way, by two other men—Benedetto Giovo, of Como, and Bono Mauro, of Bergamo. Cesariano himself returned to Milan, where he turned to architectural projects and undertook a seven-year quest for justice that finally succeeded when a Milanese court compensated him in 1528. As we know from the recently discovered autograph manuscript, he also continued to work on his own commentary to books 9 and 10. Vasari reported that he died "more of an animal than a person," having lost his sanity in the dispute. Yet his documented activities as an active architect, engineer, and author after 1528 suggest that such a report may be false.[38]

In life, Cesariano combined an itinerant practice in art and architecture with authorship focused on his Vitruvian commentary. His writing reflects his views that reason and hands-on practice are closely related. He dramatizes his life as an author and architect with unusual vividness, representing himself as escaping from poverty and his wicked stepmother into the seats of powerful and wealthy immortals with the help of his commentaries and his compass. Nevertheless, in this "up from poverty" narrative of his own life he never turns away from handwork and practice; instead he makes it inherently important to reason and intellectual activity and essentially linked to the praiseworthy practice of authorship itself.

After the 1521 publication of Cesariano's edition of *De architectura*, numerous other writers in many parts of Europe produced Vitruvian translations and commentaries as well as independent architectural treatises. Architectural authors often advocated combining the skill of handwork with learning, which included knowledge of the *De architectura* and other classical texts, and facility in mathematics. Collaboration between learned men and skilled practitioners often underlay the production of architectural books. Collaborations included expeditions to measure and study ancient buildings and ruins, study of Vitruvius's text, and finally, the actual design and construction of buildings.[39]

Although the *De architectura* became available in vernacular languages, often accompanied by commentaries, the Latin text still presented formidable difficulties, partly because of its technical vocabulary. In Rome, a group of scholars began to address some of these textual problems. The group, made up of members of the Roman Accademia della Virtù, led by Claudio Tolmei, turned to architectural interests in 1540–41. As we know from a detailed letter from Tolmei to a potential patron, the goal of the academy was to examine Roman ruins carefully and compare them with the prescriptions in the Vitruvian text. Tolmei explains that because almost all of the arts are composed of both theory and practice, "it is necessary in order to come to some excellence, not only to speculate but also to put into work." However, the members of the academy were not able to fabricate at the moment and thus had turned to thinking about ancient constructed things. "Hence," he continues, "joining the precepts of the writers with examples and notices that are pulled from works, thus they endeavor as best they can to turn the eyes from one part to the other."[40]

In addition to comparing ancient artifacts with the prescriptions of Vitruvius, a goal of the academy was to publish a Latin edition of the *De architectura* with commentaries. Although the edition was never achieved, the French humanist Guillaume Philander, who lived in Rome during these years, completed his Latin commentary in 1544 as an initial part of the project. His work included a digression on the five architectural orders, based significantly on Sebastiano Serlio's treatise on the architectural orders, published in 1537. Yet unlike Serlio's treatise, to be discussed below, Philander's annotations focus not on actual structures but on detailed philological considerations.[41]

Serlio had studied and measured ancient ruins during a sojourn in Rome, and he had been instrumental in arousing Philander's interest in Vitruvius. The son of a leatherworker, he had trained as a painter. He worked as a painter in Pesaro and in Rome in the circle of the architect Bramante, perhaps in the Vatican workshops. Having moved to Venice by 1527, he associated with the Venetian circle of the poet and humanist Pietro Aretino, with whom he became good friends, the painter Titian, and the architect Jacopo Sansovino. In Venice he devoted himself primarily to creating his illustrated treatise *On Architecture*, which he published in installments.[42]

He first published book 4, on the architectural orders, in 1537, dedicating it to the duke of Ferrara, Ercole II d'Este, who financed the publication. Serlio saw his authorship in terms of openness. God allots divine influence according to the capacity of each soul to receive it, he writes, and thus each soul receives a greater or lesser quantity of grace. Although his own talent is small, he has not ignored what God was pleased to give him, nor has he "kept it buried, hidden in my garden." Rather, he wants to bring what he understands of "this noble art of architecture out into the open." He does this, not for praise, but to "inspire the Souls of those who could benefit a great deal if they did not hide their treasures in laziness." He also wants to teach those who have the ability to add beauty to the world.[43]

Serlio aims to teach a wide range of his contemporaries, especially practitioners. He has formulated "some rules concerning architecture" so that not only exalted intellects but every average person might understand it. As Deborah Howard notes, Serlio was "a pioneer, providing the first architectural treatise of the Renaissance to convey its content chiefly by visual means."[44] His drawings are such that architects and builders did not need to travel to particular buildings to see what they looked like. Instead they could find images in his treatises of whole buildings and many building elements, such as capitals, friezes, gates, and windows, which they could readily use in their own designs.

Serlio dedicated his second book on ancient Roman buildings (book 3 in the completed treatise) to the French king Francis I, thereby acquiring patronage in the French court. Moving to Paris in 1541, he dedicated his subsequent books—on geometry, perspective, and temples—to Francis I and to Francis's sister, Marguerite of Navarre. These books were published in bilingual editions, with the French translation prepared by Jean Martin. A sixth book on domestic architecture, dedicated to Henry II, remained unpublished until modern times. Book 7, on accidents and unusual situations in building, was purchased from Serlio by Jacopo Strada, an antiquary and art dealer, who published it in 1575 in Frankfort on the Main. Strada also published the *Extraordinary Book of Architecture*, on rusticated doorways. Finally, Serlio prepared a treatise on Roman fortifications that was never published. Most of Serlio's books, however, were immensely popular. They were published in numerous translations and, much to his annoyance, in purloined editions. Yet after the deaths of his two patrons, Francis I and Marguerite of Navarre, he was replaced at court by the French architect Philibert de L'Orme. Serlio then moved to Lyons, where he lived out the rest of his life in poverty.[45]

In the Veneto, where Serlio had written and published his first architectural book in 1537, extensive building design and construction and the study of Vitruvius led to fruitful collaborations between elite patrons and skilled practitioners. Perhaps the most notable is that between Daniele Barbaro, a learned humanist from an eminent Venetian family, and Andrea Palladio, an architect who had been trained as a stonecutter. Palladio designed the Villa Maser for the two Barbaro brothers, Daniele and Marc Antonio. Daniele Barbaro himself wrote a Vitruvian commentary illustrated by Palladio, published in Italian in 1556 and then in a revised and expanded edition in both Latin and Italian in 1567. Palladio published his own architectural treatise, *The Four Books on Architecture*, in 1570.[46]

Barbaro appreciated skilled handwork as well as learning. In his 1556 commentary, following Vitruvius, he insists that "practice . . . is necessary, discourse is necessary—the discourse is as the father, the fabrica as the mother of architecture." Moreover, if anyone believes he could be an architect "with craftsmanship alone," or "with discourse alone," he is mistaken. Such an architect "would be considered an imperfect, indeed a monstrous thing." The architect without practical skill will be subjected to derision: "And pray, if one had knowledge only, and wanted to usurp the name of architect, would he not be subjected to the insults of the skilled? Would not every manual worker be able to reproach you and to say to you: 'What do you [*tu*] do?'" Yet if the architect were a skilled craftsman with no theoretical knowledge, he would suffer a different sort of humiliation: "On the other hand, if one is believed to be worthy of such a great name through having a humble practice and much experience, would not an intelligent and learned man be able to silence him, demanding an account and reasons of the things done?"[47]

Vitruvius's expression of gratitude to his parents for training in both technical and literary disciplines prompts Barbaro's further comment on theory and practice: "Not only must the architect devote himself with ardent desire to the understanding of letters but he must take great pleasure in knowing how technical things work, in investigating them, and in making them, so that his understanding does not remain dead and useless."[48] In a discussion of dials and clocks Barbaro stresses his own practical experience. He warns those who think these things difficult that they cannot understand them well "without making a test." Nor can they blame obscure writing "because in every experience there is a difficulty where there has not been practice." Barbaro insists that he himself has understood these things, "and this much more through making and experimenting, than through reading."[49]

Barbaro delights in tools and machines. The understanding and manufacture of machines is a "beautiful, useful, and wonderful practice." Who, he asks, could "not look with astonishment" upon the ability of one man "helped by a small tool" to easily lift "an immense weight," or to lift a rock seemingly as heavy as a mountain "with skillfully wound weak ropes"? Who does not read "with wonder" about the things made by Archimedes? The construction of machines includes an intellectual aspect. The "mechanical science or art" is "the demonstrated reason of the way of making machines." It is not subject to the meaning "the common man" gives it, "calling every mechanical art low." This is because what "is made first in the mind" and "then governs works with skill" is said to be "of machination and discourse."[50]

Barbaro defends the open, written transmission of knowledge. He praises the ancients, who have discovered "the very subtle causes of high things" and who are worthy of "celestial honors" because they looked "to the common good" rather than to their own advantage, bestowing benefits not only upon their own age or upon any single era but upon all ages perpetually.[51] Barbaro justifies open authorship, especially as it concerns his discussion of fortifications. He says that some believe that such information should be kept secret, available only to princes and republics. Others lament that the Italian method of fortifications would become known outside of Italy. Barbaro insists that people with such opinions "by themselves descend to the depths of those who, being men, are willing to fail in the service of humanity." They are ungrateful, since they themselves have learned "many fine things from people of various countries." Further, they do not understand that "malicious people" can teach everything about Italian fortifications without writing anything down. "To those that truly praise secrecy," he insists, "I would say that that which pertains to the preservation of men must not be held secret."[52] Barbaro believes that military technology, specifically fortification, should be openly explained.

Praising those who are willing to teach, Barbaro emphasizes the debts he has incurred in his own education. "I have sought to learn from everyone: I am a debtor to everyone who has helped me." That help, he specifies, has come especially from his collaborator, Andrea Palladio. Palladio himself started out as a stonemason. He was born Andrea di Pietro della Gondola, the son of a millworker. At the age of thirteen he was apprenticed to a stone carver in Padua, but three years later he broke the contract and escaped to Vicenza. For the next fourteen years he worked as an apprentice and assistant to Giovanni da Pedemuro and Girolamo Pittoni, the leading sculptors and stone carvers of Vicenza. At the age of thirty Palladio was called to work on a loggia being built at the villa of Count Giangiorgo Trissino. Trissino was a distinguished humanist who was interested in architectural theory and antiquities and who housed a number of young nobles in his villa as his students. He decided to include the thirty-year-old stonecutter, Andrea, among them and eventually bestowed the classical name Palladio upon him. He gave Palladio readings that pertained to architecture, engineering, ancient topography, and military science.[53]

In 1528 Trissino moved to Padua, where he stayed for three years. During that period, Palladio spent time there as well. In Padua Palladio met Alvise Cornaro, a wealthy patron, agriculturalist, scholar, and dilettante in architecture. Cornaro's writings included a short treatise on architecture. Palladio learned from Cornaro and also from the architects and painters he met in Padua. He saw the drawings and writings of Serlio, whom he met before Serlio moved to France. Palladio joined Trissino on his first trip to Rome in 1541. Thereafter, he frequently traveled to Rome and throughout Italy, studying both ancient and modern buildings. On one of his trips to Rome he was accompanied by Daniele Barbaro. In 1555 Palladio became a member of the Accademia Olimpica in Vicenza, an academy made up of scholars, mathematicians, several nobles, and himself, the sole member with an artisanal background.[54]

In *I quattro libri dell'architettura (The Four Books on Architecture)* Palladio states that he has approached the subject through the careful study of Vitruvius and through the observation and diligent measurement of ancient monuments. Both modes of study are pervasively evident in his treatise, which is notable for its detailed, precise descriptions. The treatise is divided into four parts. The first concerns principles of architecture, the orders, and elements such as chimneys and stairs. The subject of the second book is private houses. Palladio describes and shows by ground plan and elevation many of the villas that he himself designed and built. In each case he mentions the noble patron and the location, describes distinctive elements, and often mentions the painters and other artists who decorated the interiors. Book 2 functions as a walking tour of Palladio's architectural designs and simultaneously as a display of the magnificent villas of his patrons. He includes discussions of atria and porches. (He was the first architect of his own time to put classic temple porches on domestic buildings.) The third book discusses public buildings and public spaces, including roads, *piazze*, and bridges, including the bridge Julius Caesar built over the Rhine and bridges of Palladio's own invention. The fourth book is devoted to ancient temples, including the

Pantheon. It includes one modern building: the famous *tempietto* in Rome by Donato Bramante, a tribute to the famous circular building and to Bramante himself as a reviver of classical forms.[55]

Palladio's treatise is very different from Daniele Barbaro's Vitruvian commentary; it is less concerned with theory and exhibits a richness of detail concerning both ancient and modern buildings. Reading Barbaro's and Palladio's treatises together, it is easy to see how they were collaborators in the best and deepest sense of the word. Palladio possessed an intimate knowledge of construction and a far-ranging familiarity with ancient buildings and their elements derived from a lifetime of careful observation and measurement. He had used these aspects of his experience in his extensive study of Vitruvius. Barbaro was a wealthy patron and learned humanist with a deep interest in Vitruvius and in the constructive arts. Barbaro and Palladio appreciated each other's knowledge and experience, which were complementary. Both wrote books on architecture. Both advocated the openness of knowledge. Palladio points to his long labor, his diligence, and his devotion in both the understanding and the practice of architecture and thanks God for his goodness if he has not worked in vain. He emphasizes that he is at the same time "greatly indebted to those who, through their own ingenious inventions and the experience they gained, have bequeathed us the rules of this art, for they opened up an easier and more direct route to the study of new things, and (thanks to them) we know of many things that would perhaps have remained hidden."[56] It is a Vitruvian sentiment that Barbaro shared.

The central interests of the French architect Philibert de L'Orme, who displaced Serlio at the French court, were surprisingly similar to those of Palladio and Barbaro: architectural design and building construction, the careful observation and measurement of ancient ruins, and authorship. Philibert, who was the son of a master mason and trained as a master mason himself, was born in Lyons. As a young man in 1533 he went to Rome, where he met Cardinal Jean du Bellay and Du Bellay's secretary, the famous humanist François Rabelais. Philibert and Rabelais became close friends. Philibert spent three years in Rome, observing, measuring, and excavating. He returned to Lyons but was soon called to Paris to build a château for Du Bellay at St. Maur-lès-Fossés. During his work on the château he met the dauphin, Henry, and his consort, Diane de Poitiers; in 1547 he was commissioned by the latter to design and build her château at Anet. When the dauphin became Henry II in 1547, he removed Serlio (who as we know went on to lead an impoverished life in Lyons) and appointed Philibert superintendent of buildings. Philibert became the leading architect of France and exerted far-reaching powers over the royal building works. He centralized its organization, organized the training of highly skilled artisans, and exerted tight control over the budget. At Henry's death in 1559, Philibert was dismissed, a result of the hostility of a particular faction at court, combined with his arrogance in the exercise of power and his tight control over the budget. Eventually, though, in 1563, he entered the service of the Queen Mother, Catherine de' Medici.[57]

During the years when he was out of favor, Philibert wrote two treatises, the *Nouvelles inventions pour bien bastir et à petits fraiz*, published in 1561, and *La premier tome de l'architecture*, published in 1567. In these treatises he points to both his knowledge of writings and his experience. In *La premier tome de l'architecture*, for example, he writes that his knowledge of architecture comes "as much by books, as by the experience that I have had of it in different places and also by various works which I have had to make and have guided in my time." Elsewhere he stresses that his information is based on practice: "I inform you that all which I propound to you and write has been tested in various places by my edict, advice, and command." He cites his own experience in many areas, including difficulties with the servants, the relatives, and the wives of patrons; the importance of using a model; and his own study and measurement of Roman buildings.[58]

Philibert repeatedly urges architects to acquire sufficient training. In *Nouvelles inventions* he admonishes, "If the architect or superior who commands the master masons and other workers is not well trained and does not quickly understand his theory and practice," his work will be "deformed and ridiculous." Such an architect will be reputed the "slave of the master mason or some worker," who will make him understand what he wants and who will not be able to repair things poorly made because of ignorance. All this will be to the "great detriment and dishonor" of patrons if not of the architect himself.[59]

He wrote the *Nouvelles inventions* for the benefit of master masons and workers as well as for architects. He hoped that "all those who make profession of the said architecture, as also all workers and others who wish to make buildings," would profit from his writings. In the *Premier tome de l'architecture*, discussing geometry, he suggests that with such instruction the master mason can carry out all kinds of work, provided he understands the practice of measures and proportions. He complains that many workers do not study enough, and he wants "fraternally to warn, admonish and pray" that the workers will "recognize and wish to study and learn what is required and necessary to their art and profession."[60]

Yet ultimately Philibert insists upon the primacy of actual experience. The task of building construction requires knowledge and understanding, "which is learned through long experience and the practice of taking several buildings into work." In order to understand the art, the architect must know all its precepts, "not so much by means of books as by long and great practice."[61] Nevertheless,

Philibert aims his own clearly written explications of architectural theory and practice at both general readers and practitioners who wish to become successful architects.

To conclude, the range of individuals who undertook architectural authorship is notable. They include artisan practitioners such as Cesariano, Serlio, and Palladio and learned humanists such as Jean Martin, Guillaume Philander, and Daniele Barbaro. In one sense this range continues the tradition of the fifteenth century, during which Alberti, a university-trained humanist, and Filarete, a workshop-trained architect, each wrote a major architectural treatise. Architecture in these centuries was a constructive practice, involving the design and construction of buildings, but also a discursive practice, involving communication about an important body of knowledge. Architecture created a discursive space, what might be called a trading zone, in which substantive communication occurred between learned humanists and patrons and artisan-trained practitioners, who became learned in their own way and wrote books.

Authorship and Diverse Arts: Piccolpasso, Vasari, Cellini, and Palissy

In the same decades in which a diverse group of authors wrote books on architecture, practitioners wrote treatises on arts such as pottery, goldsmithing, and sculpture. Individuals undertook such writings in the context of patronage relationships, and they also found cause for writing in the growing admiration for the virtuosity of skilled artisans. Craft skill and an appreciation for beautiful objects encouraged a view of particular crafts as repositories of knowledge, thus encouraging their written explication in treatises dedicated to patrons. The printing press provided additional motivation, creating the conditions for a more general readership.

Cipriano Piccolpasso's *I tre libri dell'arte del vasaio (The Three Books of the Potter's Art)* provided a comprehensive description of pottery manufacture. The illustrated treatise exists in a single manuscript copy that was almost certainly meant to be printed. Piccolpasso was from the famous pottery town of Castel Durante, near Urbino, where his younger brother Fabio was the master of a majolica workshop. Their father served the armed forces of the dukes of Urbino in various capacities; their mother was the daughter of an apothecary. Cipriano himself acquired a humanist education and trained in the military arts. In his youth he served as a page to Cesare Riaro, bishop of Malaga and patriarch of Alexandria. Riaro took him to Padua, and from there he frequently visited Venice, a city he greatly admired. When Riaro died in 1541, Piccolpasso returned to Castel Durante. His known occupations involved for the most part military service to the dukes and to the papacy, which controlled Umbria. He served the papacy in Perugia, for example, as the supply officer *(providetore)* at the fort; there and elsewhere he also carried out various tasks as a military engineer and surveyor. In addition to his treatise on pottery, he wrote a topographical survey of the cities and lands of Umbria.[62]

Although Piccolpasso never worked professionally as a potter, it is evident from his detailed treatise that he had a thorough knowledge of the art of pottery, which he may have gained in his brother's workshop. Castel Durante was one of the flourishing pottery towns of central Italy, its wares famed throughout Europe. Piccolpasso addresses his treatise to the duke of Urbino, Guidobaldo II, who himself avidly promoted the craft. In his later treatise Piccolpasso mentions that he actually wrote the treatise at the request of the French cardinal François de Tournon, who in the course of various diplomatic and papal missions spent a year in Castel Durante, from September 1556 to September 1557. Tournon was a humanist who collected a major library, established two colleges to provide boys with a humanist education, and was interested in natural history and the mechanical arts. His encouragement of Piccolpasso undoubtedly also reflected his interest in encouraging the majolica industry in the region of Tournon and Lyons, where Italian majolica workers had arrived in the early 1550s. In the mid-sixteenth century Italian majolica workers were given privileges in many parts of Europe, often in exchange for training apprentices from the region. Yet these industries did not achieve the quality of the best majolica of central Italy, in part because they did not have access to the same materials. Piccolpasso's detailed descriptions and instructions would have been of great interest to itinerant potters and patrons alike.[63]

Piccolpasso probably completed his treatise on pottery about 1557 or 1558. Even though only one copy exists, the translators of the recent edition nevertheless provide substantial evidence that the manuscript "is a fair copy prepared for the press" and that he intended to have it published in Venice. Perhaps he did not find the necessary support for publication after the death of Cardinal Tournon in 1562. Piccolpasso clearly states his intention to openly explicate the secrets of the potter's art on the title page. His book treats "not only of the practice" of pottery "but in brief of all its secrets. A matter that up to this very day has always been kept concealed."[64]

In his prologue to readers, Piccolpasso amplifies his goal of open explication: "I have set myself faithfully," he writes, "to show forth all the secrets of the art of the potter." He defends himself against his detractors, including those who say "it is wrong to publish what has now been concealed for so many years." Piccolpasso defends himself by referring to Biringuccio's *Pirotechnia* as a model. He continues that he is not like those who tell the art but keep certain small but essential secrets until they are on their deathbed, at which time they tell their oldest sons. He, in contrast, tells all. Further, "to those who deem me presumptuous in publishing these secrets I answer that it is better that many should know a good thing than that a few should keep it hidden." As a result of his writing openly, the art of pottery, which "has full often remained among persons of small account," now "will circulate in courts, among lofty spirits and speculative minds."[65]

True to his word, Piccolpasso discusses all aspects of pottery, from clays and other materials to the construction of pottery wheels and the sticks and irons for aid in turning. For making molds he refers the reader to Biringuccio's *Pirotechnia,* although for mixing colors he tells readers that "this worshipful gentleman has been deceived." He demonstrates how to construct and fire kilns and how to make mills for grinding colors, and he discusses glazing and painting. He supplements his text with careful illustrations. He writes for a far broader readership than just craftsmen. He cautions that readers should not wonder at or jest concerning his writings about clay. For this art "will not always remain solely and forever in the hands of skilled masters." Rather, he insists, it will go abroad, "not merely from the craft, but from Italy." Whoever tries the craft of pottery will find that it is just as beautiful and sells for just as high a price as pottery from Italy. In these other lands the book will remind others of the happy state of Guidobaldo II of Urbino.[66] Piccolpasso thus appropriates the craft secrets of potters for the glory of the dukes of Urbino and presumably for the benefit of the economy of Lyons.

Piccolpasso believes that the practice of pottery will be of interest to nobles as well as artisans. Alfonso, the duke of Ferrara, for example, "took it as his relaxation to have a pottery kiln made for himself in a place near his palace." Then he set out "of his own accord to experiment concerning these matters, through which he discovered the greatest excellence of the potter's art, yet without laying aside his royal thoughts and his care for his people." Alfonso was greater even than Caesar, for if Caesar could write, dictate, and read at the same time, Alfonso, as duke, father, friend, and brother, could rule, succor, and defend many peoples, increase ducal majesty, and "at the same time likewise he was able to expound all the arts." Piccolpasso concludes that the "making of earthen pots" will not diminish "the greatness and worth of so excellent a prince."[67] His own treatise further augments the status of that art, already well-known for its fine products, by explaining its methods and reasons.

Far better known than Piccolpasso are two of his contemporaries, the painter and architect Giorgio Vasari and the goldsmith and sculptor Benvenuto Cellini. Both Vasari and Cellini spent their lives as practitioners, both depended on patronage, and both explained their crafts in writing. In addition Cellini portrayed his own life through his wonderfully vivid autobiography, while Vasari depicted the lives of more than a hundred artists in his monumental *Lives of the Artists*, modeled after the *Lives of Eminent Philosophers*, by the ancient author Diogenes Laertius. The biographical essay had become an important humanist literary form in the fifteenth century. A particularly relevant precedent had been Manetti's biography of Brunelleschi. By the mid-sixteenth century, painters, sculptors, and architects, as ingenious creators of unique objects, had become ideal subjects for autobiography and biography and also for portraits.[68]

Vasari came from a family of potters in the Tuscan city of Arezzo, south of Florence. Learning the craft of painting in workshops in Arezzo and then in Florence, he led a highly successful, peripatetic career as a painter for patrons in Venice, Naples, Rome, and most importantly Florence. Eventually he lavishly decorated his own house in Arezzo, perhaps inspired by the richly decorated houses in Mantua belonging to the painters Andrea Mantegna and Giulio Romano. In 1554 he came into the service of Duke Cosimo I di Medici of Florence, where he remained for the rest of his life. He undertook major architectural commissions as well as fresco paintings. His commissions included the remodeling and decoration of the Palazzo Vecchio as a ducal residence and the remodeling of the Uffizi Palace to house the administrative offices of Cosimo's state.[69]

Vasari began to write biographies of artists in the 1540s. The first edition of his *Lives*, dedicated to Duke Cosimo, appeared in 1550 in Florence. More than a thousand pages long, it includes a preface, a technical introduction to architecture, sculpture, and painting that focuses on materials and techniques, and biographies of 133 painters, sculptors, and architects. A second, enlarged edition, published in 1568, includes 30 additional biographies as well as a portrait of each artist at the beginning of his biography. Vasari attempts to itemize the works of the artists and critically to appraise them. He also explicates a view of the history of art (strikingly similar to Ghiberti's) according to which the perfection of ancient art declined to a low point in the German, or "Gothic," era, after which a rebirth occurred in the fourteenth century, in the time of Cimabue and Giotto. In Vasari's view, this rebirth was based on the imitation of nature and proceeded upward in three stages, reaching a pinnacle in the work of Raphael and Michelangelo in his own century. Within the biographies of individual artists he emphasizes individual style, particular works of art, personality, and character. For

Vasari, the crux of art involved *disegno*, comprising first the idea in the mind of the artist and then the drawing. Vasari was the first collector of drawings. His large collection, the Libro dei disegni, probably included at least seven volumes, now dispersed, of carefully mounted drawings dating from the fourteenth century to his own time.[70]

Benvenuto Cellini also spent the last part of his life in Florence, where he executed his famous life-size statue of Perseus for Cosimo di Medici. Ultimately, however, Cellini was far less successful than Vasari (whom he seems to have despised) in acquiring Medici patronage. Cellini was from a mid-level Florentine family. His grandfather was a mason, and his father was a skilled carpenter and maker of musical instruments, as well as a musician who played the recorder. Cellini trained as a goldsmith and spent his life working as a goldsmith, a creator of coins and medals, and a sculptor in Rome and Florence and in the court of Francis I in France. From his famous autobiography we can surmise that Cellini led a violent and unstable life. Involved in frequent brawls, he murdered two men, the first a man who had killed his brother, the second the goldsmith Poppeo de' Capitaneis for bad-mouthing him at the papal court in Rome. Cellini was accused of stealing the materials of his patrons at least three times, and he was accused of sodomy twice. He fathered at least eight children by several different women. Partly because in his work as a goldsmith he used precious materials such as silver, gold, and gems, few of the objects he created are extant. One of them is the famous salt cellar that he made for Francis I, now in Vienna. Despite the spectacular virtuosity of this object, he was forced to leave Paris after being accused of stealing part of the silver given to him to make the king's candlesticks. Returning to Florence, he became a client of Cosimo I, where he constructed the bronze Perseus. After he was accused of stealing some of the bronze from the statue, his commissions dried up, especially after Vasari's arrival in 1554.[71]

Cellini composed his treatises on goldsmithing and sculpture in 1565 and had them printed at his own expense in 1568. Ten years earlier, in 1558, he had begun to write his autobiography while under house arrest for the crime of sodomy. He tells us that he began writing in his own hand. Soon tiring of the mechanics of writing, he dictated most of the work (which was unfinished at his death) to an assistant, Michele di Goro Vesti, while he himself worked on a marble crucifix meant to be placed on his own tomb. Cellini believed that artistic talent and genius, including his own, were intrinsically valuable gifts from God. He wrote to impart the story of his life, the details of his own personality and genius, and the particulars of his many works to contemporaries and to posterity.[72]

In his writings Cellini frequently points to his own talent and ingenuity as well as to his own practice. In his treatise *On Goldsmithing* he tells readers that he writes because he knows how much people like hearing of new things and because readers will be moved to pity and anger on his behalf. He does "what no one had done before," that is, "to write about those loveliest secrets and wondrous methods of the great art of goldsmithing." He begins with a cryptic history of goldsmithing from the time of Brunelleschi, Donatello, and Ghiberti and proceeds to a detailed account of various techniques. In a chapter on filigree work he admits the difficulty of explaining a craft in writing—"which [i.e., the soldering of filigree wires] one is not able to teach with writing, but a good part of it is told with live words and with experience; nevertheless we will follow our way of reasoning."[73]

Cellini intersperses detailed explanations of technique with discussions of his patrons and commissions. In one episode he relates how he explained to the admiring king of France, Francis I, how to make a particular kind of bowl with filigree work. He reports another conversation with the king concerning the technique of making life-size silver statues, and he reports as well Francis's great admiration and delight. In a discussion of his work for papal patrons in Rome he scolds both ignorant jewelers and the princes who encourage them.[74]

Cellini insists throughout on the priority of practice over theory. Discussing the settings for rubies, he writes: "I consider that practice always has come before theory in every craft, and that rules of theory, in which your skillful craftsman is accomplished, are always grafted on to practice afterwards." Here and elsewhere he emphasizes his own practical experience, his ingenious inventions of technique and creations, and his novel ideas. Before describing the fabrication of his famous saltcellar, he writes that he will use examples that he has made himself so that readers "can know that I have taught these true experiences not having stolen them from the labors of others." Cellini delights in describing his own innovative techniques. Occasionally he shows how his own methods represent an improvement on those of the ancients.[75]

He openly discusses his techniques in order to display his own ingenuity but also because he delights in the techniques themselves. For example, he proudly reveals a method he has developed of strengthening clay by mixing it with cloth scraps and letting the mixture decompose for four months. "And note that this is a wonderful secret, that has not ever been used." He warns that the clay will appear too fatty "to those who have not made such experiment," but he assures readers that it is not. Similarly, in his treatise on sculpture he describes large-scale bronze casting in great detail, including the construction of furnaces. In a description of how to measure for the construction of colossal statues, knowing that the standard method will not work, he describes his own invention: "I found another rule with this, which was made by me myself, nor ever understood by others, born from my great studies, thus I generously teach it to those who would have good work at heart."[76]

A far more ambiguous attitude toward revealing secrets is found in the writings of the French glass painter and potter Bernard Palissy. Palissy was an artisan who initially worked as an itinerant *peintre-vitrier*, one who paints and installs stained-glass windows. He also worked as a surveyor, constructed grottoes in the gardens of noble patrons, and spent decades as a potter experimenting with new glazes, especially white tin glazes. A Protestant who was twice arrested as a heretic, he died in the Bastille in 1590.[77]

Palissy's work of fabrication led to an interest in the nature of materials and in certain aspects of natural philosophy, including the nature of rocks, fossils, soils, and clays and the principles of rivers and springs. He authored three works: a pamphlet on a grotto that he had constructed, a book of essays, *Receptes véritable*, and then his mature work, the *Discours admirables (Admirable Discourses)*, published in 1580. The latter takes the form of a somewhat contentious dialogue between *Théorique* (Theory) and *Practique* (Practice), the latter representing Palissy's own views. Topics include waters and springs, rocks, fossils, salts, alchemy and metals, the use of marl (a kind of natural fertilizer) in agriculture, and a remarkable description of Palissy's experiments involving glazes.[78]

Palissy dedicates the *Admirable Discourses* to his longtime friend and patron Antoine de Pons, a nobleman who was the governor of the province of Saintonge, where Palissy himself lived for many years. In the dedication Palissy explains in a way that is notably similar to Serlio's why he writes openly. He notes that it is written "that we should take care not to abuse the gifts of God nor to hide talent in the ground" and, further, that "the dolt concealing his foolishness is better than the sage hiding his knowledge." Therefore, he brings to light the things "which it has pleased God to make me understand according to the measure in which he has been pleased to endow me, in order to benefit posterity." Palissy is harshly critical of other books, including those of the followers of Paracelsus and those of many alchemists, which have caused many to lose both time and wealth. "Such pernicious books," he writes, "have led me to scratch the earth during forty years, and to search its bowels, in order to know the things it produces within itself." By these means he has found grace before God, "who has revealed to me secrets which until now have remained unknown to men, even the most learned."[79]

Palissy defends himself against those who claim that knowledge of nature is impossible without Latin and against those who object to his boldness in writing against so many famous and ancient philosophers. He warns that others will say that he is a poor workman and thereby will try to make his writings appear harmful. To counter criticism and legitimate his specific claims, Palissy adopts a novel procedure: he informs his readers that he has set up a cabinet in his dwelling in which he has placed many things "from the bowels of the earth" that "give reliable evidence" of what he says. Palissy's ultimate proof of his statements requires that the reader come to his dwelling and actually look at objects that he has gathered from the earth. The last section of the *Admirable Discourses* provides a list of the labels indicating the precise contents of the cabinet.[80]

Warning against reliance on theory, Palissy counsels the reader "not to dull your mind with sciences written in the study by an imaginative or biased theory or taken from some book written from imagination by those who have practiced nothing." He cautions against those who claim that theory begets practice, including those who claim that one must imagine what to do in the mind before carrying it out in practice. If theory did precede practice, alchemists would do wonderful things without searching for fifty years, and war leaders would never lose a battle that they had planned. Those who rely on theory in this way "could not make a shoe, not even the heel of a boot, even if they had all the theory in the world." If you were to sail to all the countries of the world, he asks, would you rely on someone who had spent years studying books on the subject, or would you sail with "a man who is expert and practical?"[81]

Palissy finds validation for his own writing through his practice. He views his cabinet of natural things as a collection of proofs of his written claims. He again invites the reader to his cabinet, "in which will be seen marvelous things which are placed there as witness and proof of my writings, arranged in order on shelves, with labels below them; in order that one may learn by himself." He insists on the primacy of firsthand experience. By looking at the objects in the cabinet, he writes, even the first few hours of the first day, "you will learn more natural philosophy about the things contained in this book, than you could learn in fifty years by reading the theories and opinions of the ancient philosophers." He envisions a highly participatory way of reading. If readers are confused, they should ask the printer where he lives. They should find him, and he will provide further demonstrations and explanations. Further, if someone wants to build a fountain and is unable to understand the intent of the author, they should come, and he will build a model.[82]

Discussing his experiments with glazes, Palissy insists that the matter cannot be understood without the benefit of actual practice. He expresses a craftsman's view of knowledge acquisition: knowledge, he says, is that which is acquired by practice and handwork. Further, he offers a craftsman's justification for secrecy, one that is at odds with the ideal of open writing expressed in his dedication to Antoine de Pons. Theory reminds Practice of his promise to teach him the art of the earth (i.e., pottery and glazing) after he has discussed the earth clays, and he expresses surprise that Practice told him to come back later so that (in Theory's view) he would forget his interest. "Do you believe," Practice replies, "that a man of sound judgement would thus wish to give away the secrets of an art that has cost dearly to the man who has invented it? As for me, I am not willing to do so unless I know a reason in it." Theory replies that Practice has no charity: "If you wish thus to keep your secret hidden, you will carry it to the grave and no one will benefit from it, and thus you will be accursed: for it is written that every man, according to the gifts he has received from God, should give to others." Thus, Theory concludes, "if you do not teach me what you know of this art, you are misusing the gifts of God."[83]

Practice retorts that his art and secrets are not like others. For example, a remedy against a plague or other disease must not be kept secret; nor should "the secrets of agriculture" be concealed, nor the hazards and dangers of navigation, nor the word of God, nor the sciences that serve the whole state. But, he argues, this is not true of his art of the earth and of many other arts. Many prized objects in the houses of nobles and princes would be less valued if they were common. Practice gives numerous examples: Certain glasses became so common that the glassmakers live more poorly than porters. The inventors of enameled buttons once sold them at a high price, but "because those who invented them did not keep their invention a secret," so many were made that they were sold for a tiny price and came to be despised, so that now no one wears them. Likewise, printing has hurt painters and clever draftsmen. He remembers seeing depictions of "Our Lady" coarsely printed after the invention of a German called Albert (he is probably referring to Dürer), which are despised because of the great number that were made. Similarly, casting hurts clever sculptors. After a sculptor has spent a long time making a figure, "it falls into the hands of some cast-maker." The cast maker will make such a large number of them that the name of the inventor will no longer be known and the figures will sell cheaply, "to the sorrow of one who carved the first piece." However, Practice concludes, "if I thought you would keep the secret of my art as jealously as it deserves, I would not hesitate to teach it to you."[84]

Finally Theory persuades Practice to discuss his experiments. Yet the conversation that ensues does not reveal glaze recipes or specific techniques of firing. Rather, it is a remarkably vivid account of the experimenter himself (namely, Palissy) and the numerous trials and tribulations that he encountered during his years of carrying out glaze experiments. Practice makes clear that to actually learn the secrets, you have to do the work yourself. Theory asks for Practice's knowledge in writing so that he can avoid the same troubles. Practice replies, "Even if I used a thousand reams of paper to write down all the accidents that have happened to me in learning this art, you must be assured that, however good a brain you may have, you will still make a thousand mistakes." The art of glazing cannot be learned from books but only from practice.[85]

Palissy himself made hundreds of experiments, during which he invented a fine white glaze and many others. Yet he also wrote books. He articulated the ideal of openness and also explicated an artisan's reasons for secrecy. He stood at a crossroads in which the commercial market for finely made objects, noble patronage, the legitimation of practical experience, and interest in the nature of the natural world met in uneasy confluence. In his writings both openness for the benefit of the public and craft secrecy find forceful expression.

Architecture and certain other arts, such as painting, sculpture, and pottery, created a middle ground of communication among skilled and learned men, a ground in part constituted and mediated by authorship. This is not to claim that social hierarchies were obliterated or even fundamentally modified for society as a whole. But it does underscore the importance of the development of an arena of discursive practice in which the productive value of certain technical arts (inherent in their ability to produce fabricated and constructed objects) was augmented by their status as knowledge-based disciplines. As such, these arts found expression in written works and served as arenas of collaboration and communication among skilled practitioners, learned humanists, and wealthy elites.

The sixteenth century was transformative with regard to certain mechanical arts. Eventually arts such as painting, sculpture, and architecture became fine arts, far afield from their mechanical origins. This transition, by no means completed by the end of the century, nevertheless proceeded rapidly, assisted by practices of authorship. Authors from diverse walks of life emphasized the value of openness for the progress of knowledge. Equally important, they often stressed the importance of handwork and experiential, skill-based knowledge, which they emphatically joined with learning and mathematics. Authors expressed such ideals in books, many of which emerged from a patronage system.

This study revises a view of that patronage system as one that encompassed a moral economy exclusively concerned with exchanges that enhanced the glory and power of the patron and remunerated the client with various material and status benefits. Rather, patrons, clients, and others were involved in exchanges of various kinds of substantive knowledge. Together, they created trading zones of knowledge exchange that would have significant effects on the culture of knowledge itself.

Epilogue

Values of Transmission and the New Sciences

THIS STUDY INVESTIGATES values concerning the possession of knowledge and its communication as phenomena that are contextually grounded and involve both written authorship and oral transmission. It addresses not just the substantive content of books but the social and cultural circumstances of their production. It also focuses on the technical arts and investigates the influence of craft and engineering traditions beyond specific activities of construction and fabrication. A brief recapitulation points to some of the ways in which this study qualifies or undermines broad generalizations—concerning craft secrecy, the influence of print, and the relative prevalence of openness or secrecy in ancient, medieval, and early modern cultures.

Greek and Roman authors wrote openly on practices such as agriculture and on technologies such as catapults and pneumatic devices. Yet such writings occupied separate conceptual categories in the ancient world. Agricultural authors, who invariably belonged to the elite classes, wrote for their peers. They produced praxis writings that promoted the hegemony of the elite classes by inculcating traditional values such as persistence, courage, and virtue. In contrast, engineer-authors produced technē writings within the context of client-patron relationships, whether in institutional settings such as the Museum of Alexandria or within more informal patronage relationships. In one such treatise, the *De architectura*, Vitruvius articulates the ideal of open, written transmission and credit to authorship. His views can be understood within the context of traditional Roman religion, reinstated by the emperor Augustus in the 20s B.C.E. Vitruvius suggested that openly transmitted writings allowed the progress of knowledge based on past accomplishments. He believed that giving credit and honor to past authors involved acts of piety, whereas both the theft of writings and criticism of dead authors were acts of impiety.

There is significant evidence for secrecy and esoteric groups in late antiquity within texts pertaining to magic and alchemy and within Neoplatonic philosophical writings. The great appeal of secret and esoteric traditions in this era points to expanded arenas of privacy as Roman civic values diminished in importance for traditional Roman elites. Vivid, detailed portrayals of esoteric groups that followed charismatic leaders are provided by Iamblichus's account of the Pythagoreans and Porphyry's description of the followers of Plotinus. As they are represented by these authors, both groups practiced secrecy, but in very different ways. The two groups also differed in their approach to authorship and the attribution of writings. Yet both rejected material and physical goals for noncorporeal spirituality. Authors such as Apuleius also ultimately rejected magic, with its worldly aims. Philosophical Neoplatonists left material interests behind as they strove to attain understanding of the noncorporeal divine.

Concerning the ordinary crafts, the view that craft secrecy existed wherever there were crafts must be abandoned. There is little evidence for craft secrecy in the ancient and early medieval worlds outside of magic and late antique temple crafts associated with the development of alchemy. During the premodern centuries the handing down of craft knowledge occurred for the most part by means of apprenticeship, whether in workshops or as part of household economies. Yet oral transmission is not synonymous with secrecy.

In the high Middle Ages, particularly in the cities, craft knowledge per se acquired much greater cultural value than it had had in previous centuries. Beginning in the thirteenth century, evidence for craft secrecy is abundant and signifies the development of proprietary attitudes toward craft knowledge, craft processes, and inventions. The emergence of patents for inventions provides further evidence for "intellectual property" attitudes. The essential context for this development was the rise of urbanism and commercial capitalism that accompanied the proliferation of artisan crafts in the high and late Middle Ages and, somewhat later, the development of habits of conspicuous consumption among elites, which further enhanced the status of certain mechanical arts.

Authors wrote books on the mechanical arts with increasing frequency beginning in the early fifteenth century. I suggest that these writings came out of a new alliance between technē and praxis. Both university-trained humanists and workshop-trained artisans wrote such books, both groups within the context of patronage. Rulers of numerous fifteenth-century cities, especially in Italy and southern Germany, lacked the legitimacy that accrued to the traditional nobility through kinship ties. Such new rulers needed to legitimize their de facto power. One avenue of legitimation involved urban redesign and major construction projects, including the lavish ornamentation of interiors. The arts were themselves enhanced when artisan practitioners and humanists produced treatises on various mechanical arts and dedicated them to patrons. These books explicated in writing and in drawings particular arts such as painting, sculpture, architecture, and engineering. Formulating their principles in treatises, authors created potentially learned disciplines out of arts previously concerned primarily with craft production and construction. They thereby enhanced the cultural value of certain arts, making them all the more useful to patricians and rulers.

The fifteenth- and sixteenth-century revival of Platonism, with its extensive study, circulation, and reinterpretation of ancient texts, also created something new in its joining of magic with philosophical, spiritual Neoplatonism for utilitarian purposes. Renaissance Neoplatonic authors did not practice secrecy consistently or in a uniform way. Some, such as Ficino, disseminated their ideas openly. Others, such as Cornelius Agrippa, advocated openness for some topics and secrecy for others. Printing, which was contiguous with the revival of Neoplatonism, facilitated the proliferation of books, including alchemical books and Neoplatonic treatises that espoused secrecy or utilized techniques of concealment such as dispersion, opaque symbolism, and other forms of mystification. Printing was entirely neutral in one sense: it widely disseminated both the values of secrecy and the values of openness.

Printing facilitated the expansion of writings on the mechanical arts in the sixteenth century. There emerged broad traditions of authorship on mining, metallurgy, painting, architecture, the military arts, and even arts such as pottery and goldsmithing. Such writings present visible manifestations of collaboration and communication between practitioners, learned humanists, other university-educated men, and ruling elites. They depict discussions between skilled practitioners and the learned either through dialogues or through references to conversations that reportedly had occurred in real life. Some writings, such as Daniele Barbaro's Vitruvian commentary, were themselves the products of collaboration between learned men and practitioners. Such communication signals the development of border areas where substantive communication occurred or was represented as occurring between individuals from diverse social strata. Without abolishing social differences, such communication across boundaries, or "trading zones," was significant for the development of experimentally and empirically based natural philosophies, in which natural philosophy and the mechanical arts were joined.[1]

This book builds on the work of scholars who used diverse theoretical models to suggest the contribution of artisanal culture to the "scientific revolution." Such scholarship includes the work of Leonardo Olschki on writings from Alberti to Galileo, which chronicled the development of "scientific-technological literature." It includes Robert Merton's *Science, Technology, and Society in Seventeenth-Century England,* especially the second half, in which Merton explores the significance of technological developments for the development of science; and Charles Webster's *Great Instauration*, which similarly investigates a multitude of technological and practical interests in seventeenth-century England. It includes Edgar Zilsel's articles of the early 1940s on the contribution of artisans to the scientific revolution; Paolo Rossi's elaboration of the Zilsel thesis in *Philosophy, Technology, and the Arts in the Early Modern Era*; A. C. Crombie's discussions of the influence of medieval technology on science and his later work on "styles of science," including the experimental style; and the work of Alex Keller and J. A. Bennett, among others, on the importance of engineering and instrumentation to the development of experimental philosophy.[2]

In significant ways the new sciences used the mechanical arts both to explore the natural world and to legitimate knowledge claims about that world. Galileo invented a military compass, about which he wrote a small book. He also successfully defended both the invention and the book in extended proprietary disputes. Some time later, after hearing about an optical instrument, he fabricated a similar one, creating a telescope. William Gilbert wrote his classic work on the magnet by reading about and thinking about the mariner's compass. Francis Bacon's *Novum organum*, written as a new approach to natural philosophy, was explicitly indebted to the mechanical arts. Later, Robert Boyle explored the nature of air using the air pump.[3] In these and numerous other examples proponents of the new experimental philosophy used devices, instruments, and machines to establish truth claims about the natural world.

This book asks how the mechanical arts came to be involved in making knowledge claims about the world in the first place. The complex answer to this question includes the development of commercial capitalism, the expansion of artisanal trades in general, the increasing cultural importance of objects, and the development of conspicuous consumption on the part of elites. Objects and the fabrication of objects gained greater cultural importance and thus came to be the focus of written treatises and accounts. Construction and fabrication became connected not just to craft know-how and the making of things but to knowledge about the world itself. This development involved a thoroughgoing, centuries-long process. Authorship on the mechanical arts in particular involved and brought about relationships and alliances between artisans, learned humanists, and elite rulers. The expansion of such authorship turned certain mechanical arts into discursive, even learned subjects, preparing them for use in investigating the world.

Many fifteenth- and sixteenth-century authors explicitly advocated openness as they openly explained topics such as mining, ore processing, artillery, and fortification. Ancient technē writings provided significant precedents. Later, the new scientific societies of the seventeenth century advocated both openness and experimentation. Yet the stated values of openness in the societies were circumscribed in various ways. For example, the *Saggi* of the Accademia del Cimento (1657–67) in Florence contains a statement explaining the practices of openness and experimentation followed by the society. However, as Biagioli points out, the image of a stable institution of experimenters sharing their experiments with visitors is considerably attenuated by the circumstance that the informal academy was created by Prince Leopold de' Medici, who called it into session and dismissed it at will; that by the time the *Saggi* had been published, the informal academy had ceased to exist; and that its membership consisted of individuals already established in the Medici court.[4]

Similarly, the statements concerning openness and experimentation presented by Henry Oldenburg in the Royal Society of London's *Philosophical Transactions* (1665) do not tell the whole story. Steven Shapin's essay on experimentation within the Royal Society underscores the point that openness could be a highly complex matter, that it could depend on differences between private and public space, on degrees of access, and on the social status of participants. It is relevant that membership in the Royal Society was highly selective and restricted largely to gentlemen. The Académie Royale de France also expressed ideals of open communication. Yet Alice Stroup's work on the Parisian Royal Academy of Sciences shows that open communication on the part of academicians was tempered by a regulation that the meetings be secret and by the exclusionary practices of the academy. The Académie Royale's practices involved a complex mixture of openness and secrecy.[5]

Throughout the seventeenth century alchemists and others practiced and advocated esoterism and secrecy, especially with regard to certain disciplines. Adepts and would-be adepts made alchemy a focus of intense interest. Alchemy as a discipline supported practices of concealment. Much alchemical writing was disseminated in manuscript, but much was also disseminated in print. Lawrence Principe shows that Robert Boyle was a devotee of alchemy and defended its practices of secrecy, in which he participated. Boyle, as is well known, also advocated open experimentation. Betty Jo Teeter Dobbs establishes the centrality of alchemy to Newton's thought.[6] These two practices—openness and secrecy—represent, not a contradiction, but separate attitudes concerning dissemination attached to separate kinds of philosophical activity.

Both open and esoteric traditions from the sixteenth century shared, however, an orientation toward manipulating the material world—what might be called a technological or utilitarian orientation. This orientation was inherited from the artisanal culture of the late Middle Ages. Yet authors of the mechanical arts often turned from the proprietary values of that culture to open authorship. The proprietary attitudes that developed within craft and artisanal cultures did not remain tied to those cultures. Such values became central to sixteenth-century priority conflicts, which were fueled by a growing appreciation for novelty and new inventions. Proprietary attitudes toward both inventions and ideas could be used to promote profit and commerce, but they could also be used to gain credit and standing in the courts, the cities, and the universities. Such attitudes first developed among artisans in the context of craft culture, but they were taken up by learned men, particularly those engaged in the new sciences.

The utilitarian orientation of both esoteric and open traditions had significant philosophical implications. In his important study of Bacon and the maker's-knowledge tradition, Antonio Pérez-Ramos provides an extended investigation of the three key ingredients of Bacon's notion of science—*forma, opus,* and *inductio.* He shows that the meanings of these terms in Bacon's thought was fundamentally different from their meaning in the prior Aristotelian tradition. As Pérez-Ramos puts it, the overarching notion of the maker's knowledge that is fundamental to Bacon's thought uses "a conceptual apparatus which is philosophically richer than the standard contrast of 'practical' versus 'theoretical' orientations." For Bacon, there is "an intimate relationship between objects of cognition and objects of construction" that is, "distinct from the utilitarian ethos imparted to knowledge *after* its constitution as natural science."[7] Bacon's philosophical integration of the maker's knowledge into a new methodology was made possible, I suggest in this study, by the changing status of the mechanical arts in the prior two centuries, changes that involved the practice of authorship. When authors transformed craft know-how into forms of discursive knowledge, they prepared it for integration into philosophical methodologies pertaining to investigation of the natural world.

The sixteenth and seventeenth centuries saw the development of numerous sites of knowledge production, some traditional and some new, including courts, universities, building sites, ancient ruins, artillery proving grounds, arsenals such as the Venetian arsenal, scientific societies, the homes of adepts, the laboratories of experimentalists of all kinds, libraries (private, semipublic, and public), colleges, and, as Rob Iliffe vividly demonstrates, London coffee houses.[8] On such sites occurred interchanges among various kinds of people, both practical and learned. It was in the context of such discursive practices, which included the practice of authorship, that experimental and empirical methodologies emerged.

Advocates of experimentation and various forms of empiricism did not find immediate acceptance for their views. The use of instruments and mechanical demonstrations to make statements about the world contradicted the presuppositions of the learned culture of Aristotelianism. One result was that advocates of the new sciences in the seventeenth century sometimes distanced themselves from artisans and from the mechanical arts; at the same time, they represented themselves as "disinterested," that is, not influenced by political or economic concerns. This is the context in which "invisible technicians," as Shapin calls them, should be seen. It is also the context of seventeenth-century remarks about low mechanicians' being unworthy of association with natural philosophy, as, for example, Pamela H. Smith shows in the case of Joachim Becher.[9] The distancing of experimental philosophers from the mechanical arts, where it does occur, does not represent the playing out of a centuries-old history of disparagement of the mechanical arts. I suggest a far more complex history in which the mechanical arts were first transformed, making them suitable for philosophical endeavors, and then appropriated by the new sciences. Once such appropriation was accomplished, the task of legitimating the new sciences required distancing them from artisanal practice and the mechanical arts per se.

I claim that seventeenth-century struggles to validate new experimental methodologies would not have occurred at all if some of the mechanical arts had not been transformed into discursive disciplines explicated in writing in the previous two centuries. An intermediate step occurred: a close association between technē and praxis. To put it in Aristotelian terms, before technē could join epistemē to become disinterested experimental philosophy, it allied itself with praxis as new political elites of the fifteenth and sixteenth centuries legitimated their power and authority by massive building programs, urban transformations, and conspicuous consumption and also patronized and sometimes themselves engaged in authorship on the mechanical arts. During this complex process certain arts were transformed, becoming culturally fit to be used in a philosophical quest concerning the nature of the world.

Empirical and experimental sciences flourished in the seventeenth century without forming a unified "science." Rather than unity, there was a complex development of diverse new sciences and methodologies that challenged traditional natural philosophy without fully replacing it. Part of the complexity of this cornucopian philosophical situation involves diverse values of transmission, including both openness and secrecy, as well as evolving attitudes of ownership and priority. Such views were influenced by a complex prior history of fabrication and authorship on the mechanical arts.

Notes

Introduction

1. Merton, "Über die vielfältigen Wurzeln"; Whitney, *Paradise Restored*.

2. Aristotle, *Nichomachean Ethics* 6.4–5, 1140a–1141a. See also Dunne, *Back to the Rough Ground*, for a discussion of the importance of the categories in Aristotle and in modern philosophy; and Habermas, *Theory and Practice*.

3. For the contested role of experiment in the seventeenth century, see esp. Dear, *Discipline and Experience*, and Shapin and Schaffer, *Leviathan and the Air-Pump*.

4. Alasdair MacIntyre's "disquieting suggestion" with regard to moral philosophy is relevant. MacIntyre suggests that whole areas of moral philosophy have been obliterated because history itself as a discipline was created out of the fragments of the breakdown of those philosophies. MacIntyre, *After Virtue*, 1–5; MacIntyre, *Three Rival Versions of Moral Enquiry*.

5. For discussions of changing disciplinary boundaries, see Kelley, *History and the Disciplines*; and Greenblatt and Gunn, *Redrawing the Boundaries*. For Skinner's work, see esp. Tully, *Meaning and Context*.

6. Baldwin, "Snakestone Experiments"; Biagioli, *Galileo, Courtier*; Findlen, "Controlling the Experiment"; Hannaway, *Chemists and the Word*; Hannaway, "Laboratory Design and the Aim of Science"; Hull, *Science as a Process*; Shackelford, "Tycho Brahe"; Shapin, "House of Experiment"; Shapin and Schaffer, *Leviathan and the Air-Pump*.

7. Williams, *Keywords*; Skinner, "Language and Social Change."

8. Merton, *Sociology of Science*, 273–77; Price, *Science since Babylon*, 117–35. Ernan McMullin reiterated the thesis of the openness of science and the secrecy of technology in McMullin, "Openness and Secrecy in Science."

9. For a cogent discussion of the ancient notion of technē, see Ferrari and Vegetti, "Science, Technology, and Medicine," esp. 200–202. For an example from architecture, see Vitruvius, *De architectura* 1.1.1, Vitruvius said that the work of the architect comes from *fabrica* (construction) and *ratiocinatio* (reason).

10. For social constructivism, see Golinski, *Making Natural Knowledge*, and see Hull, "Openness and Secrecy in Science," and Hull, *Science as a Process*. Both Findlen, "Controlling the Experiment"; and Biagioli, *Galileo, Courtier*, associate openness with display connected with patronage. See also Shapin, "House of Experiment."

11. Long, "Openness and Empiricism"; Long, "Openness of Knowledge."

12. Bok, *Secrets*, 4–14.

13. Paolo Rossi and William Eamon point out that in early modern Italy the metaphor of the *venatio*, or the hunt, was frequently used with reference to the "secrets of nature." Rossi, *Philosophy, Technology, and the Arts*, 42; Eamon, "Court, Academy, and Printing House," 25–28; Eamon, *Science and the Secrets of Nature*, 11.

14. See, e.g., Minnis, *Medieval Theory of Authorship*.

15. Quint, *Origin and Originality*, esp. 1–30; Dunn, *Pretexts of Authority*; Loewenstein, "Script in the Marketplace"; Kernan, *Printing Technology, Letters, and Samuel Johnson*.

16. Barthes, "Death of the Author"; Foucault, "What Is an Author?" For a useful discussion of the postmodern view of the author, see Rosenau, *Post-modernism and the Social Sciences*, 25–41. For discussions of the "death" of the author, see esp. Kernan, *Death of Literature*. As Seán Burke remarks, Roland Barthes "does not so much destroy the Author-God, but participates in its construction. He must create a king worthy of the killing" (Burke, *Death and Return of the Author*, 26).

17. Scholars who posit the eighteenth-century development of modern authorship and its connection to copyright include Hesse, "Enlightenment Epistemology"; Nesbit, "What Was an Author?"; Rose, "Author as Proprietor"; Rose, *Authors and Owners*; and Woodmansee, "Genius and the Copyright." Dissenters from the view that connects authorship and copyright include Ginsburg, "Tale of Two Copyrights"; Saunders, *Authorship and Copyright*; and Saunders and Hunter, "Lessons from the 'Literatory.'"

18. Hathaway, "Compilatio." For an example of alchemical pseudoauthorship, see Newman, *"Summa Perfectionis" of Pseudo-Geber*, 57–108, discussed in detail in chapter 5.

19. Biagioli, *Galileo, Courtier*, esp. 149–57; Chartier, *Order of Books*, esp. 25–59. Biagioli emphasizes the ambiguous situation that such patronage represented for authors such as Galileo and points to Galileo's *bricolage* skills to negotiate to his own advantage. Chartier discusses the complexity of authorship and the "author-function" in the early modern period. He describes the patron-author in the context of the court but also points to authors who asserted their presence and even originality in their textual productions long before the eighteenth century.

20. See esp. Kemp, "From 'Mimesis' to 'Fantasia'"; and Kemp, "'Super-Artist' as Genius." For Brunelleschi's patent, see Prager, "Brunelleschi's Patent"; and see Cole, "Titian and the Idea of Originality."

21. Saunders, *Authorship and Copyright*, 79–81. For an introduction to intellectual-property law, see esp. Boyle, *Shamans, Software, and Spleens*; Kaplan, *Unhurried View of Copyright*, 38–78; and Kuflik, "Moral Foundations of Intellectual Property Rights," esp. 236–39. For a study of a culture for which the concept of intellectual property was foreign, see Alford, *To Steal a Book Is an Elegant Offense*. For discussions of property in ancient law, see Kenney, "Books and Readers in the Roman World," 19; Noyes, *Institution of Property*, 131–220; and Schlatter, *Private Property*, 9–32.

22. Kaplan, *Unhurried View of Copyright*, 38–78.

23. Long, "Invention, Authorship, 'Intellectual Property' and the Origin of Patents," esp. 870–81.

24 For the number of books published with a privilege, see Armstrong, *Before Copyright*, 78. For early modern authors' copyright, see Grendler, "Printing and Censorship"; and Febvre and Martin, *Coming of the Book*, 159–66

25. See Armstrong, *Before Copyright*, 78–99, for grounds for seeking a privilege.

26. See Chavasse, "First Known Author's Copyright"; Feather, "Authors, Publishers, and Politicians", Franceschelli, "Le origini e lo svolgimento del diritto industriale", Franceschelli, "Il primo privilegio in materio di stampe"; Grendler, "Printing and Censorship," 33–34, Grendler, *Roman Inquisition and the Venetian Press*, 69–181, 233–52; and Rose, *Authors and Owners*, 9–30

27 For Pliny's discussion of first inventors and discoverers, as well as other marvels, see Pliny, *Naturalis historia* 7.16–60. The tradition of identifying inventors was in evidence in the fifth century B.C.E. and continued into the early modern period. See esp. Thraede, "Das Lob des Erfinders"; and French, *Ancient Natural History*, 196–255.

28 Copenhaver, "Historiography of Discovery." Robert K Merton's early discussion of such disputes imposed an idealized view of modern science onto the sixteenth century; see Merton, "Priorities in Scientific Discovery."

29. Gingerich and Westman, *Wittich Connection;* Rosen, *Three Imperial Mathematicians*; Jardine, *Birth of History and Philosophy of Science;* Biagioli, *Galileo, Courtier*, Iliffe, "In the Warehouse."

30. "This is the truth of historically effected consciousness. It is the historically experienced consciousness that, by renouncing the chimera of perfect enlightenment, is open to the experience of history " Gadamer, *Truth and Method*, 377–78

31. See Spiegel, *Past as Text*, 49–50. "Mediation is an active process that constructs its objects in precisely the sense that poststructuralism conceives of the social construction of reality in and through language Rather than functioning as a middle term relating two disjunct phenomenal orders from which it stands apart, mediation is intrinsic to the existence and operation of the reality that it actively produces In studying history, then, what we study are the mediatory practices of past epochs which, then as now, constructed all being and consciousness."

Chapter 1
Open Authorship within Ancient Traditions of *Technē* and *Praxis*

1. For the low status of handwork in antiquity, see Burford, *Craftsmen in Greek and Roman Society;* Geoghegan, *Attitude towards Labor;* and Rossi, *Philosophy, Technology, and the Arts*, esp. 13–15. For a view that stresses the varying status of craftwork in antiquity, see esp. Mondolfo, *Polis, lavoro e tecnica.*

2. Rhodes, "Athenian Revolution," emphasizes the disagreements and unknown aspects of the reform, see also Ober, *Mass and Elite in Democratic Athens*

3 Rhodes, "Athenian Revolution," 85–87.

4. For an introduction to the sophists, see Kerford, *Sophistic Movement;* and Kerford, *Sophists and Their Legacy*. On Protagoras, see Schiappa, *Protagoras and Logos*, esp. 39–63, 89–102, 117–33

5. For Protagoras's associations with Athens and Pericles, see esp. Morrison, "Place of Protagoras in Athenian Public Life"; Placido, "El pensamiento de Protágoras y la Atenas de Pericles"; Placido, "Protágoras y Pericles"; and Schiappa, *Protagoras and Logos*, esp. 175–79, quotation on 199. For the association of the sophists with democracy, see Müller, "Sophistique et démocratie."

6. For the rhetorical manuals, see Plato, *Phaedrus* 266c–278e; and Plato, *Protagoras* (trans. Taylor) 318d5–319a5, where Protagoras says that when a student comes to him he will learn "the proper management of one's own affairs, how best to run one's household, and the management of public affairs." Modern scholarship includes Fuhrmann, *Das systematische Lehrbuch*; Kennedy, "Earliest Rhetorical Handbooks"; and Meyer, *Geschichte des Altertums*, vol. 4, pt. 1, 890–909. For the *hoplomachoi*, see Wheeler, "*Hoplomachia* and Greek Dances in Arms"; and Wheeler, "*Hoplomachoi* and Vegetius' Spartan Drillmasters."

7. For sophist handbooks, see esp. Fuhrmann, *Das systematische Lehrbuch*, 122–44; Kennedy, "Earliest Rhetorical Handbooks"; and Turner, *Athenian Books*, esp. 19–20.

8. I have used the translation from Conacher, *Aeschylus' "Prometheus Bound,"* 82–97, quotation on 84. For Xenophanes and the extant fragments, see esp. Barnes, *Presocratic Philosophers*, esp. 82–99, 136–43; Kirk, Raven, and Schofield, *Presocratic Philosophers*, 163–80; and McKirahan, *Philosophy before Socrates*, 59–68. See also Cole, *Democritus and the Sources of Greek Anthropology*; and O'Brien, "Xenophanes, Aeschylus, and the Doctrine of Primeval Brutishness."

9. For Anaxagoras and Democritus, see McKirahan, *Philosophy before Socrates*, 196–231, 303–43; Barnes, *Presocratic Philosophers*, esp. 318–77; and Kirk, Raven, and Schofield, *Presocratic Philosophers*, 352–57, 402–33. For Democritus's writings, see Diogenes Laertius, *Lives of Eminent Philosophers*, 1:443–63; see also Cole, *Democritus and the Sources of Greek Anthropology*, 56–59.

10. Corso, *Monumenti Periclei*; Kagan, *Pericles of Athens*, 95–106; Meiggs, *Athenian Empire*; Rhodes, *Athenian Empire*; Rhodes, "Delian League to 449 B.C.," 34–61, 535–39.

11. Kagan, *Pericles of Athens*, 152–67; Knell, *Perikleische Baukunst*; Mark, "The Gods on the East Frieze"; Pollitt, *Art and Experience*, 64–110. See also Pollitt, "Art: Archaic to Classical," esp. 180–83; Wycherley, "Rebuilding in Athens and Attica"; and Ostwald, "Athens as a Cultural Centre." For Pheidias, see esp. G. Lippold, "Pheidias 2) Sohn des Carmides"; and Gross, "Pheidias." For the quotation see Plutarch, *Life of Pericles* (trans. Perrin) 13.1.

12. Rumpf, "Agatharchos, 2. Des Eudemos Sohn aus Samos"; Rumpf, "Classical and Post-Classical Greek Painting," esp. 13; Pfuhl, *Malerei und Zeichnung*, 2:665–67. For ancient perspective drawing, in the development of which Agatharchus played an important role, see White, *Perspective*.

13. For Iktinos, see esp. Dinsmoor, *Architecture of Ancient Greece*, 148–49, 154–79; and Gross, "Iktinos."

14. For Polyclitus, see Gross, "Polykleitos, 5. P. Von Argos"; and Pollitt, *Art and Experience*, 105–10.

15. For Hippodamus, see esp. Castagnoli, *Orthogonal Town Planning*, esp. 65–72, Falciai, *Ippodamo di Mileto*; Gerkan, *Griechische Stadteanlagen*, 42–61; Martin, *L'urbanisme*, esp. 103–6; and Wycherley, *How the Greeks Built Cities*, 15–35

16. For the persecution of Pericles' friends, see Kerford, *Sophistic Movement*, 21–23 and (for Protagoras) 43. Schiappa, *Protagoras and Logos*, 143–45, 205–6, argues that Protagoras's condemnation for impiety and subsequent exile is doubtful. For Pheidias's troubles, see esp. Frost, "Pericles and Dracontides." For Anaxagoras's condemnation for impiety, see Kirk, Raven, and Schofield, *Presocratic Philosophers*, 352–55

17. Sealey, *History of the Greek City States*, 369–498, Hammond, *History of Greece*, esp. 533–95. For the rise of Macedonia, see Hammond, Griffith, and Walbank, *History of Macedonia*, esp. 2:203–698 and 3.1–94; and see Hamilton, *Alexander the Great*.

18 Aineias the Tactician, *How to Survive under Siege*, 34–42, this recent English translation by Whitehead includes an astute commentary and introduction For the Latin text, see Illinois Greek Club, *Aeneas Tacticus, Asclepiodotus, Onasander*, 27–215. See also Gabba, "Technologia militare antica", and Bettalli, "Enea Tattico e l'insegnamento."

19 Aineias the Tactician, *How to Survive under Siege*, see 9.2 and 10 3 with commentary (52–53, 116–17) for plotters and the certainty that they will exist in the city-state, 18.1–20 5 (65–69, 147–53) for gates, locks, and bolts; and 31.1–35 (84–90 and 183–93) for cryptography and secret messages. For the military and political context of the treatise, see Bengtson, "Die griechische Polis bei Aeneas", and Gabba, "Technologia militare antica."

20 See Anderson, *Xenophon*; Delebecque, *Essai sur la vie de Xénophon*; Nickel, *Xenophon*; and for his probable membership in the Athenian cavalry, Bugh, *Horsemen of Athens*, 128–29. For an introduction to Xenophon's writings, see Sandbach, "Xenophon," 478–80, 788–89. For the *Oeconomicus*, see Xenophon, *Memorabilia and Oeconomicus*, and for the treatises on hunting, cavalry commandment, and horsemanship, see Xenophon, *Scripta minora*. For the uses of authorship and literacy to promote social hegemony in the ancient world, see esp. Graff, *Legacies of Literacy*, 11–12

21 See Xenophon, *Oeconomicus* (trans Marchant) 4.1–3, for the banausic arts. Later Xenophon has the interlocutor, Socrates, note that if husbandmen and craftsmen were divided into two separate groups and asked to defend the territory, husbandmen would be likely to fight in order to protect their land, whereas artisans would refuse (6.4–8).

22. See ibid , 5.1–17, where additional benefits are also listed

23. Ibid , 6.13–17, 11 12–14.10.

24. Ibid., 15.4–13, 18.9–10, 19.17–19

25. Aristotle, *Nichomachean Ethics* 6.3–5, 1139b14–1141a8

26. A cogent account of Alexander's career is Hamilton, *Alexander the Great*. For the military "revolution" of the fourth century B.C.E., see Long and Roland, "Military Secrecy," 271–72

27. Marsden, "Macedonian Military Machinery," esp 218–19, Jahns, *Geschichte der Kriegswissenschaften*, 1:36–37. For Polyidus, see Ziegler, "Polyidos, 6." For Charias, see Hultsch, "Charias, 11." For Diades, see Fabricius, "Diades 2) Mechaniker"; and Kroll, "Diades Nr. 2." For the recruiting effort of Dionysius I, see esp. Marsden, *Greek and*

Roman Artillery: Historical Development, 48–49; for the high level of skill required to construct and operate the catapult, see ibid., 67–68, 73–83.

28. For a brief introduction to the Hellenistic states, see Price, "History of the Hellenistic Period." For the Library, see esp. Delia, "From Romance to Rhetoric," quotation on 1457; Fraser, *Ptolemaic Alexandria*, 1:305–35, 2:462–94; and Wendel and Göber, "Das griechisch-römische Altertum," 62–82.

29. Schürmann, *Griechische Mechanik und Antike Gesellschaft*, esp. 1–32, 60–92.

30. See esp. Drachmann, "Ctesibius (Ktesibios)"; Drachmann, *Ktesibios, Philon, and Heron*, 1–21; and Gille, *Les mécaniciens grecs*, 84–102.

31. Lawrence, *Greek Aims in Fortification*, 73. For Philo's writings and context, see Drachmann, "Philo of Byzantium"; Fraser, *Ptolemaic Alexandria*, 1:428–34, 2.619–20 nn. 425, 426; Gille, *Les mécaniciens grecs*, 103–21; and Schürmann, *Griechische Mechanik und Antike Gesellschaft*, esp. 7–8. See also Garlan, *Recherches de poliorcétique grecque*, 279–404, the Greek text and a French translation of the sections on fortresses and besieging and defending towns (bks. 7 and 8, traditionally bk. 5), and 282 for his lost treatise on cryptography. For the text and English translation of the *Belopoeika* with commentary, see Marsden, *Greek and Roman Artillery: Technical Treatises*, 105–84. For the political and military context of these writings, see esp. Gabba, "Scienza e potere nel mondo ellenistico"; and Garlan, "Cités, armées et stratégie."

32. Philo of Byzantium, *Pneumatica* 1. See also Drachmann, *Ktesibios, Philon, and Heron*, 41–73. The name Aristo is misconstrued as Muristom in Arabic translations.

33. For the Greek text and an English translation, see Marsden, *Greek and Roman Artillery: Technical Treatises*, 105–84, 106–9 for the portion of the text referred to here. For Polyclitus, see above, n. 14; and Stewart, "Canon of Polykleitos."

34. Long and Roland, "Military Secrecy"; Schürmann, *Griechische Mechanik und Antike Gesellschaft*, [illegible]; Delia, "From Romance to Rhetoric," [illegible].

35. Fraser, *Ptolemaic Alexandria*, 1:325. See also Grafton, *Forgers and Critics*; and Speyer, *Die literarische Fälschung*.

36. Fraser, *Ptolemaic Alexandria*, 1:447–58.

37. Ibid., 1.459–61; see also Fraser, "Aristophanes of Byzantion," 119 n. 7. Porphyry made his remarks in his no longer extant *On Literary Theft*, which is cited by Eusebius in *Praeparatio Evangelica* 10.3. For general accounts of ancient plagiarism, see Ziegler, "Plagiat," *RE*, where col. 1979 deals with Aristophanes' book on Menander; Ziegler, "Plagiat," *KP*; and Stemplinger, *Das Plagiat*.

38. Vitruvius, *De architectura* 7 pref.4–7.

39. Fraser, "Aristophanes of Byzantion," 115–22. Fraser suggests that the immediate source for the story is Varro, behind which he argues for a Pergamene source of the second century B.C.E.

40. For Augustus, see Bowman, Champlin, and Lintott, *Cambridge Ancient History*, vol. 10. For Roman patronage, see Gold, *Literary and Artistic Patronage*; Rawson, *Intellectual Life*, and the studies pertaining to Rome in Wallace-Hadrill, *Patronage in Ancient Society*.

41 Marsden, *Greek and Roman Artillery. Technical Treatises*, 4–5.

42. Schneider, "Griechische Poliorketiker, III: Athenaios uber Maschinen," 8–13; Rochas D'Aiglun, "Traduction du traité des machines d'Athénée," 800. See also Cichorius, *Römische Studien*, 271–79.

43. Schneider, "Griechische Poliorketiker, III. Athenaios über Maschinen," 10–13.

44. Ibid., 30–33.

45 Vitruvius, *De architectura* 1 pref 2–3, 1.1–18. For an introduction to the scholarship on the treatise, see esp Gros, "Vitruve", Callebat, "La prose du 'De Architectura'", and Baldwin, "Date, Identity, and Career of Vitruvius " On Vitruvius's architectural theory, see Knell, *Vitruvs Architekturtheorie*, and Rawson, *Intellectual Life*, 185–200. On his place in scientific and technical culture, see Tabarroni, "Vitruvio nella storia della scienza e della tecnica." For Varro's *Nine Disciplines*, see Grimal, "Encyclopédies antiques," 470–72.

46 Vitruvius, *De architectura* (trans. Granger) 1.1.1–3, 15.

47. Ibid., 2 1.1–7 For Vitruvius's role in the anthropological tradition and its positive view of craftwork, see esp. Cole, *Democritus and the Sources of Greek Anthropology*, 60–69, 193–95.

48. Vitruvius, *De architectura* (trans Granger) 3.pref.1–3 As told by Plato in the *Apology* 20E–22E, the story concerns Socrates' disbelief of the oracle's pronouncement that he was the wisest of all men, as well as his search among politicians, poets, and craftsmen to find someone wiser than himself. Socrates decided that although they all believed they were wise, they were not because they did not understand how little they really knew. He concludes that the oracle's decree means that human wisdom was of little or no value and that true wisdom was a property of God Cicero, *Academica* 1.4, adds that the oracle also recognized Socrates' wisdom in turning away from the unknowable mysteries of nature to moral philosophy Vitruvius's addition of windows and his interpretation of Socrates' wisdom as his recognition of the need for openness so that men could judge one another's knowledge and craft competence is strikingly different from these versions. For one possible source, see Ferri, *Vitruvio*, 90 n 1, which suggests that Vitruvius probably attributed to Socrates a story derived from a myth (reported by Lucian, *Hermotimus* 20) in which Athena, Poseidon, and Hephaestus quarrel about who is the best artist. To resolve the quarrel, they have a contest in which Poseidon makes a bull, Athena designs a house, and Hephaestus constructs a man. Momus, the judge, criticizes Hephaestus because the man does not have windows in his chest so that everyone can see his desires and thoughts and whether he is lying or telling the truth

49. Vitruvius, *De architectura* (trans Granger) 3.pref 1

50. Vitruvius, *De architectura* 3.pref.1–3 (my translation).

51. Ibid 7.pref.1.

52. Vitruvius, *De architectura* (trans. Morgan) 7 pref 10; for the condemnation of literary theft, see 7.pref.3–7. On Aristophanes, see Fraser, *Ptolemaic Alexandria*, 1:459–61; and Fraser, "Aristophanes of Byzantion "

53. Vitruvius, *De architectura* (trans. Granger) 7.pref.8–9. See Fraser, "Aristophanes of Byzantion," 121–22, for possible sources for the story of Zoilus

54. Vitruvius, *De architectura* 7.pref 3, where Vitruvius describes literary thieves as those who "impio more vixerunt" (lived in an impious way). See also Ogilvie, *Romans and Their Gods*, 101

55. For Roman religion in the early Principate, see Price, "Place of Religion."

56. Drachmann, "Hero of Alexandria " Hero's writings, except for the *Belopoeika*, are published with German translations in *Heronis Alexandrini opera*. Discussions of his writings include Gille, *Les mécaniciens grecs*, 122–44; and Landels, *Engineering*, 199–208 Drachmann, *Mechanical Technology*, 19–140, provides a detailed discussion of the *Mechanics* and English translations of many passages

57. See Drachmann, *Ktesibios, Philon, and Heron*, 77–161, for a detailed commentary on this work and its possible sources, including the treatise by Philo. For the Greek text and German translation, see Hero of Alexandria, *Pneumatica* 1 pref.1–20; see also Hall's introduction to Hero of Alexandria, *Pneumatics of Hero*, xi–xii.

58. Hero of Alexandria, *Automata* 20.1–3, 5; and see Schürmann, *Griechische Mechanik und Antike Gesellschaft*, 190–201. For a discussion of this passage from the point of view of Philo's work, see Philo of Byzantium, *Pneumatica* (ed and trans. Prager), 16.

59. Marsden, *Greek and Roman Artillery: Technical Treatises*, 1–2, 19.

60. For a general introduction to Roman authorship, see Kenney, "Books and Readers in the Roman World."

61. Cato, *De agri cultura* pref.1–4, translation in Hooper and Ash, *Marcus Porcius Cato on Agriculture, Marcus Terentius Varro on Agriculture*. See Astin, *Cato the Censor*, esp. 182–210 for a discussion of his writings and 240–66 for the context of *De agri cultura*; see also Gratwick, "Prose Literature."

62. White, "Roman Agricultural Writers," 440–58

63 Cicero, *De officiis* 1.150–51, translation in Cicero, *On Duties*; see also Rawson, *Intellectual Life*, 84–88.

64. For a summary of Varro's life and works see esp. Dahlmann, "Varroniana"; and Horsfall, "Prose and Mime." For the *Lingua latina*, see esp. Rawson, *Intellectual Life*, 117–31. I have used the edition of Varro's treatise on agriculture contained in Hooper and Ash, *Marcus Porcius Cato on Agriculture, Marcus Terentius Varro on Agriculture* See also Martin, *Recherches sur les agronomes latins*; and Skydsgaard, *Varro the Scholar*.

65. Varro, *De rerum rusticarum* 1 1.1–4. See White, "Roman Agricultural Writers," 482–94, for a discussion of the composition of the treatise and Varro's motivation for writing it.

66 Varro, *De rerum rusticarum* 1.1.4–11 to 1.2.1–2, 1.49.2–3, 3 1.1–6

67. Ibid., 1.3–1.4.1.

68. Ibid., 1.5.1–2.

69. For the background of encyclopedism, see Grimal, "Encyclopédies antiques."

70 For Frontinus's life and work, see Goodyear, "Technical Writing," 672 73, 903.

71. For Pliny's life, see Reynolds, "Elder Pliny and His Times"; and Syme, "Pliny the Procurator " Pliny's history of the German wars is discussed in Sallmann, "Der Traum des Historikers "

72 For an introduction to Columella, see esp. Goodyear, "Technical Writing," 668–70, 901; Martin, "État présent des études sur Columelle"; Martin, *Recherches sur les agronomes latins*, 289–373; and Baldwin, "Columella's Sources." The text I have used is Columella, *On Agriculture*.

73. Columella, *Rei rusticae* 2 1.1–7, 4.1.1, 5.1 1–4, 10.pref 1–5

74. Ibid., 1.pref.1–3.

75. Ibid., 1.pref 13–17. For the influence of Stoicism, see esp. Hoven, *Work in Ancient and Medieval Thought*, esp 21–49

76 Serbat, "Pline l'Ancien"; Wallace-Hadrill, "Pliny the Elder and Man's Unnatural History"; French, *Ancient Natural History*, 196–206, for a discussion of Pliny's view of God and nature; Beagon, *Roman Nature*. For the *Naturalis historia* I have used Pliny, *Natural History*, ed. and trans. Rackham et al. For the preface, see Koves-Zulauf, "Die Vorrede der plinianischen 'Naturgeschichte.'"

77. "C Plinius Baebio Macro Suo S.," *Epistulae* 3.5.1–20, in Pliny the Younger, *Letters and Panegyricus*; see also Sherwin-White, *Letters of Pliny*, 215–25.

78. Pliny, *Naturalis historia* pref.17–19. Ferraro, "Il numero delle fonti," emphasizes that Pliny is using numbers symbolically in specifying the number of volumes and facts.

79. Pliny, *Naturalis historia* pref.21–23.

80. Frontinus, *Stratagematon* 1.pref., 3.pref. I have used the translation in Frontinus, *Stratagems and The Aqueducts of Rome*; I have also consulted the more recent German edition, *Kriegslisten*.

81 Frontinus, *De aquis urbis Romae* 1.1–13. I have used the translation in Frontinus, *Stratagems and The Aqueducts of Rome*. A recent German translation of the text—Frontinus Gesellschaft e V , *Sextus Iulius Frontinus, Curator Aquarum*—includes essays on Roman water technology and extensive illustrations. See also Evans, *Water Distribution*, which includes an English translation of the text and detailed discussions of the aqueducts

82 Frontinus, *De aquis urbis Romae* 2.64–75 Hodge, "How Did Frontinus Measure the Quinaria?" shows that Frontinus actually made the measurements and suggests that he or his engineers may have invented the measuring technique that he used

83. Skydsgaard, *Varro the Scholar*, 101–6.

84. Kenney, "Books and Readers in the Roman World, [illegible]9–20.

85 Thomas, *Oral Tradition and Written Record*; Harris, *Ancient Literacy*, esp. 175–284

86 Veyne, "Roman Empire"

Chapter 2
Secrecy and Esoteric Knowledge in Late Antiquity

1. A good general introduction to these topics is Luck, *Arcana Mundi* In an important study of miracles Howard Clark Kee provides a model for a contextually based approach to these diverse practices as he criticizes the universalizing generalizations of scholars such as Mircea Eliade and Karl Jung. See Kee, *Miracle in the Early Christian World*,

2–3, 21–31 Examples of the methodologies he criticizes are found in Eliade, *Myth of the Eternal Return*; and Jung, *Psychology and Alchemy* For the syncretism of some of these traditions, see Scarborough, "Hermetic and Related Texts"

2. See Brown, "Sorcery, Demons, and the Rise of Christianity"; and for private life and the nature of marriage, Veyne, "Roman Empire," esp. 33–49, and Brown, "Late Antiquity," esp. 95–115

3. For Roman traditional religion, see esp. Ogilvie, *Romans and Their Gods*. For an introduction to mystery religions in the Roman empire, see Burkert, *Ancient Mystery Cults*; and Turcan, *Les cultes orientaux*.

4. Gordon, "Aelian's Peony" And see Aune, "Magic in Early Christianity"; Dodds, *Greeks and the Irrational*; and Lloyd, *Magic, Reason, and Experience*. For an important analysis of the often inappropriate distinctions between religion, science, and magic in the history of anthropology, see Tambiah, *Magic, Science, Religion*.

5. Betz, *Greek Magical Papyri*, xli–liii

6. Ibid , xlvi–xlvii

7. *Papyri graecae magicae* 1.1–42; translations are from Betz, *Greek Magical Papyri*, and all brackets and parentheses appear therein

8. Ibid.

9. Ibid , 1.42–132. *Keryx* means "herald," either a real or an ideal person of priestly or holy status.

10. Ibid., 4.2519; see Betz's n. 317 for other examples of addressing an apprentice as "son"

11 Ibid , 12.401–44. See also Scarborough, "Hermetic and Related Texts," 33–34

12. *Papyri graecae magicae* 1.248, 2.53, 5 370–446.

13. Betz, *Greek Magical Papyri*, xli. See also MacMulten, *Enemies of the Roman Order*, 124–27, 130–36.

14 For a short summary of Apuleius's life and literary production, see Walsh, "Apuleius." Apuleius, *Golden Ass*, contains an especially valuable introduction and notes with reference to much recent scholarship; and see Gersh, *Middle Platonism and Neoplatonism*, 1:215–328, for a detailed discussion of Apuleius's philosophy.

15 Apuleius, *Apologia and Florida of Apuleius of Madaura*.

16. Apuleius, *Opuscules philosophiques (Du dieu de Socrate, Platon et sa doctrine, Du monde)*, which includes an excellent introduction to ancient demonology and its relationship to middle Platonism; and see Dillon, *Middle Platonists*, esp. 184–230, 306–38

17 Apuleius, *Metamorphoses* (ed. and trans. Hanson)

18 Ibid., 2 1–2.

19. Ibid , 3.19–3.25, quotation at 3.19.

20. Ibid., 10.20–33 and 10 34–35 for his escape, bk. 11, the "Isis book," for his transformation into an initiate in the cult of Isis. For an edition and translation of the Isis book with extended commentary, see Apuleius, *Isis-Book (Metamorphoses, Book XI)*. Critics disagree concerning the eleventh book of the *Metamorphoses* and its divergence in tone and subject matter from the prior ten books. John J Winkler, who explicates the complex narratological character of the work as a whole, suggests that Apuleius was ambivalent toward

the Isis cult. Other scholars insist on the unity of the novel and suggest that Apuleius intended in the final book to attract readers to the Isis mysteries. See Winkler, *Auctor and Actor;* and for a contrasting view, see Schlam, *Metamorphoses of Apuleius,* esp. 113–25. See also Shumate, *Crisis and Conversion,* a detailed reading of the entire *Metamorphoses* as a conversion narrative.

21. Apuleius, *Metamorphoses* 11.5–21

22 Ibid., 11.22–23.

23. Ibid , 11 23.

24. Ibid., 11.24.

25 The dynamics of the Isis cult described by Apuleius in many ways confirms Georg Simmel's analysis of the dynamics of esoteric societies, in which secret societies are dependent for their identity upon the wider culture of which they are a part and provide the matrix from which they distinguish themselves. Simmel emphasizes the importance of hierarchy, ritual, and secrecy itself in the initiation of individuals into the group. Simmel, "Sociology of Secrecy and of Secret Societies," esp. 477–84.

26. This development is argued by Smith, *Map Is Not Territory,* 172–89, primarily on the basis of the autobiography of a magician, Thessalos A similar argument is made by Brown, "Rise and Function of the Holy Man," which focuses on holy men of the fourth and fifth centuries.

27 Iamblichus, *On the Pythagorean Way of Life* Recent studies of Iamblichus's philosophy and his religious and intellectual milieu include Finamore, *Iamblichus and the Theory of the Vehicle of the Soul;* Shaw, *Theurgy and the Soul;* and Stäcker, *Die Stellung der Theurgie in der Lehre Jamblichs.*

28. See Iamblichus, *On the Pythagorean Way of Life,* intro p. 29, and 56–57 (*De vita pythagorica* 6 31–32).

29. Ibid , 58–67 (*De vita pythagorica* 7.33–34, 8 35–44).

30 Ibid., 68–71 (*De vita pythagorica* 9 45–47).

31. Ibid., 78–81 (*De vita pythagorica* 11.54–57).

32 Ibid., 96–97 (*De vita pythagorica* 17.71–72).

33 Ibid., 96–99 (*De vita pythagorica* 17.72–74).

34. Ibid., 98–101 (*De vita pythagorica* 17.75–76).

35. Ibid., 126–29 (*De vita pythagorica* 23.103–5).

36. Ibid., 128–29 (*De vita pythagorica* 23.105)

37. Ibid., 172–73, 202–3 (*De vita pythagorica* 29.158, 31.198–99)

38. Porphyry, "On the Life of Plotinus." For an introduction to Plotinus, see Armstrong, "Plotinus."

39. Porphyry, "On the Life of Plotinus," 3.1–36.

40 Ibid., chs. 1, 3, 7, 9.

41 Ibid., 9, 11.

42. Ibid., 3, 4.

43 Ibid., 5–6, 18, which includes Porphyry's list of Plotinus's treatises in the order that he sent them.

44. Ibid., 8.
45. Ibid., 17.
46. Ibid., 24, 21, 6.
47. Fowden, *Egyptian Hermes*, esp. 1–74, 97–104. The standard Greek edition and a French translation are found in Nock and Festugière, *Corpus Hermeticum*. An English translation with useful commentary is Copenhaver, *Hermetica*. Significant studies include Festugière, *La révélation d'Hermès Trismégiste*. Gasparro, "La gnosi ermetica come iniziazione e mistero," which analyzes the Hermetic writings in terms of initiation into the gnosis of divine mysteries; and Grese, "Magic in Hellenistic Hermeticism," which discusses the overlap between Hermetic texts concerning technical magic and the "philosophical *Hermetica*."
48. Copenhaver, *Hermetica*, "⟨Discourse⟩ of Hermes Trismegistus: Poimandres" (*Corpus Hermeticum* 1.1–5).
49. Ibid. (*Corpus Hermeticum* 1.6–14).
50. Ibid. (*Corpus Hermeticum* 1.16–21).
51. Ibid., discourse 13, "A Secret Dialogue of Hermes" (*Corpus Hermeticum* 13.1–16).
52. Ibid. (*Corpus Hermeticum* 13.16–22), and see also pp. 180–83 for a useful discussion of the scholarship concerning how the dialogue may have been used.
53. See Fowden, *Egyptian Hermes*, esp. 192–93. The Nag Hammadi Library is a collection of texts discovered in Upper Egypt in 1945 that included two Hermetic texts. One of them, "The Prayer of Thanksgiving" (6.7), in Robinson, *Nag Hammadi Library in English*, contains evidence for Hermetic cultic practices (321–38, and for the prayer, 328–29). The prayer ends with the statement, "When they had said these things in the prayer, they embraced each other and they went to eat their holy food, which has no blood in it," indicating a ritual embrace or kiss and a cultic meal. Douglas Parrott, one of the translators of the Nag Hammadi texts, says that such texts "may well have been used in the context of small groups devoted to secret knowledge and mystical experience, in which those who were more advanced would teach and direct neophytes, and in which certain cultic acts were engaged in (prayers and hymns are found throughout the Hermetic corpus). The tractates would have served as the basis for discussion and as texts for individual meditation" (322). For an edition and detailed study of the Hermetic texts in the Nag Hammadi collection with French translations, see Mahé, *Hermès en Haute-Égypte*.
54. Luck, *Arcana Mundi*, 361. For early alchemy, see Eliade, *Forge and the Crucible*; Holmyard, *Alchemy*, Lindsay, *Origins of Alchemy*; Merkur, "Study of Spiritual Alchemy", and Stillman, *Story of Alchemy and Early Chemistry*. The edition of early alchemical texts, Berthelot and Ruelle, *Collection des anciens alchimistes grecs*, is in the process of being replaced by new editions, the completed volumes of which are cited below.
55. See Robert Halleux, *Les alchimistes grecs*, 12, for the possibility of copying by the same scribe. There are also English translations: Caley, "Leyden Papyrus X"; and Caley, "Stockholm Papyrus." For the relationships between the chemical papyri and the *Greek Magical Papyri*, see Fowden, *Egyptian Hermes*, 168–72.

56 On rare occasions a source is mentioned; for instance, a recipe for purple dye is cited as being from the book of Africanus (the *Cestes* of Julius Africanus is an encyclopedia written around 230 C.E.). See Halleux, *Les alchimistes grecs*, 6–17 and *Papyrus holmiensis*, recipe 116. *Papyrus holmiensis* comprises pp. 110–51 of Halleux, *Les alchimistes grecs*.

57. Halleux, *Les alchimistes grecs*, 24–30.

58. See ibid., 28, for this and other examples of amateur usages; the example cited is *Papyrus leidensis*, recipe 68, line 378 (*Papyrus leidensis* comprises pp. 81–109 of Halleux, *Les alchimistes grecs*); and see *Papyrus holmiensis*, recipes 108–13, for dyers

59 Ibid , 26 and *Papyrus leidensis*, recipe 39; Caley, "Leyden Papyrus X," 1156, recipe 40 (which translation I use).

60. Halleux, *Les alchimistes grecs*, 11 and *Papyrus leidensis*, recipe 88, Caley, "Leyden Papyrus X," 1161 (recipe 90).

61. *Papyrus holmiensis*, recipe 106; Caley, "Stockholm Papyrus," 993 (recipe 101)

62. See Berthelot and Ruelle, *Collection des anciens alchimistes grecs*, 2.41–53 (for the Greek text), 3.44–45 (for a French translation); and Stillman, *Story of Alchemy and Early Chemistry*, 154–57, which includes a partial translation, which I have used See also Lindsay, *Origins of Alchemy*, 90–130. An important recent discussion of the text, and its authorship and origins is Hershbell, "Democritus and the Beginnings of Greek Alchemy"

63. Berthelot and Ruelle, *Collection des anciens alchimistes grecs*, 3.44–45; Stillman, *Story of Alchemy and Early Chemistry*, 154–57.

64 See Zosimos of Panopolis, *Memoires authentiques*, in *Les alchimistes grecs*, 168–69 n. 2.

65 See Zosimos of Panopolis, *Les alchimistes grecs*, esp. xi–xviii, for a discussion of his life and times; 197 n. 1 (*Memoires authentiques* 8.1) for the ways of addressing Theosebeia that suggest a society or confraternity of alchemists, 26 n 2 for the discussion of the term *structor*; and 27 (*Memoires authentiques* 8.2) for the "Jewish books." Mertens suggests that the extremely rare Greek term *strouktorion* derived from the Latin *structor*, meaning "constructor" or "architect," and referred to the slave organizer of a banquet She suggests that the use of the term contains a connotation of snobbery that points to the social context of the alchemists. See also Lindsay, *Origins of Alchemy*, 323–57 For an edition and English translation of one part of the *Memoirs*, see Zosimos of Panopolis, *On the Letter Omega*.

66. Zosimos of Panopolis, *Memoires authentiques*, 1–2, 8, 66–67 n. 17, 113–14 n 99; and see Lindsay, *Origins of Alchemy*, 325–27

67 Zosimos of Panopolis, *Memoires authentiques*, 8, 113–14 n. 99

68 Ibid., 9. Mertens points out that such medical books with crosshatch illustrations must have once existed but are no longer extant.

69. Ibid., 14–15, 23–25, 17–21, 30–33 (*Memoires authentiques* 3.1–2 on the *tribicos* and tubes, 7 1–6 on the *kerotakis*, 4.2–4, 9.1–4 on the fabrication of waters, including "divine water," and the use of eggs in the process). For a detailed discussion of alchemical apparatus, including the *tribicos* and the *kerotakis*, see Holmyard, *Alchemy*, 43–59

70. Zosimos of Panopolis, *Memoires authentiques*, esp. 6–8 and 110 n. 95, which treats the deceitful and secret actions of the evil demon.

71 Ibid , 9–10

72. Ibid., 16–17, 140 nn. 4 and 5, a cogent discussion of the passage.

73. Zosimos of Panopolis, *Memoires authentiques*, 18–19 (4.2).

Chapter 3
Handing Down Craft Knowledge

1 See esp. Schick and Toth, *Making Silent Stones Speak*; and Renfrew and Bahn, *Archaeology*, esp. 295–334.

2. For Neolithic agriculture and crafts, see Renfrew and Bahn, *Archaeology*, 253–94; Fagan, *People of the Earth*, esp. 225–339; and Wenke, *Patterns in Prehistory*, esp. 225–76. For artisanal work in the ancient Near East, see esp. Zaccagnini, "Le tecniche e le scienze"; and for that in Greece and Rome, Burford, "Crafts and Craftsmen."

3. See, e.g , Burford, *Greek Temple Builders at Epidauros*, esp. 198, which suggests that the Epidaurans were "unused to specialists" and more familiar with "occupations basic to maintaining a decent standard of living, stone masonry and woodwork."

4 See Burford, *Craftsmen in Greek and Roman Society*, 82–91, for a range of evidence of transmission through families, Kümmel, *Familie, Beruf und Amt*, esp. 20–48, 161–62; Meiggs, *Roman Ostia*, 323; Weisberg, *Guild Structure and Political Allegiance*, esp. 77–85, although Weisberg's hypothesis concerning the existence of independent guilds has not found general acceptance, and Westermann, "Apprentice Contracts," esp. 304–5

5. Finley, "Technical Innovation and Economic Progress " Although Finley's point concerning anachronistic readings of ancient cultures is well taken, his discussion concerning the lack of technological innovation in the ancient world has been thoroughly criticized. See Greene, "Technological Innovation and Economic Progress."

6. Long and Roland, "Military Secrecy," esp. 266–68.

7. Dilley, "Secrets and Skills "

8. For ancient craft associations, see San Nicolò, "Guilds: In Antiquity"; Boak, "Guilds: Late Roman and Byzantine"; and Epstein, *Wage Labor and Guilds*, 10–49.

9. San Nicolò, "Guilds· In Antiquity." For the opposite point of view, see Mendelsohn, "Gilds in Babylonia and Assyria," where Mendelsohn discusses what he considered to be evidence of independent guilds in ancient Assyria at the time of Hammurabi (1792–1750 B.C.E) and thereafter, and Weisberg, *Guild Structure and Political Allegiance*, esp. 86–105. This view has not been accepted. See Renger, "Notes on the Goldsmiths, Jewelers, and Carpenters"; and Zaccagnini, "Patterns of Mobility," 261.

10. See Stöckle, "Berufsvereine." For Greek guilds, see Poland, *Geschichte des griechischen Vereinswesens*, esp. 116–27; Tod, *Sidelights on Greek History*, 71–96, and Ziebarth, *Das Griechische Vereinswesen*, esp. 74–90. For Egyptian guilds, see San Nicolò, *Ägyptisches Vereinswesen*, 1.66–206. For Roman guilds, see esp. Duff, *Personality in Roman Private Law*, 95–158; Kornemann, "Collegium"; Robertis, *Storia delle corporazioni*; Ruggini, "Le associazioni professionali"; and Waltzing, *Étude historique sur les corporations*.

11. See Waltzing, *Étude historique sur les corporations*, 1.61–154, and for the suppression of the Roman collegia in 64 B.C.E., 1:90–113. See also Robertis, *Il diritto associativo*; and Treggiari, *Roman Freedmen*, 168–77. For Numa's creation of the guilds, see Gabba, "*Collegia* of Numa."

12. Epstein, *Wage Labor and Guilds*, 10–49.

13. Waltzing, *Étude historique sur les corporations*, 2:6–348, which outlines this development in great detail; Sirks, *Food for Rome*. See also Boak, *Manpower Shortage*, 65–84, Robertis, *Storia delle corporazioni*, 2:93–234, and Dill, *Roman Society*, 227–81.

14. Epstein, *Wage Labor and Guilds*, 44–49. An important source for the Byzantine guilds is the book of regulations of the eparch (a market inspector) from the reign of Leo VI ("Leo the Wise"), 880–912. See Boak, "Book of the Prefect"; and Vryonis, "Byzantine *Demokratia*."

15. For the status of artisanal work, labor, and commerce (relevant because many artisans also sold their products from their shops) in antiquity, see esp. Bradley, *Slaves and Masters*, Burford, *Craftsmen in Greek and Roman Society*, 184–218, D'Arms, *Commerce and Social Standing*; Robertis, *Lavoro e lavoratori*; Geoghegan, *Attitude towards Labor*, Mondolfo, *Polis, lavoro e tecnica*; and Treggiari, *Roman Freedmen*. A controversial thesis concerning ancient valuation of Greek pottery is relevant to the issue of the status of ancient artisans. Michael Vickers and David Gill argue in *Artful Crafts* that ancient pottery was not considered valuable or precious by the ancients despite the skill with which it was made and decorated. They suggest that pottery was often fabricated in imitation of gold, silver, bronze, or ivory objects and that objects made of precious materials were valued for their materials as much as for the skill with which they were produced. The controversial aspects of their thesis (see, e.g., Pollitt, "Gold Pastoral!") do not detract from their fundamental insight that artisanal skill was viewed very differently in antiquity than it is in the modern world. For a study that attempts to get at the artisan's own point of view, see Joshel, *Work, Identity, and Legal Status*.

16. Plutarch, *Life of Pericles* (trans. Perrin) 1–2.

17. Epstein, *Wage Labor and Guilds*, 12–13, Joshel, *Work, Identity, and Legal Status*, esp. 113–22.

18. Oppenheim et al., *Glass and Glassmaking*, 4–6.

19. Ibid., 6.

20. Gadd and Campbell Thompson, "Middle-Babylonian Chemical Text"; and Campbell Thompson, *Dictionary*, xii.

21. Oppenheim et al., *Glass and Glassmaking*, 59–65.

22. Ibid., 22–59, 80–86.

23. See ibid., 32–33, on tablet B, for the ritual instructions and 51–53, on tablet L, whose ritual instructions include reference to the dead masters.

24. See ibid., 50–51, on recipe I, and, for the second recipe (Q), 53–54.

25. Oppenheim, "Mesopotamia in the Early History of Alchemy," esp. 37 (quotation) and 42–43 n. 3 (on coinage).

26. See Halleux, *Les alchimistes grecs*, esp. 54–62, for a cogent discussion of recipe literature, and 154–66, for papyrus recipe fragments. For Pliny's remarks on writings con-

cerning the coloration of stones, see Pliny, *Naturalis historia* 37.24.90–91, 37.75 197. See also Festugière, *La révélation d'Hermès Trismégiste*, 1:220; and for Dioscorides, Riddle, *Dioscorides*.

27 Oppenheim et al , *Glass and Glassmaking*, 4.

28. For an introduction to craft writings in the Latin West, see Bischoff, "Die Uberlieferung der technischen Literatur " For the *Compositiones variae*, see Hedfors, *Compositiones ad tingenda*, which includes a German translation and extensive commentary, Burnam, *Classical Technology*, an English translation; and Schiaparelli, *Il codice* 490, a photographic reproduction of the manuscript. Essential studies include Alexander, "Medieval Recipes," esp. 40–42; Johnson, *Compositiones Variae*; and Svennung, *Compositiones Lucenses*

29 See Smith and Hawthorne, *Mappae Clavicula*, 4 (for the Reichenau catalog entry). Smith and Hawthorne's edition contains a useful introduction and bibliography, an English translation, and photographic reproductions of two of the most complete manuscripts (the Sélestat and Phillipps-Corning manuscripts); and for a transcription of the latter, see Phillipps, "Letter from Sir Thomas Phillipps." A succinct recent introduction to this and related writings is Davis-Weyer, "Panel and Wall Painting"; see also Roosen-Runge, *Farbgebung und Technik*.

30. For an English translation, see Smith and Hawthorne, *Mappae Clavicula*, 28. However, I have used the version of this preface provided in Halleux and Meyvaert, "Les origines de la 'Mappae clavicula,'" 14–15, which contains significant emendations, and have translated into English from this text. "⟨M⟩ultis et mirabilibus in Hermetis libris conscriptis", "heresim"; "clausis domibus/sine clavi impossibile est facile potiri his quae in/domibus sunt, ita et sine isto commentario omnis scriptura/quae in sacris libris conscripta est clausum et tenebrosum sensum efficiet eius qui legerit"; "pium et iustum/ sensum . . . et isto conservare."

31. "Invenies quomodo se habeat, sanctum laudabile que secretum." Smith and Hawthorne, *Mappae Clavicula*, 31; Phillipps, "Letter from Sir Thomas Phillipps," 195.

32. "Absconde sanctum, et nulli tradendum secretum, neque alicui dederis propheta." Smith and Hawthorne, *Mappae Clavicula*, 32; Phillipps, "Letter from Sir Thomas Phillipps," 196.

33. "Ut nichil honerosius sit dictum, absconde confectionem " Smith and Hawthorne, *Mappae Clavicula*, 35, Phillipps, "Letter from Sir Thomas Phillipps," 201.

34 Smith and Hawthorne, *Mappae Clavicula*, 15; Halleux and Meyvaert, "Les origines de la 'Mappae clavicula,'" 25

35. Merrifield, *Original Treatises*, 1:166–203. Essential studies include Alexander, "Medieval Recipes," esp. 35–36; Giry, "Notice sur un traité"; Richards, "New Manuscript of Heraclius"; and Roosen-Runge, *Farbgebung und Technik*.

36 Merrifield, *Original Treatises*, 1 182–83

37 See ibid., 1:188–89, for the quotations from Pliny; and Giry, "Notice sur un traité," 225–26

38 Theophilus, *On Divers Arts*, xxx–xxxi. Roger of Helmarshausen's authorship is probable, although it has not been proven. It is suggested by attributions in late-seventeenth-

century manuscripts and by comparison of several objects fabricated by Roger with instructions in the treatise. For the standard edition of the Latin text and an English translation, see Theophilus, *Various Arts* A critical review of both of these editions is Thompson, "Theophilus Presbyter", see also White, "Theophilus Redivivus." For a recent German translation, with detailed commentary on book 3, on metalwork, see Brepohl, *Theophilus Presbyter.*

39. Theophilus, *Various Arts,* 1; Van Engen, "Theophilus Presbyter and Rupert of Deutz"; White, "Theophilus Redivivus."

40. Theophilus, *Various Arts,* 1.

41. Ibid., 36–37.

42. Ibid., 61–63.

43. Ibid

44. Ibid., 2. See also Post, Giocarinis, and Kay, "Medieval Heritage of a Humanistic Ideal"

45 Ovitt, *Restoration of Perfection;* Van Engen, "Theophilus Presbyter and Rupert of Deutz," 162; White, "Medieval Engineering", Whitney, *Paradise Restored,* Hoven, *Work in Ancient and Medieval Thought.*

46. For urbanism and urban commercial and artisanal culture, see esp. Martines, *Power and Imagination,* Miller and Hatcher, *Medieval England;* Miskimin, *Economy of Early Renaissance Europe,* 73–115; and Swanson, *Medieval Artisans.* A still valuable introduction to the medieval guilds is Thrupp, "Gilds." See also Barral i Altet, *Artistes, artisans,* Epstein, *Wage Labor and Guilds,* esp. 50–101; and Rosser, "Crafts, Guilds and the Negotiation of Work," all of which contain much further bibliography. For English crafts and guilds, see Blair and Ramsey, *English Medieval Industries.* For German guilds, see esp. Wissell, *Des alten Handwerks Recht;* and Cordt, *Die Gilden.*

47 See Epstein, *Wage Labor and Guilds;* Epstein, "Craft Guilds, Apprenticeship, and Technological Change"; and for the relationship of the guilds to the development of political thought, Black, *Guilds and Civil Society.* See also Gustafsson, "Rise and Economic Behaviour of Medieval Craft Guilds," which emphasizes quality control as central to the origin and function of the guilds.

48. See esp. Epstein, *Wage Labor and Guilds,* 102–256; and Thrupp, "Gilds"

49. Mackenney, *Tradesmen and Traders,* 9 (quotation) and, for the Venetian guilds' relationship to the government, 1–43; see also Romano, *Patricians and Popolani,* 65–90. A study that emphasizes the material basis of Venetian glassmaking is McCray, *Glassmaking in Renaissance Venice.* For the Venetian guild statutes, see Monticolo and Besta, *I capitolari delle arti veneziane.* For the confraternities, see esp. Sbriziolo, "Per la storia delle confraternite" For the Giustizia Vecchia, see Monticolo, *L'Ufficio della Giustizia Vecchia*

50. For the capitulary of the glassmakers, see Monticolo and Besta, *I capitolari delle arti veneziane,* 2:61–98. For Venetian glasswork, see esp. Brunello, *Arti e mestieri,* 17–30; Gasparetto, *Il vetro di Murano;* McCray, *Glassmaking in Renaissance Venice;* Mentasti, "Tecnica del vetro"; Mentasti, *Il vetro veneziano;* and Zecchin, *Vetro e vetrai di Murano.*

51. Monticolo and Besta, *I capitolari delle arti veneziane,* 2:64, ch. 2, 2:67, ch 11; 2:70, ch 23.

52. Ibid., 2:62–64, ch. 1; 2·85, ch. 72; and 2:86, ch. 75 designate the feast days that the fieroli were required to celebrate, whereas 2:89–91, chs. 81–82, and 2:96–97, ch. 93, refer to the seven-month work and five-month selling periods to which the glassworkers were compelled to conform.

53. Two of the most important election rules limited the term of office for the gastaldus to only one year and forbade padroni di fornace to run for office in either the arte or the scuola. Ibid., 2.75, ch. 41 (November 1265), 2:76–77, chap. 45.

54. Adjudication of most disputes and enforcement of the regulations were the responsibility of the gastaldus and the judices. Many chapters of the capitulary concern these functions. See ibid , 2.67, chs. 12–14, 2:68–70, chs. 17–22; 2:72, ch. 30; 2:83, ch. 63, 2:92, ch. 84.

55. Ibid., 2:65, ch. 6; 2:71, chs. 24–25; 2:74, ch 38; 2·82, ch. 57; 2:83, chs. 61, 62, 64; 2:85, chs 69–70. The regulations specify the minimum age of apprentices (eight years) and prohibit attempts to persuade another master's apprentice to work for oneself.

56. Ibid., 2:67–68, ch 15, and 2:79, ch. 52 (concerning stolen goods); 2:73, ch 34 (concerning the sale of imperfect products); 2:86, ch 73 (concerning the sale of non-Venetian glass in Venice).

57. The wood to be used in the furnace was specified first as alder or willow, later as alder only Not more than three or four boche were allowed in the main furnace. Ibid., 2:65, ch 7, and 2:94–95, chs. 88, 90 (concerning wood); 2:65, ch. 5, 2·91–92, ch. 83, and 2:92–93, ch. 86 (concerning furnace "mouths"). See also 2:95–96, chs. 91–92, for attempts to regulate the use of *fuliginus*, the potassium carbonate that northern glassmakers used instead of the sodium carbonate used in Venetian glass, for which see Gasparetto, *Il vetro di Murano*, 239, 247.

58. For glassmaking technology in both northern and southern Europe in the late medieval period, see Charleston and Angus-Butterworth, "Glass."

59. Monticolo and Besta, *I capitolari delle arti veneziane*, 2:66, ch. 8· "unusquisque de arte predicta qui exierit extra Venecias occassione exercendi dictam artem", "gastaldus non debeat accipere sacramentum ab hominibus qui istius artis causa exierit extra Venecias sine licencia iusticiariorum." The fine was ten Venetian denarii.

60. Ibid., 2.79, ch. 51. The fine was raised to five libras.

61. Ibid., 2:88–89, ch 80· "desertantur et extrinseci elevantur."

62. Ibid., 2·92, ch. 85.

63. Ibid., 2:97, ch. 94.

64. Ibid., 2:97–98, ch. 95

65. A transcription of Gian Antonio's account appears in Cicogna, *Delle inscrizioni veneziane*, 466–71; the quotation, "primo autore e inventore de'colori variamente mescolati nel vetro," is on 467. See also Mentasti, *Il vetro veneziano*, 48–50, and Zecchin, "Giorgio Ballarin." Mentasti, "Tecnica del vetro," discusses the increasingly complex technology of fifteenth-century glassmaking, the secrets of which are still being discovered.

66. I have used the terms *patent* and *privilege* interchangeably throughout. Rather than suggesting either traditional medieval or modern usages by these terms, I use them

to describe a conceptual development and practice which occurred before the emergence of precise terminology to describe it. The best general introduction to early European patents is Silberstein, *Erfindungsschutz und merkantilistische Gewerbeprivilegien*. For Venetian patents, see Mandich, "Le privative industriali", and Mandich, "Primi riconoscimenti veneziani." For English patents, see MacLeod, *Inventing the Industrial Revolution;* see also Braunstein, "A l'origine des privileges d'invention." For further bibliography, see Long, "Invention, Authorship," 875–81.

67 See Frumkin, "Early History," 50–54. For a detailed study of Venetian glassmaking in the Netherlands beginning in the sixteenth century, see Schuermans, "Verres 'façon de Venise.'" As MacLeod has demonstrated in "Accident or Design?" Venetian glassmaking influenced the patent system well into the seventeenth century.

68. See esp. Mandich, "Le privative industriali," 524–25, which emphasizes the requirement of utility to the state.

69. Mandich, "Primi riconoscimenti veneziani," 105–6: "si aliquis medicus vellet facere aliquam medicinam suam secreto, teneatur eam facere modo scilicet de melioribus rebus, et teneant omnis in credentia, et jurent omnes stationarii non intromittere se de predictis."

70 Ibid., 106–10, 107 for the windmill.

71 Ibid., 115–16, 149–50. For an English translation of the document, see Mandich, "Venetian Origins," 379 n 6. I have used this translation with minor changes.

72. See Mandich, "Primi riconoscimenti veneziani," 116–55, for a detailed discussion of patents and related provisions enacted by the senate from the time of Petri's patent to 1547.

73. For the text of the law and a discussion, see Mandich, "Le privative industriali," 518–19: "accutissimi ingegni"; "apti ad excogitar et trovar varij ingegnosi artificij"; "auctor et inventor" For further discussion and an English translation of the law, see esp Mandich, "Venetian Patents," 176–77; and Phillips, "English Patent."

74. Mandich, "Le privative industriali," 537–47, lists 109 Venetian patents awarded between 1475 and 1549

75 For an introduction to Brunelleschi, see Battisti, *Brunelleschi* For his perspective studies, see esp. Tsuji, "Brunelleschi and the Camera Obscura"; and for his machines and construction, Prager and Scaglia, *Brunelleschi*, 1–109; Scaglia, "Drawings of Machines for Architecture", and Settle, "Brunelleschi's Horizontal Arches." A convenient summary of the available documents and some of the older scholarship is Hyman, *Brunelleschi in Perspective*. For the cupola, see Saalman, *Filippo Brunelleschi: The Cupola*; see also Saalman, *Filippo Brunelleschi· The Buildings*. Smith, *Architecture in the Culture of Early Humanism*, esp. 19–39, argues that Brunelleschi was admired in his lifetime especially for the dome construction and for the machines that he invented and designed for that project. For Brunelleschi's own participation in Florentine governing councils, see Zervas, "Filippo Brunelleschi's Political Career."

76. See Hyman, *Brunelleschi in Perspective*, 34–35, for Brunelleschi's conflict with the stonemasons and woodworker's guild. Many details of Brunelleschi's life, including his

conflicts with Ghiberti, are known from the biography of his younger contemporary Antonio di Tuccio Manetti. See Manetti, *Life of Brunelleschi*, esp. 83–89, for his conflicts with Ghiberti.

77. Brunelleschi's patent is the most famous from the fifteenth century, although contrary to recent claims by several scholars, it was by no means the first. For the original document, see Gaye, *Carteggio inedito*, 1:547–49. See also Prager, "Brunelleschi's Patent," esp. 109–10, for a translation of the document (which I have used with some minor changes); and Prager and Scaglia, *Brunelleschi*, 111–23.

78. Gaye, *Carteggio inedito*, 1:547–49; Prager, "Brunelleschi's Patent," 109–10; Prager and Scaglia, *Brunelleschi*, 111–23.

79. See Prager and Scaglia, *Brunelleschi*, 118 for translations of the two sonnets and 143–44 for the Italian.

80. See Prager, "A Manuscript of Taccola," 138–42, for a transcription of the Latin text and an English translation; and Prager and Scaglia, *Mariano Taccola and His Book*, 11–13. See also Taccola, *De ingeneis*, 1.134–36.

81. Prager and Scaglia, *Mariano Taccola and His Book*, 11–12.

82. For Brunelleschi's father, see Manetti, *Life of Brunelleschi*, 36–38.

83. Hyman, *Brunelleschi in Perspective*, 24; translation slightly altered.

84. Smith, *Architecture in the Culture of Early Humanism*, 19–39; and Kemp, "From 'Mimesis' to 'Fantasia.'"

Chapter 4
Authorship on the Mechanical Arts in the Last Scribal Age

1. See esp. Crossgrove, *Die deutsche Sachliteratur*, 103–44; Eis, *Mittelalterliche Fachliteratur*, Galluzzi, *Prima di Leonardo*; Gille, *Engineers of the Renaissance*, 55–170; Hall, "Der Meister sol auch kennen schreiben und lesen"; Hall, *Technological Illustrations*, 121–33; and Jähns, *Geschichte der Kriegswissenschaften*, 1:241–43.

2. Smith *Architecture in the Culture of Early Humanism*, 50–51.

3. For the relationships between political power, the decorative arts, and architecture in the Renaissance, see esp. Kaufmann, "Editor's Statement", Woods-Marsden, "Images of Castles"; Goldthwaite, *Building of Renaissance Florence*; Goldthwaite, *Wealth and the Demand for Art*, 212–42, Kent and Simons with Eade, *Patronage, Art, and Society*; Kempers, *Painting, Power, and Patronage*; Martines, *Power and Imagination*; Starn and Partridge, *Arts of Power*; Rosenberg, *Art and Politics*, and Thomson, *Renaissance Architecture*.

4. For the abacus tradition see Goldthwaite, "Schools and Teachers"; and Grendler, *Schooling in Renaissance Italy*.

5. Hall, "Giovanni de' Dondi and Guido da Vigevano," quotation on 141; Pesenti, "Dondi Dall'Orologio, Giovanni", Aiken, "Truth in Images"; Hall, "Guido's *Texaurus*."

6. See Kyeser, *Bellifortis*, an edition and German translation that includes a facsimile reproduction of a manuscript at Göttingen dated 1405 (2 Cod. philos. 63 cim., Univer-

sitätsbibliothek, Gottingen); Battisti and Battisti, *Le macchine cifrate;* and for physicians' involvement in technology, White, "Medieval Engineering."

7 Kyeser, *Bellifortis,* 1:4, 53 (2: fols 3a, 85a). "diem impium"; "perfusa amaritudine spiritus"; "inaudite audacie principis Sygismundi Regis ermofrodati ungarie suorumque dominorum et regnicolarum repentinam fugam"; "perfugum atque furibundum / Fallacem nequam quia rem non diligit equam." For Sigismund and the complex political system of the empire, see esp. Baum, *Kaiser Sigismund,* Mályusz, *Kaiser Sigismund in Ungarn;* and Wefers, *Das politische System Kaiser Sigmunds.* For the battle of Nicopolis, see Atiya, *Crusade of Nicopolis.*

8. Kyeser, *Bellifortis.* For Ruprecht see Thorbecke, "Ruprecht III."

9. Kyeser, *Bellifortis,* vols. 1, pp. 3, 15, and 2, fol. 11b: "sicud celum ornatur sideribus, sic alemania prefulget disciplinis liberalibus, honestatur mechaniis diversisque artibus adornatur." Quarg, the editor and translator, suggests that *Meufaton* refers to Philip of Macedonia's conquest of Methone in 353 B.C.E. during which the Macedonian king lost an eye What Kyeser makes clear is the almost magical properties of the weapon, which caused enemies to flee. For the war carriage, see ibid., 1:17, where the descriptive verse begins, "Hoc instrumentum ab allexandro repertum" (This instrument was invented by Alexander); and see vol. 2, fol. 16a, for the illustration

10 Ibid., 1:100–117 (2. fols. 135a–139b); for the epicedium, "Sit mea tibi anima altissimo coniuncta," see 1·100 (2: fol. 135a). See also White, "Kyeser's 'Bellifortis,'" quotation on 438.

11. Kyeser, *Bellifortis,* vols. 1, pp 104–5, and 2, fol. 137a–b.

12. For Fontana's life and writings, see esp. Clagett, "Life and Works"; and Battisti and Battisti, *Le macchine cifrate,* 39–41. For Fontana's relationship to earlier writings and technical illustrations, see esp. Aiken, "Truth in Images", and Prager, "Fontana on Fountains."

13. Battisti and Battisti, *Le macchine cifrate.*

14. Ibid., 36–37 for the quotation (my translation), 141–58 for a transcription of the text of the *Secretum,* and 39; see also Clagett, "Life and Works," 16–17. The manuscript of the *Bellicorum instrumentorum liber* is Cod. icon. 242, Bayerische Staatsbibliothek, Munich.

15. Battisti and Battisti, *Le macchine cifrate,* 96 (fol. 62v) (my translation). See Prager, "Fontana on Fountains," for a detailed discussion of his interest in fountains and his sources.

16. For a summary of Taccola's career, see Taccola, *De ingeneis,* 1.11–15, and Prager and Scaglia, *Mariano Taccola and His Book,* 1–21. Most of the relevant documents are published in Milanesi, *Documenti* For the role of the stimatore, see Adams, "Life and Times of Pietro dell'Abaco."

17. For a facsimile of the first two books (Codex Latinus Monacensis 197, pt 2, Bayerische Staatsbibliothek, Munich), with transcription, translation, and notes, see Taccola, *De ingeneis.* See also Prager and Scaglia, *Mariano Taccola and His Book.* For a facsimile edition of books 3 and 4, see Taccola, *Liber tertius de ingeneis,* which includes transcriptions of documents pertaining to Taccola. For *De machinis,* see Taccola, *De machinis,* which includes a facsimile edition of Codex Latinus Monacensis 28,800, Bayerische

Staatsbibliothek, Munich, Taccola, *De rebus militaribus*, which includes a facsimile of Codex Parisinus Latinus 7239, Bibliothèque Nationale, Paris. For an introduction to Taccola and fifteenth-century Sienese culture, see Galluzzi, "Le macchine senesi." For the ravaging by mercenary companies, see Caferro, *Mercenary Companies*.

18. Prager and Scaglia, *Mariano Taccola and His Book*, 144–45.

19. For evidence of Taccola's presence in the Veneto, see Milanesi, *Documenti*, 299–300; Prager and Scaglia, *Mariano Taccola and His Book*, 16–17, 20; Taccola, *De ingeneis*, 1:116 (bk. 2, fol. 82r); and Taccola, *De machinis*, 1:20–23, 29–58 (for a detailed discussion of the manuscripts).

20. Taccola, *De machinis*, 1:18, 59–60.

21. Plutarch, *Life of Marcellus* (trans. Perrin) 17.4.

22. Taccola, *De ingeneis*, 1.64 and 2:61 (fol. 31r): "Scio quid facio super piscem nata(ta)ntem. Et in ore premo spumam expluentem oleum, ut baiuletur a pisce equitans eum. Et intus habet quod ipsum portat et quod ab se ipso portatur. Et nota quod in capitulis de precipitantibus et troquentibus sine me aliquid ad prefectionem fieri non potest. Et velate locutus sum. Quod diutius cum labore acquisivi non cito sciatur." I have modified the translation slightly.

23. Hall, *Technological Illustrations*, 121–33; Jähns, *Geschichte der Kriegswissenschaften*, 1:241–443. For the history of gunpowder and gunpowder artillery, see Hall, *Weapons and Warfare*; Schmidtchen, *Bombarden, Befestigungen, Büchsenmeister*, Egg, *Der Tiroler Geschützguss*; Egg, *Das Handwerk der Uhr*, 183–201, and Kramer, *Berthold Schwarz*.

24. Cod. germ. 600, Bayerische Staatsbibliothek, Munich. For a detailed description, see Schneider, *Die deutschen Handschriften*, 227. See also Hassenstein, *Das Feuerwerkbuch*, 48, 56.

25. For a reprint of the 1529 printed edition and a translation into modern German, see Hassenstein, *Das Feuerwerkbuch*, see 15–16 for the original version of the quotations, "diener / die als from[m] unnd fest leut seyen" and "weissleut," and 41–43 for the modern German. See also Hall, "Der Meister sol auch kennen schreiben und lessen," 49–51; Jähns, *Geschichte der Kriegswissenschaften*, 1:392–408; and Kramer, *Berthold Schwarz*.

26. Hassenstein, *Das Feuerwerkbuch*, 16: "Und darumb wann der stuck sovil sind die darzügehored / die ein yetlicher gütter püchsenmaister künden soll / und die ein mayster on die geschrift in seinem sinne nit gedencken kan."

27. Ibid., 16–17, 43–45.

28. Ibid., 17–18, 45–47: "sol er got eren und vor allen dingen vor auge[n] haben"; "sich auch bescheidenlich mit der welt halten / mitt der er dann wandelt"; "ain endlicher unverzagter man sein"; "sich tröstlich halten"; "d[er] maister sol auch künden schreiben un[d] lesen." See Partington, *History of Greek Fire and Gunpowder*, 91–97, for the legend of "Black Berthold," also known as Berthold Schwartz; and see Kramer, *Berthold Schwarz*, 121–22, arguing that he was a real person.

29. Hassenstein, *Das Feuerwerkbuch*, 31, 69. "diezu in bringen mochtenn durch der kunst weissheit / radt und hülf / sy iren feinden widerstan."

30. Hall, *Technological Illustrations*, 11–25, 118–33, which includes reference to other bibliography, Jähns, *Geschichte der Kriegswissenschaften*, 1.243–443.

31. HS 25,801, the *Kriegsmaschinen,* in the Germanisches Nationalmuseum, in Nuremberg, is an early-fifteenth-century manuscript that illustrates cannon resting on carriages with their elevating arches for aiming, fire bombs, ladders, ships, and other military machines, with explanatory text in verse.

32. Hall, *Technological Illustrations,* a facsimile edition of the "Anonymous of the Hussite Wars" including a transcription and English translation of text and a technological commentary.

33 Bossert and Storck, *Das mittelalterliche Hausbuch.*

34. Examples of his authorship include Cod. germ. 734, Bayerische Staatsbibliothek, Munich, fols. 60v–71r, of which fol. 60v is signed "Johann Formschneider, buchsen meister un[d] gutter aben teurer"; MS 1949/258, a group of similar drawings found in fragments, in the Bibliothek des Deutschen Museums, Munich; and HS 719 (formerly Kr. 300), Germanisches Nationalmuseum, Nuremberg, among others. See esp. Hall, *Technological Illustrations,* 21–22, 127–29, for the items listed above

35. The version of Mercz's treatise that I have used is Cod germ. 599, fols. 66r–101v, in the Bayerische Staatsbibliotek, Munich. For a description, see Schneider, *Die deutschen Handschriften,* 225–26. The location of the copy mentioned in earlier sources—Cod. ms. 3, 1471, in the private Liechtenstein collection, Vienna—has been unknown since the end of World War II. For Mercz, see Gessler, "Merz (Mercz), Martin"; Sarton, *Introduction to the History of Science,* 1553, and Schmidtchen, *Bombarden, Befestigungen, Büchsenmeister* 153, 182–83. For Friedrich of Siegreiche, see Gruneisen, "Friedrich I. der Siegreiche"

36 Mönch's treatise is Cod. palat. germ 126, in the Universitätsbibliothek, Heidelberg. See Gessler, "Mönch, Philipp", Hall, *Technological Illustrations,* 23, 125; Jähns, *Geschichte der Kriegswissenschaften,* 1:271; and Wegener, *Beschreibendes Verzeichnis,* 99–101.

37 Freysleben was a master gunner at Innsbruck for the emperor Maximilian I between ca. 1493 and 1509. See esp. Egg, "From the Beginning," 26–31; Egg, *Der Tiroler Geschützguss,* 49–94; and Schmidtchen, *Bombarden, Befestigungen, Büchsenmeister,* 172–74.

38. For introductions to humanism, including extensive further bibliography, see esp. Rabil, *Renaissance Humanism;* and Kraye, *Cambridge Companion to Renaissance Humanism.*

39 For Alberti's life and work, see esp. Borsi, *Leon Battista Alberti;* Grayson and Argan, "Alberti, Leon Battista"; Gadol, *Leon Battista Alberti;* and Tavernor, *On Alberti.* For the Alberti family's exile, see esp. Baxendale, "Exile in Practice" For developments in the visual arts, see Hartt, *History of Italian Renaissance Art,* Edgerton, *Renaissance Rediscovery Of Linear Perspective,* and White, *Birth and Rebirth of Pictorial Space.*

40 For Alberti's own practice of painting and sculpture, see Alberti, *"On Painting" and "On Sculpture,"* 143–54. For his architectural works, see Borsi, *Leon Battista Alberti;* and Grayson and Argan, "Alberti, Leon Battista." See also Aiken, "Leon Battista Alberti's System", Smith, *Architecture in the Culture of Early Humanism,* 19–39 (for his relationship to Brunelleschi) and 98–129 (for Pienza), and for his involvement in Nicholas V's Rome, Westfall, *In This Most Perfect Paradise,* and an important article that both extends and revises Westfall's thesis, Tafuri, "Cives esse non licere"

41. For the Latin text and English translation, see Alberti, "*On Painting*" *and* "*On Sculpture*," 31–116. For the quotation, see Alberti, *On Painting* (trans. Grayson), 12, 37. For the Italian text printed next to the Latin, see Alberti, *De pictura*. The Italian text, which differs slightly from the Latin, has also been translated: see Alberti, *On Painting* (trans. Spencer), 18–20, where Spencer discusses Alberti's use of Cicero's *De amicitia* 5.19.

42. Alberti, *On Painting* (trans. Grayson), quotation on 87. For a comparison of the *De pictura* with Quintillian's *Institutio oratoria*, see Wright, "Alberti's *De pictura*", see also Jarzombek, "Structural Problematic." On *istoria* see esp. Greenstein, "Alberti on Historia."

43. Alberti, *On Painting* (trans. Grayson), 34–35; Smith, *Architecture in the Culture of Early Humanism*, 19–39.

44. Alberti may have intended to dedicate a copy of the work to Federico da Montefeltro, military captain and lord of Urbino. The printed version of the treatise, which appeared in 1485, after Alberti's death, contains a dedication by the humanist Poliziano to Piero di Medici of Florence. See Alberti, *L'architettura (De re aedificatoria)*; and Alberti, *On the Art of Building*. For an introduction to the *De re aedificatoria*, see esp. Wittkower, *Architectural Principles in the Age of Humanism*, 1–56. See also Smith, *Architecture in the Culture of Early Humanism*, 19–39; and Oppel, "Priority of the Architect," quotations on 251, 262.

45. Alberti, *On the Art of Building*, 3–4; and Alberti, *L'architettura (De re aedificatoria)*, 1.6–13.

46. Alberti, *On the Art of Building*, 136 and n. 43 for *De navis*, 135 for war machines.

47. Ibid., 4–5.

48. Ibid., 37. For an illuminating discussion of Alberti's civic and moral positions as well as his debt to Cicero's *De officiis*, see Onians, *Bearers of Meaning*, 147–57.

49. Alberti, *On the Art of Building*, 7.

50. Ibid., 7, 155. See also Krautheimer, "Alberti and Vitruvius."

51. Valturio, *Elenchus et index rerum militarium*. For a partial facsimile reproduction of one manuscript at the Archivio Storico AMMA at Turin, Lat., cart., sec. XV, 142 cc.—and useful articles concerning the illustrations, see Bassignana, *Le macchine di Valturio*, which includes a list of manuscript copies and further bibliography. See also Galluzzi, *Prima di Leonardo*, 199–201; and Edgerton, *Heritage of Giotto's Geometry*, 145–46. For Sigismund Malatesta, see esp. Jones, *Malatesta of Rimini*, 176–239, and Tabanelli, *Sigismondo Pandolfo Malatesta*. For his building projects, see Pietramellara and Turchini, *Castel Sismondo*.

52. The fundamental study of the building is Ricci, *Il Tempio Malatestiano*. A detailed reevaluation of the building history is Hope, "Early History of the Tempio Malatestiano," 58–59 for the papal bulls, 91 for Piero della Francesca, 86–87 and 118–19 for Roberto Valturio, 95 ff. for Alberti, and 59–62, 66–68, 94–96, and 151–54, for Sigismund's personal involvement in the project. For Piero della Francesca's work in Rimini, see Lavin, *Piero della Francesca a Rimini*.

53. See Valturio, *Elenchus et index rerum militarium*, n.p.; and see Bassignana, *Le macchine di Valturio*, 171 for a list of manuscripts, 172 for printed editions.

54. The printed edition contains eighty-two woodcuts of military machines and de-

vices, possibly made by Matteo de' Pasti, a sculptor in the court of Sigismund. The illustrations indicate Valturio's acquaintance with Conrad Kyeser's *Bellifortis* For Valturio's life, see Klemm, "Valturio, Roberto", and see Bassignana, *Le macchine di Valturio,* 13–16, for Matteo de' Pasti's possible role.

55. Valturio, *Elenchus et index rerum militarium,* "Ad magnanimum et illustrem Heroa sigismundu[m] pandulfum," n.p "veluti sues in luto grunientes"; "veluti apum fuci et improbissimi Sicopha[n]te", "corpore p[ro]pter umbram molli nullo unq[ue] vulnere admisso", "q[ui] et castris nutritus sit. Et exercitus maxi[m]os duxerit se[m]per invictus"; "difficiles monstruosi· lucifugi", "multa sibi arroga[n]tes", "inepte cornicantes."

56. Ibid · "no[n] tanq[ue] imitatorem verum etiam tanq[ue] defloratorem [et] veterum co[m]pillatorem publicumq[u]e furem", "Comicorum summus"; "merito non sua sed aliena pro suis edentem"; "animat[a]e divin[a]eq[u]e potius eloque[n]ti[a]e"; "i[n] tollera[n]da superbia"; "totis voluminibus lacerase traditur: q[ue] i[n] philosophia paululu[m] a se suisq[u]e insanis opinionibus discordaret", "zeno . . maledicus."

57. Ibid : "Non mirum tibi Sigismunde nec cuipiam videri debet. si contra me pusillum homine[m] et indoctum· scholastici quidam et circunforan. rabul[a]e latrent atq[u]e deseviant"; "iurgiis"; "omnino nova inquisitione", "ut non erudiendi ostentandiq[u]e novi aliquid"; "de perdita . . . clarorum virorum monumenta", "maior[um] demonstratione vestigiorum "

58 Valturio, *Elenchus et index rerum militarium* In the copy I used (the first printed edition, published in Verona in 1472, now no. 218 in the Rosenwald Collection, Library of Congress), the index and lists of ancient sources are bound at the end of the text Some later printed editions, such as the 1532 edition, published in Paris by Christian Wechelus, have omitted this section altogether.

59. Ibid , "Ad magnanimum et illustrem Heroa sigismundu[m] pandulfum," n.p.. "concilium caelestium terrestriumq[u]e insedibus i[m]mortalitatis"; "litterar[um] custodiae."

60 Ibid., bk. 1, ch. 3· "Non vulgarem, non crassam et barbaram", "sed pr[a]eclarum et accuratam illam cum rerum multarum scientia coniunctam"; "italic[a]e magnificenti[a]e vera admiratio "

61. Jones, *Malatesta of Rimini,* 176–79, which suggests that Sigismund's bad reputation was due to the success of hostile testimony rather than to his character per se.

62. Cennini, *Craftsman's Handbook*; and see Kemp, *Behind the Picture,* 84–90.

63. The best introduction to Ghiberti's life and works is Krautheimer with Krautheimer-Hess, *Lorenzo Ghiberti.* The standard edition of the *Commentarii* is Ghiberti, *Lorenzo Ghibertis Denkwürdigkeiten;* see 1.3–4 for the excerpt from Athenaeus A detailed commentary and German translation of Ghiberti's third commentary is Ghiberti, *Der dritte Kommentar Lorenzo Ghibertis.*

64. Ghiberti, *Lorenzo Ghibertis Denkwurdigkeiten,* includes a transcription of the sole existing manuscript—II, I,333 (formerly Magl. cl. XVII.33), at the Biblioteca Nazionale Centrale, Florence—and an extensive commentary. The single manuscript copy of Ghiberti's treatise, which is imperfect and is not Ghiberti's autograph, dates from about the mid-fifteenth century. For an Italian edition in which Ghiberti's often difficult Italian

has been modernized, see Ghiberti, *I commentari* (ed. Morisani). See Krautheimer and Krautheimer-Hesse, *Lorenzo Ghiberti*, 306–14, and for the edition of Athenaeus, 308.

65. Ghiberti, *Lorenzo Ghibertis Denkwürdigkeiten*, 1·4 (*I commentarii* 1 2): "L'iscultura et pictura è scientia di più discipline et di varij amaestramenti ornata"; "con certa meditatione"; "per materia et ragionamenti "

66. Ibid., 1:4–6 (*I commentarii* 1.2): "per proportione d'astutia et di ragione"; "con magno animo"; "non sia arrogante"; "agevole et humile et fedele et sança avaritia " Ghiberti is paraphrasing Vitruvius, *De architectura* 1.1.2.

67. See ibid., 1.8–9 (*I commentarii* 1.3), for Ghiberti's paraphrases of Vitruvius, *De architectura* 7.pref.1 and 3.pref.1, and 33–51 for the second commentary. The second commentary has been translated by Holt, *Documentary History of Art*, 1:152–67.

68. For Francesco Sforza and Milan under his rule, see esp. Catalano, *Francesco Sforza;* and Chittolini, *Gli Sforza.* For Filarete's life, see Romanini, "Averlino (Averulino), Antonio, detto Filarete." For his treatise, see Filarete, *Trattato di architettura;* and an English translation, Filarete, *Filarete's Treatise on Architecture.* Filarete dedicated one copy to Francesco Sforza and a later copy to Piero di Medici of Florence.

69 Filarete, *Filarete's Treatise on Architecture*, 1·48–65.

70. Ibid., 1:79, 232–33 (for the furnaces), 258–59 (plaster), 277 (his treatise on engines), 317 (where he notes that two books of his treatise on agriculture are complete), and see bks. 21, 22, and 23 of his treatise for his discussions of drawing, perspective, and painting. See also Spencer, "Filarete's Description "

71. Filarete, *Filarete's Treatise on Architecture*, 1.228.

72. Ibid., 1·231, 254–55

73. Kemp, *Science of Art*, 27–35, quotation on 27. For Piero's life and works, see esp. Battisti, *Piero della Francesca*, which includes a summary of documents and sources (2:213–46); Bertelli, *Piero della Francesca*, esp. 7–50; Poggetto, *Piero e Urbino;* and Lavin, *Piero della Francesca and His Legacy;* and for an astute study of the treatises, Davis, *Piero della Francesca's Mathematical Treatises.* For the treatises, see Piero della Francesca, *Trattato d'abaco;* Piero della Francesca, *De prospectiva pingendi;* and Mancini, "L'opera 'De corporibus regularibus.'"

74. Bertelli, *Piero della Francesca*, 7–50, 170 (chronology).

75. For Federico da Montefeltro and his court at Urbino, see esp. Baiardi, Chittolini, and Floriani, *Federico di Montefeltro;* Tommasoli, *La vita di Federico da Montefeltro;* and Rotondi, *Ducal Palace of Urbino* For the frieze on the facade of the ducal palace, see esp. Pezzini, *Il fregio;* and Manno, "Architettura e arti meccaniche nel fregio." The copy of Valturio's treatise that was in Federico's library is now MS Urb. Lat. 281, Biblioteca Apostolica Vaticana, Vatican City.

76. Piero della Francesca, *De prospectiva pingendi*, 64–66: "il vedere, cioè l'ochio"; "la forma de la cosa veduta"; "la distantia da l'ochio a la cosa veduta"; "le linee che se partano da l'estremità de la cosa e vanno a l'ochio"; "il termine che è intra l'ochio e la cosa veduta dove se intende ponere le cose"; "tractare de prospectiva con dimostrationi le quali voglio sieno comprese da l'ochio"; "una cosa tanto picholina quanto è posibile ad ochio comprendere."

77. Ibid., 66. "omne quantità se rapresenta socto angolo nell'ochio." See also Field, "Mathematics and the Craft of Painting," 73–95; Field, "Piero della Francesca's Treatment"; and Kemp, *Science of Art*, 28, 30–35.

78. See Mancini, "L'opera 'De corporibus regularibus,'" 449 n. 2, which notes that the prefatory letters are missing from extant copies of the *De perspectiva pingendi*. The treatise was translated into Latin by Matteo da Borgo, who probably also translated the *Libellus*, on the regular solids. See Davis, *Piero della Francesca's Mathematical Treatises*, 54.

79. For Francesco di Giorgio, see esp. Dechert, "Military Architecture of Francesco di Giorgio", Fiore and Tafuri, *Francesco di Giorgio architetto;* Toledano, *Francesco di Giorgio Martini;* and Scaglia, *Francesco di Giorgio*. The dates and chronology of Francesco's notebooks and treatises are subjects of ongoing scholarly debate. See esp. Betts, "On the Chronology of Francesco di Giorgio's Treatises"; and Kolb, "Francesco di Giorgio Material."

80. Scaglia, *Francesco di Giorgio*, 43–50 for the manuscript, Codex 197 B 21 [MS Harley 3281], in the British Library, London, and 101–4, which describes the copy of that manuscript that includes the dedication to Federico, Codex serie militare 383 (Cod. 14856, 14876, 14896 D[otazione] C[orona]), Biblioteca Reale, Turin.

81. Francesco di Giorgio, *Trattati di architettura ingegneria e arte militare*, 2.425, for Francesco's tribute to Federico.

82. See Millon, "Architectural Theory of Francesco di Giorgio."

83. Francesco di Giorgio, *Trattati di architettura ingegneria e arte militare*, 1:36: "prespicace e singulare ingegno e invenzione", "è solo una sottile immaginazione concetta in nella mente la quale in nell'op[e]ra si manifesta"; "d'ogni e ciascuna cosa non si può la ragione assegnare, perchè lo ingegno consiste più in nella mente e in nello intelletto dell'architettore che in iscrittura o disegno, e molte cose accade in fatto le quali l'architetto overo op[e]ratore mai pensò"; "pratico e sciente", "arroganti e presentuoisi"; "i quali nelli errori fondati sono" "per forza della lingnia [*sic*]."

84. Ibid., 1:75, 142: "molte cose all'animo dell'architetto paia facile"; "Io per me delle invenzioni che qui demostrate seranno, d'assai buona parte, in me non confidando, spirienza ho veduta."

85. Ibid., 2:307–9 (noxious winds), 309–10 (types of marble), 316 (best kind of lime), 358–59 (searching for water).

86. Ibid., 2:297: "tutto quello che in questa mia operetta si contiene"; "di mia invenzione", "di più autentici libri", "del mio debile ingegno invenzioni"; "piacere o vero utilità"; "saranno a ciascuno manifeste, le quali per molte età sono state occulte."

87. Ibid., 2:492–93: "di non volere manifestare alcuna mia macchina"; "con grande mia spesa di esperienzia e grave incomodo"; "impedimento dell'altre cure utili"; "la fatiga sprezzano della invenzione"; "delle fatighe aliene"; "questo vizio nelli tempi nostri abbonda."

88. See esp. Kemp, *Leonardo da Vinci;* Galluzzi, "Career of a Technologist"; and Pasquale, "Leonardo, Brunelleschi, and the Machinery of the Construction Site."

89. Galluzzi, "Career of a Technologist." For Leonardo's letter to Ludovico, see Leonardo da Vinci, *Notebooks*, 2:395–98 (no. 1340).

90. For the complex tradition of Leonardo's notebooks, see Galluzzi, "Career of a Technologist," 41–48; Marinoni, "I manoscritti di Leonardo da Vinci"; Pedretti, *Leonardo da*

Vinci on Painting, 252–59; and Zwijnenberg, *Writings and Drawings of Leonardo da Vinci*. For Leonardo and Francesco di Giorgio, see Scaglia, "Leonardo da Vinci e Francesco di Giorgio."

91. See Marinoni, "The Writer."

92. Leonardo da Vinci, *Madrid Codices*, 3:29–31.

93. Leonardo da Vinci, *Il Codice Atlantico di Leonardo da Vinci*, vol. 1, fol. 77v, Leonardo da Vinci, *Il Codice Atlantico della Biblioteca Ambrosiana di Milano*, 1:57, Pedretti, *Codex Atlanticus of Leonardo da Vinci*, pt. 1, 57, whose dating for the sheet I accept; Leonardo da Vinci, *Madrid Codices*, vol. 1, fols. 111v–112r, and 4:283–86.

94. For Leonardo's brilliantly innovative illustrative techniques, see Ackerman, "Involvement of Artists in Renaissance Science," esp. 102–11; Galluzzi, "Career of a Technologist," 96–109; Galluzzi, "Leonardo da Vinci"; and Kemp, *Leonardo da Vinci*, esp. 54–57. Francesco di Giorgio was an important predecessor in such illustration (see Kemp, "La diminutione di ciascun piano"). For the submarine, see Leonardo da Vinci, *Notebooks*, 2:274 (no. 1114).

95. Pedretti, *Leonardo da Vinci on Painting*, 9.

96. Ibid., esp. 10–15.

97. See esp. Marinoni, "The Writer."

98. See Reti, "Elements of Machines," esp. 272. For Leonardo's library see Leonardo, *Madrid Codices*, 3:91–108. For his difficulty with contemporary methods of printing, see Reti, "Leonardo da Vinci and the Graphic Arts." Leonardo describes his invention of relief etching in Codex Madrid II (see Leonardo, *Madrid Codices*, vol. 2, fol. 119r, and 5:255–56).

99. Davis, *Piero della Francesca's Mathematical Treatises*, 98–123.

Chapter 5
Secrecy and the Esoteric Traditions of the Renaissance

1. Overviews include Muller-Jahncke, "Von Ficino zu Agrippa"; Shumaker, *Occult Sciences in the Renaissance*; and Webster, *From Paracelsus to Newton*. For an astute reassessment, see Copenhaver, "Natural Magic, Hermetism, and Occultism."

2. See Newman, "Occult and the Manifest," which points to further bibliography; and Daston and Park, *Wonders and the Orders of Nature*, esp. 127–33. For occult qualities in Neoplatonic cosmology, see Ficino, *Théologie platonicienne*.

3. For early Western alchemy, see esp. Holmyard, *Alchemy*, 105–52; Halleux, *Les textes alchimiques*, esp. 43–45, for a discussion of the components of alchemy; Crisciani and Pereira, *L'arte del sole e della luna*, 3–105, and Pereira, *L'oro dei filosofi*, esp. 1–83, which includes an excellent discussion of the historiography of Western alchemy.

4. Crisciani and Pereira, *L'arte del sole e della luna*, 77–94; Halleux, *Les textes alchimiques*, 97–100.

5. From this point of view, it is a fundamental distortion of the early historiography of alchemy to see it as an early phase of the history of chemistry, a protochemistry that reached its full development in the "chemical revolution" of the eighteenth century. On this point, see esp. Crisciani and Pereira, *L'arte del sole e della luna*, 95–97.

6. See Newman, "Technology and Alchemical Debate," 427; and Newman, "*Summa Perfectionis*" *of Pseudo-Geber*, esp. 1–56

7. See Newman, "*Summa Perfectionis*" *of Pseudo-Geber*, 57–58 for a concise summary of the text, 249–785 for a Latin edition and English translation

8. Ibid., 57–108, quotation on 61, provides a complete account of the scholarly debates. For Berthelot's discussion, see Berthelot, *La chimie au moyen âge*, 1:336–50; and see Kraus, *Jābir Ibn Hayyān*.

9. Newman, "*Summa Perfectionis*" *of Pseudo-Geber*, 630–32, 784–85.

10. For the pseudo-Lullist alchemical corpus, see Pereira, *Alchemical Corpus*; and Pereira, "*Medicina* in the Alchemical Writings." For John of Rupescissa, see esp. Halleux, "Les ouvrages alchimiques de Jean de Rupescissa." For Arnald of Villanova, see Manselli et al., "2. A[rnald] v. Villanova"; and Benton, "Arnald of Villanova." For a succinct summary of the development of Western alchemy, including its acquisition of vitalism and prophesy, see Newman, *Gehennical Fire*, 1–114

11. See Obrist, *Les débuts de l'imagerie alchimique*, and Newman, *Gehennical Fire*, 115–35, on which my account is dependent. For *Decknamen*, see esp. Ruska and Wiedemann, "Beiträge zur Geschichte der Naturwissenschaften."

12. See esp. Pereira, *L'oro dei filosofi*, 1–83; and Crisciani and Pereira, *L'arte del sole e della luna*, 3–105. And see Moran, *Alchemical World of the German Court*.

13. For a brief discussion of medieval and Islamic Platonism, see Rees, "Platonism and the Platonic Tradition." See also Hankins, *Plato in the Italian Renaissance*.

14. Kristeller, *Marsilio Ficino and His Work*, 5–8 For Ficino's work of synthesizing Christian and Platonic thought, see esp. Allen, *Plato's Third Eye*. The foundational study of Ficino's thought is Kristeller, *Philosophy of Marsilio Ficino*. Valuable recent discussions include Bono, *Word of God and the Languages of Man*, 26–47; and Daston and Park, *Wonders and the Order of Nature*, esp. 144–45, 161–64. For Ficino's writings, see Ficino, *Opera omnia*, and Kristeller, *Supplementum Ficinianum*.

15. For Ficino's life and work, see esp Marcel, *Marsile Ficin*; Field, *Origins of the Platonic Academy*, esp 175–201; Garfagnini, *Marsilio Ficino e il ritorno di Platone*; and Hankins, *Plato in the Italian Renaissance*, 1:267–359. For Ficino's influence through personal contacts with visitors from northern Europe, see esp. Spitz, "*Theologia Platonica* in the Religious Thought"

16. My account depends on Hankins, *Plato in the Italian Renaissance*, 1:15–16. And see Garin, *L'età nuova*, 263–92.

17. Field, *Origins of the Platonic Academy*, esp 176–77, 198–201, Hankins, *Plato in the Italian Renaissance*, 17; Hankins, "Cosimo De' Medici and the 'Platonic Academy'", and Hankins, "Myth of the Platonic Academy of Florence."

18. Hankins, *Plato in the Italian Renaissance*, 17.

19. Kristeller, *Philosophy of Marsilio Ficino*, 15 (first quotation), 26–29, where Kristeller cites the *Theologia Platonica*, bk. 17, ch. 1

20. Ficino, *Théologie platonicienne*, bk. 17, chs. 1–4.

21. See Kristeller, *Philosophy of Marsilio Ficino*, 106–10, and Vasoli, "La *prisca theologia*." See also Ficino, *Théologie platonicienne*,, 11–16, for a succinct summary of this

complex treatise and Ficino's doctrine of the soul within it, and bk. 1, ch. 3, for Ficino's initial discussion of the rational soul; the quotation is on 137 (bk. 3, ch. 2), and see bk. 6 for a lengthy description of the nature of the soul and the reasons for its immortality. Trinkaus, *In Our Image and Likeness*, 2:461–504, provides a cogent discussion of Ficino's doctrine of the soul in which he notes Ficino's delight in the human capacity for material production and the mechanical arts.

22. Ficino, *Three Books on Life*, 102–5 (dedication to Lorenzo), 164–65 (dedication to Valori), 236–39 (dedication to the king of Hungary). For a cogent discussion of the Neoplatonic influences on the *De vita triplici* that emphasizes Porphyry's *De sacrificio*, see Copenhaver, "Hermes Trismegistus, Proclus."

23. For Kaske's observation, see Ficino, *Three Books on Life*, 28. For a discussion of the Neoplatonic sources of Ficino's theory of magic in the *De vita triplici*, see esp. Zambelli, "Platone, Ficino e la magia"; Copenhaver, "Iamblichus, Synesius"; and Copenhaver, "Renaissance Magic and Neoplatonic Philosophy."

24. Ficino, *Three Books on Life*, 238–39.

25. Ibid., 242–43, 278–81 (for rings).

26. Ibid., 294–95, 298–99. See also Pingree, *Picatrix*.

27. Ficino, *Three Books on Life*, 397–99. For a discussion of good and bad magic in Ficino, see Walker, *Spiritual and Demonic Magic*, esp. 36–53.

28. For Pico, see esp. Trinkaus, "Giovanni Pico della Mirandola"; and Grafton, *Commerce with the Classics*, 93–134. For Ficino's influence on Lodovico Lazzarelli, see Kristeller, "Marsilio Ficino e Lodovico Lazzarelli." For Symphorien Champier, see Copenhaver, *Symphorien Champier*. For Reuchlin, see Spitz, "*Theologia Platonica* in the Religious Thought"; Zika, "Reuchlin's *De Verbo Mirifico*", and Scheible, "Johann Reuchlin."

29. For Agrippa's life, see Nauert, *Agrippa and the Crisis of Renaissance Thought*; Prost, *Les sciences et les arts occultes*; and Poel, *Cornelius Agrippa*, 15–49. For Agrippa's relationship to the religious movements of his time, see Zambelli, "Magic and Radical Reformation"; and Kuhlow, *Die Imitatio Christi*. For Ficino's influence on Agrippa's thought, see Walker, *Spiritual and Demonic Magic*, 90–96.

30. Agrippa, *De incertitudine et vanitate scientiarum*, Agrippa, *Of the Vanitie and Uncertaintie of Artes and Sciences*. Essential studies include Poel, *Cornelius Agrippa*, esp. 116–52; and Keefer, "Agrippa's Dilemma."

31 Poel, *Cornelius Agrippa*, esp. 2–4 and 82, on Agrippa's reputation.

32 Ibid., 50

33. Agrippa, *De occulta philosophia libri tres* (ed. Compagni), 1–53, which includes a valuable discussion of Agrippa's sources. Translations from the Latin are my own For a reprint edition of the first printed edition of 1533, a photographic reproduction of the 1510 manuscript, and an extended commentary, see Agrippa, *De occulta philosophia* (ed. Nowotny). For Agrippa's thought in the context of medieval and fifteenth-century attitudes toward witchcraft, demonology, and magic, see Müller-Jahncke, "Agrippa von Nettesheim"; Zambelli, "Scholastic and Humanist Views"; and Zambelli, "Cornelius Agrippa." For the influence of Reuchlin, see Zika, "Reuchlin's *De Verbo Mirifico*."

34. Agrippa, *De occulta philosophia libri tres* (ed. Compagni), "Ad Lectorem," 65–66: "obliquae opinionis mente languidi, multi etiam maligni et ingenium nostrum ingrati"; "temeraria sua ignorantia"; "in deteriorem partem"; "ad avertendos malos eventus, ad destruendum maleficia, ad curandos morbos, ad exterminanda phantasmata, ad conservandam vitae, honoris, fortunae dexteritatem", "sine Dei offensa, sine religionis iniuria."

35. Ibid., 66.

36. Ibid., 66–67: "viro arcanarum rerum admodum industrio", "interceptum opus, priusquam illi summam manum impossuissem"; "truncum et impolitum"; "impatientius nescio, an impudentius"; "minus periculi fore si libri isti paulo castigatiores mea manu prodirent, quam si laceri per incondita fragmenta invulgarentur per manus aliorum." For Trithemius, see esp. Müller-Jahncke, "Johannes Trithemius"; and Brann, *Abbot Trithemius*.

37. See Agrippa, *De occulta philosophia libri tres* (ed. Compagni), 68–75, for Agrippa's letter to Trithemius and the latter's reply. Both letters were published in the 1533 edition. For the quotation, see "Ioannes Tritemius . . . Agrippae ab Netteszheym salutem et charitatem" (72): "Unum hoc tamen te monemus custodire praeceptum, ut vulgaria vulgaribus, altiora vero et arcana altioribus atque secretis tantum communices amicis. da foenum bovi, saccarum psitaco tantum—intellige mentem, ne boum calcibus (ut plerisque contingit) subiiciaris." For Trithemius's *Steganographia* and accusations against it, see Zambelli, "Scholastic and Humanist Views," 133–36, and Shumaker, *Renaissance Curiosa*, 91–131.

38. See Agrippa, *De occulta philosophia libri tres* (ed. Compagni), 7–10, for a succinct summary of these events, upon which this account depends.

39. See also Nauert, *Agrippa and the Crisis of Renaissance Thought*, esp. 104–15; and Poel, *Cornelius Agrippa*, 41–49. For the political and religious context of the Cologne printing of *De occulta philosophia* and the controversies surrounding it, see Zika, "Agrippa of Nettesheim and His Appeal." For Agrippa's patron, Hermann von Wied, see Decot, "Hermann von Wied."

40. Agrippa, *De occulta philosophia libri tres* (ed. Compagni) 1.1: "elementaris, coelestis, et intellectualis"; "quisque inferior a superiori regatur ac suarum virium suscipiat influxum ita ut ipse Archetypus et summus Opifex per angelos, coelos, stellas, elementa, animalia, plantas, metalla, lapides."

41. Ibid., 1.14 on world spirit, 1.15: "a vita spirituque mundi per ipsos stellarum radios proficiscuntur, ⟨quae a nobis non aliter quam experientia et coniecturis indagari possunt⟩", 1.22–1.32.

42. Ibid., 1.33: "totius universi completissima imago, in seipsa omnem coelestem continens harmoniam"; "omnium stellarum coelestiumque influxuum signacula characteresque"; "efficaciora"; "a coelesti natura minus sunt remota", "partim ratione, partim experientia"; "in naturae thesauris delitescunt occlusa."

43. Ibid., bk. 2, "Amplissimo Domino Principi illustrissime Hermanna ab Wyda. . .": "coelestis magiae mysteria intimamus, patefactis omnibus atque monstratis quae de iis perita rerum prodit antiquitas"; "foribus patefactis haec e vinculis emitterentur." For the dedicatory letter to bk. 3, "Amplissimo Domino, Principi illustrissimo Hermanno ab Wyda. . . ," see pp. 399–401: "sine fictione didici" (Agrippa cites from The Wisdom of

Solomon 7:13); "quae vetustatis squalore obsita et oblivionis caligine velut cimmeriis tenebris hucusque involuta iacuerunt ad lucem exponontes."

44. Ibid , 2 2–15, binary chart at 2 5.

45. Ibid., 2.20: "de rebus occultis futurisque"; "in illis numeris lateant occulta aliqua mysteria a paucis intellecta"; "non casu sed certa (licet nobis ignota) ratione."

46 Ibid., 2.60: "Coelestes animae virtutes suas corporibus coelestibus influunt, quae deinde illas huic sensibili mundo transmittunt."

47. Ibid., 3 2: "tam sacrum dogma intra secreta religiosi pectoris tui penetralia silentio tegito et constanti taciturnitate celato", "irreligiosae mentis"; "Dei et naturae sacramenta involvere et variis aenigmatibus obtegere", "Dicerem si dicere liceret, cognosceres si liceret audire; sed parem noxam contraherent aures et linguae temerariae curiositatis."

48. Ibid., 3.2–3: "publicis committere literis"; "Oportet igitur magicum operatorem"; "in omnia operatione effectum impedit atque disturbat."

49. Ibid., 3.65 "quaedam cum ordine, quaedam sine ordine scripta sunt, quaedam per fragmenta tradita sunt, quaedam etiam occultata et investigatione intelligentium relicta", "prudentes et intelligentes"; "pravos vero et incredulos."

50 Ibid., 3.65: "Vos igitur, doctrinae et sapientiae filii, perquirite in hoc libro colligendo nostram dispersam intentionem quam in diversis locis proposuimus et quod occultatum est a nobis in uno loco, manifestum fecimus illud in alio, ut sapientibus vobis patefiat", "arcana multis aenigmatibus abscondita."

51. See Zambelli, "Umanesimo magico-astrologico," esp 153–55. For the secret society, see Nauert, *Agrippa and the Crisis of Renaissance Thought*, 17–20; and Agrippa, *Epistolarum ad familiares* 1, no 11, Landulphus to Agrippa, Lyons, 4 February 1509, in *Opera*, vol. 2 (Nauert's translation on p. 18). For Agrippa's friendship with Brennonius, see Poel, *Cornelius Agrippa*, 27; and Agrippa, *Epistolarum ad familiares* 2, nos. 43–47, 49, 50–57, 59, 61, 3, nos 5–6, 8, 60–62; 4, nos. 20, 26, 27. For his oath of secrecy concerning alchemy, see Agrippa, *Of the Vanitie and Uncertaintie of Artes and Sciences*, 330. Muller-Jahncke, "Attitude of Agrippe von Nettesheim," stresses Agrippa's skepticism toward alchemy for gold production and his use of alchemical methods such as distillation and sublimation to prepare medicinals. But see Compagni, "Dispersa Intentio," which argues that alchemy is central to Agrippa's thought.

52. For Aquapendente, see Poel, *Cornelius Agrippa*, 42, Agrippa, *Three Books of Occult Philosophy*, 679–80, and for the Latin, Agrippa, *Epistolarum ad familiares* 5.14, Agrippa to Aurelius ab Aqua-pendente, Augustinian, 24 September 1527: "Qui sunt duces tui, quos sequeris?"; "mera aenigmata"; "si quis sine perito fidoq[ue] magistro, sola librorum lectione possit assequi, nisi fuerit divino numine illustratus, quod datur paucissimis", "Clave tamen operis solis amicissimis reservata "

53 Nauert, *Agrippa and the Crisis of Renaissance Thought*, 14 (military and diplomatic service for Maximilian), 16–17 (the military adventure in Spain and its pyrotechnic element), 98 (treatise on engines of war), 114 (treatise on mining and supervision of imperial mines)

54. Zika, "Agrippa of Nettesheim and His Appeal," 172.

55. For Paracelsus's life and work, see esp. Sudhoff, *Paracelsus;* Pagel, *Paracelsus;* and Pagel, "Paracelsus als 'Naturmystiker.'" Recent studies include Blumlein, *Naturerfahrung und Welterkenntnis;* Gause, *Paracelsus;* and Weeks, *Paracelsus.* For his relationship to medieval alchemy, see Ganzenmüller, "Paracelsus und die Alchemie." For Paracelsus's influence on later thinkers and on the development of Paracelsianism, see esp. Debus, *Chemical Philosophy.*

56 Pagel, "Paracelsus, Theophrastus Philippus Aureolus Bombastus von Hohenheim," 304–5. For his Basel sojourn, see also Sudhoff, *Paracelsus,* 24–49

57. Paracelsus, *Reply to Certain Calumniations,* 24–29

58 See ibid., 11, where Paracelsus mentions being hindered from publishing his writings For the modern edition of the writings, which when completed will comprise twenty-eight volumes, see Paracelsus, *Sämtliche Werke,* pt 1, *Medizinische, naturwissenschaftliche und philosophische Schriften,* and the as yet incomplete pt. 2, *Theologische und religionsphilosophische Schriften.* For Paracelsus's relationships with printers, see esp. Baron, "Paracelsus und sein Drucker."

59. Weeks, *Paracelsus,* 1–19. See also Webster, "Paracelsus on Natural and Popular Magic"; Webster also minimizes the erudite Neoplatonic tradition in Paracelsus's thought in favor of popular magic and religion.

60. See Pagel, *Paracelsus,* 50–202; Bianchi, "Visible and the Invisible"; Bianchi, *Signatura rerum,* 31–86; Goldammer, "Magie bei Paracelsus"; and Newman, "Occult and the Manifest," 185–90

61. Golinski, "Chemistry in the Scientific Revolution," esp 372–75 See also Eis, *Vor und nach Paracelsus,* 51–73.

62 Paracelsus, *Decem libri Archidoxis,* 93–94 "lieben filii"; "unser elend und verlasenheit"; "zu grossem reichtumb"; "mit vil liegens"; "an gewisser end und practiken", "dan mysterium naturae", "wie ein liecht in einer finsternus." Weeks, *Paracelsus,* 106–9, emphasizes the theological content of these passages.

63. Paracelsus, *Decem libri Archidoxis,* 95–96: "uns zu einem memorial"; "wir uns enthalten wollen und alein mit uns, den unsern reden, den selbigen verstendig gnug geschriben, und schreiben das nit in die commun der volkern"; "wir wollen unser sinn und gedanken, herz und gemut den surden nit zeigen noch geben, und beschliessen also mit einer guten mauren und mit einem schlussel"; "vor solchen idioten [nit] behüt wurde sein die dan aller künsten feind sind"; "von dem brauch der andern all"; "den gozen nit ein freudaffen einstossen, aber nichts desto minder den unsern gnug verstanden wird "

64. Ibid., 118–200. For detailed explications of Paracelsus's philosophico-medical system and his sources, see esp. Blümlein, *Naturerfahrung und Welterkenntnis,* Pagel, *Paracelsus,* and Weeks, *Paracelsus,* esp. 129–84.

65. See Paracelsus, *Decem libri Archidoxis,* 118 for the preliminary definition of *quintessence,* "die nature, kraft, tugent und arznei"; and 136. "sol sich des niemants verwundern unserer kurzen hant und federn aus ursachen"; "die arbeit, die dorauf geschehen wird und sol, zeiget unser schreiben grüntlich und klar an"; "in verdrossenheit tragen"; "das die arbeit und ubung solchs alles wol anzeigt "

66. Paracelsus, *Liber de longa vita:* "den gemeinen arzten verborgen und unwissend" (221–22); "die uns noch die natur nicht erkennen"; "wollen wir unser process zugeschriben haben und den selbigen genugsam entdekt" (245).

67. Paracelsus, *On the Miners' Sickness,* quotations on 77 and 79.

68. Ibid., 79–80.

69. Benzenhöffer, "Zum Brief des Johannes Oporinus," which includes a transcription (62–63) of the oldest and most recently discovered copy of the letter. Five copies of the letter, including this one but not the original, have been discovered to date. I have used the transcription published here as the basis for my discussion: "erat totis diebus ac noctibus dum ego ipsi . . . convixi ebrietati et crapulae deditus, ut vix unam aut alteram horam sobrium eum reperire licuerit"; "inter nobilies rusticos"; "tanquam alter Aesculapius"; "reversus domum dictare mihi aliquod Philosophiae solebat"; "ita sibi pulchre coherere videbantur, ut a maxime sobrio melius fieri potuisse non viderentur." See also Weeks, *Paracelsus,* 3–4.

70. Benzenhofer, "Zum Brief des Johannes Oporinus," 62–63: "Semper habebat officinam suam carbonariam paratam perpetuis ignibus"; "se vaticinari quaedam simulabat et arcanorum quorundam cognitionem prae se ferebat; ut clam aliquid"; "e mortuis vivos reddere posse"; "quemque scripturae nucleum recte eruisse, sed circa corticem et quasi membranam tantum haerere."

71. For Bruno's life, see esp. Aquilecchia, "Bruno, Giordano"; Yates, *Giordano Bruno and the Hermetic Tradition,* Gosselin, "Bruno's 'French Connection'"; and Gosselin, "Fra Giordano Bruno's Catholic Passion." On Bruno's inquisitional interrogation, see Firpo, *Il processo di Giordano Bruno.*

72. Yates, *Giordano Bruno and the Hermetic Tradition;* León-Jones, *Giordano Bruno.*

73. Bruno, *La Cena de le Ceneri / The Ash Wednesday Supper,* includes an insightful introduction, to which this paragraph is indebted.

74. See ibid., 126–27, for the description of the Lord's Supper.

75. Ibid., 85. For Bruno's cosmology, see esp. Michel, *Cosmology of Giordano Bruno;* and Granada, "Thomas Digges, Giordano Bruno."

76. Bruno, *La Cena de le Ceneri / The Ash Wednesday Supper,* 86–88.

77. Ibid., 89–90.

78. Ibid., 92 and 99–100.

79. Ibid., 96.

Chapter 6
Openness and Authorship I

1. Eisenstein, *Printing Press as an Agent of Change,* esp. 2:520–74.

2. Kellenbenz, *Rise of the European Economy,* esp. 79–88, 106–18; Nef, "Mining and Metallurgy"; Pounds, "Mining."

3. For mining, see esp. Suhling, *Aufschliessen, Gewinnen und Fördern;* and Wilsdorf and Quellmalz, *Bergwerke und Huttenanlagen.* For mining authorship, see Koch, *Geschichte und Entwicklung,* 8–59; and Long, "Openness of Knowledge."

4. See Darmstaedter, *Berg-, Probir- und Kunstbüchlein,* 13–24. For the *Bergbüchlein,* see Pieper, *Ulrich Rulein von Calw;* and for English translations and further discussion, Sisco and Smith, *Bergwerk- und Probierbüchlein,* quotation on 70.

5. Sisco and Smith, *Bergwerk- und Probierbüchlein,* 17–19.

6 See Darmstaedter, *Berg-, Probir- und Kunstbüchlein,* esp 25–36, and Sisco and Smith, *Bergwerk- und Probierbüchlein,* 157–78 for the editions, 179–90 for technical content. For an edition with the dedication to Knobloch, see *Probir buch / leyn tzu Gotes lob;* the quotation in the text reads, "auss erfarnheit der schrifft und selbst versuchung." For Duchess Elizabeth's mining activities, see Boyce, *Mines of the Upper Harz,* 20–22

7 Biringuccio, *De la pirotechnia;* Biringuccio, *Pirotechnia of Vannoccio Biringuccio* See also Brunello, "Vannoccio Biringuccio."

8. For Biringuccio's admiring reference to Pandolfo's iron plants, see Biringuccio, *Pirotechnia of Vannoccio Biringuccio,* 63 For biographical information, see esp. ibid., ix–x; Tucci, "Biringucci (Berniguccio), Vannoccio", and Biringuccio, *De la pirotechnia,* xxxv–lix

9 Tucci, "Biringucci (Berniguccio), Vannoccio"; Biringuccio, *De la pirotechnia,* xxxv–lix; Biringuccio, *Pirotechnia of Vannoccio Biringuccio,* ix–x.

10 The printer Curtio Navo, of Venice, dedicated the first edition "to the very magnificent Messer Bernadino di Moncelesi of Salo." For the quotations, see Biringuccio, *Pirotechnia of Vannoccio Biringuccio,* 329, 119.

11. Ibid , 49–52. Biringuccio appears to have been fighting not only avaricious and indolent rulers but also ancient prohibitions against mining. See esp. Merchant, *Death of Nature,* 29–41.

12. See Biringuccio, *Pirotechnia of Vannoccio Biringuccio,* 20–21 and, for another example in which Biringuccio cites courage and persistence in excavation after the discovery of gold by a washerwoman in Hungary, 33–34

13. Ibid , 28

14. Ibid., 35–43, quotation on 43 For further discussion of Biringuccio's anti-alchemical views, see Rossi, *Philosophy, Technology, and the Arts,* 43–46.

15. Biringuccio, *Pirotechnia of Vannoccio Biringuccio,* 41.

16. Ibid , 241.

17. Ibid , 323 (metal melting), 364 (goldsmith techniques), 371 (ironwork), 373 (tarsia work), 384–85 (amalgamation).

18. See Winkelmann, *Schwazer Bergbuch,* v–viii, for a useful introduction. See also Berninger, *Das Buch vom Bergbau;* and Kirnbauer, *400 Jahre Schwazer Bergbuch.*

19. Egg, "Ludwig Lassl and Jorg Kolber."

20 Ibid.

21 Winkelmann, *Schwazer Bergbuch,* 10–12.

22. The influence of Agricola's family connections on his writing is pointed out in Stimmel, "Die Familie Schutz," which suggests (377) that Agricola probably described his father-in-law's copper smelter in detail in *De re metallica,* bk 11. For Agricola's life, see esp. Wilsdorf, *Georg Agricola und sein Zeit;* for his family background, see 82–98. For *De re metallica,* see Agricola, *De re metallica;* and Agricola, *De re metallica libri XII.*

23. Wilsdorf, *Georg Agricola und sein Zeit*, esp. 99–275; Agricola, *De re metallica*, vi–xii; Suhling, "Giorgius Agricola und der Bergbau," esp. 157–60.

24. The edition I have used is Agricola, *Georgii Agri- / COLAE MEDICI* / BERMANNUS. I have also consulted Agricola, *Bermannus oder über den Bergbau*, 295 and 312, for the careers of Heinrich von Könneritz and Plateanus's and those men's relationship to Agricola; and Agricola, *Bermannus (Le mineur)*. For Plateanus's role in securing the support of Erasmus, see Wilsdorf, *Georg Agricola*, 184–88; see also Kaemmel, "Plateanus, Petrus P."

25 Ibid., "Evolvi charissimi iuvenes . . . ," by Erasmus, 3 ("valles illas & colles, & fodinas & machinas"), and "Nobili et clarissimo viro Henrico A. Conritz . . . ," by Petrus Plateanus, 5–6 ("quam illi, qui vel arteis vel naturae arcana, per se aliosque inventa, literis ad posteritate[m] transmittunt"; "ad abstrusissima quaeque natura[e] penetrare"). Erasmus's comment makes one wonder whether he actually read the work since it contains no description of a machine.

26. Ibid., 13–14: "gens omnium doctissima"; "turpe nobis sit res nostras per socordiam & ignaviam nostram etiam nunc tenebris quasi obrutas esse & sua luce carere."

27. Ibid., 100· "ea quae magno labore invenit aliis facillime & diligentissime explanat, ac minime qui non paucis mos est pessimus, invidentia quadam ta[n]quam mysteria & arcana celat."

28. Ibid. 16–17, 27–28. "quae certa erant, incerta"; "bene sperant omnes & foeliciter saepius p[rae]cedit, nemo vero animo qui abiecto & timido fuit, unqua[m] re[m] fecit, aut etia[m] faciet."

29. Agricola, *De Natura Fossilium*, 1–2 Agricola follows Pliny, *Naturalis historia*, pref.20–24 and bk. 1. For Agricola as a humanist, see Hannaway, "Georgius Agricola as Humanist"; and Suhling, "Georgius Agricola und der Bergbau."

30. Agricola, *De re metallica*, 1–24.

31 Ibid , xxvi–xxix.

32. Beierlein, *Lazarus Ercker*, Hubicki, "Ercker (also Erckner or Erckel), Lazarus"

33 Ercker, *Drei Schriften*, 9–214.

34. Bornhardt, *Geschichte des Rammelsberger Bergbaues*, 147–54; Boyce, *Mines of the Upper Harz*, 23–65; Henschke, *Landesherrschaft und Bergbauwirtschaft*, 24–26 and passim (see "Personenregister," s.v. "Heinrich der Jungere, Herzog von Braunschweig-Wolfenbüttel"); Schmidt, "Heinrich der Jungere"

35. Beierlein, *Lazarus Ercker*, 19–24. For an introduction to the *Münzbuch* and a transcription of the text, see Ercker, *Drei Schriften*, 267–326. For Julius's mining activities, see Kraschewski, *Wirtschaftspolitik*, 151–65 For Julius's *Instrumentenbuch*, see Spies, ed. *Technik der Steingewinnung*; Spies, "Werkzeuge, Geräte und Maschinen"; and Moran, "German Prince-Practitioners," 261–62.

36. Ercker, *Drei Schriften*, 284

37. Ibid., 269 and 284 (quotation): "diese meine arbeit nicht vor Jeden komen lassen, uff das es eine schöne Kunst, wie bieshero bleibe."

38. Beierlein, *Lazarus Ercker*, 24–34; Hubicki, "Ercker (also Erckner or Erckel), Lazarus"

39. Ercker, *Beschreibung der allervornehmsten mineralischen Erze;* Ercker, *Treatise on Ores and Assaying,* quotation on 3–4

40 For the great influence of Ercker's treatise, see Armstrong and Lukens, "Lazarus Ercker and His 'Probierbuch.'" For Ercker's career in Bohemia, see Beierlein, *Lazarus Ercker,* 32–55, and for editions and translations of the work, 68–97.

41. For a general discussion of the decline of mining and the confusion and widespread fraud in the mints, see Janssen, *History of the German People,* 70–106 Miskimin, *Economy of Later Renaissance Europe,* 35–43, discusses the decline of mining in relation to price inflation. Production statistics documenting the decline are occasionally available; see, e.g , Westermann, *Das Eislebener Garkupfer,* 313–15. For the assaying books see Schreittmann, *Probierbuchlin,* Fachs, *Probier Büchlein,* and Zimmermann, *Probierbüch;* and for a more detailed discussion, Long, "Openness of Knowledge," 341–46.

42. Hall, *Weapons and Warfare,* esp. 90–95, 105–33, and 157–200.

43 See esp. Argiolas, *Armi ed eserciti;* Hale, *War and Society;* Hale, "Early Development of the Bastion"; Hale, *Renaissance Fortification;* Hall, *Weapons and Warfare,* esp 158–64, Parker, *Military Revolution;* and Tallett, *War and Society.*

44 Hale, "Printing and Military Culture of Renaissance Venice " For treatises published throughout Europe, see also Cockle, *Bibliography of Military Books;* Jahns, *Geschichte der Kriegswissenschaften,* 1:445–865; and Pollak, *Military Architecture.*

45 For an introduction to Florentine politics in the late fifteenth century, see Butters, *Governors and Government,* 1–46 For Machiavelli and republicanism, see esp. the articles in Bock, Skinner, and Viroli, *Machiavelli and Republicanism* A convenient summary of Machiavelli's activities pertaining to the army of the republic is Machiavelli, *The Art of War,* esp. ix–xviii.

46 See esp. Butters, *Governors and Government,* 47–165; Gilbert, *Machiavelli and Guicciardini;* Rubinstein, "Beginnings of Niccolò Machiavelli's Career", Rubinstein, "Machiavelli and the World of Florentine Politics"; and Stephens, *Fall of the Florentine Republic* For Machiavelli's military ideas, see Mallett, "Theory and Practice of Warfare", and Vismara, "Il pensiero militare "

47. For Machiavelli's military responsibilities, see Machiavelli, *Art of War,* xiv–xv. See also Butters, *Governors and Government,* 104–6, 111–39; Gilbert, "Bernardo Rucellai and the Orti Oricellari"; Gilbert, "Machiavelli"; and Anglo, "Machiavelli as a Military Authority." For Machiavelli's ancient sources, see Burd, "Le fonti letterarie di Machiavelli nell''Arte della guerra.'"

48 Colish, "Machiavelli's *Art of War.*" A different view of Machiavelli's use of Colonna is expressed by Verrier, "Machiavelli e Fabrizio Colonna."

49. See Cockle, *Bibliography of Military Books,* 197 (no 765) I have used the first Venetian edition, Della Valle, *Vallo libro continente appertenentie ad capitanii;* this edition measures six inches by four, the perfect size for a uniform pocket. For what is known of Della Valle's life, see Muccillo, "Della Valle, Battista (Giovanni Battista) " For mercenary armies, see Mallett, *Mercenaries and Their Masters.* For women in early modern armies, see Hacker, "Women and Military Institutions in Early Modern Europe."

50. Della Valle, *Vallo libro continente appertenentie ad capitanii*, "ALO EXCELLENTISSIMO ET MOLto Strenuo cavaliero S. He[n]rico Pandone, Conte de Venafra," fol. 1r–v: "le sententie, ragione, & precepti militanti"; "rude ingegno"; "excellentia . . . immaginati"; "referiti & exquisitamente imparati", "quelli precepti curiosamente, & attentamente ausultando in la mia debil mente & memoria, como ad uno duro marmoro imprimeva"; "con experientia, & longo exercitio."

51. Ibid., fol. 2v. "solamente per authorita, & imitatione de altri authori, & non per propria exercitatione"; "Ma io elquale da mei teneri, & giovenil anni in gli ecercitii de larme me son io exercitato"; "quello che per longa experientia ho experto, & provato con continue fatiche, sudori, & pericoli"; "con basso, inculto & trivial parlar, & ad tutthomo cognito."

52. Ibid., fol. 3r–v: "fulminante gelosia."

53. Ibid., fols. 4r–19r. For Battista's advocacy of earthworks, see Croix, "Literature on Fortification," 38.

54. Ibid., fols. 20r–29r (bk. 2), 29v–53r (bk. 3), and 53v–72r (bk. 4). Della Valle acquired his material on dueling from the 1471 treatise on the subject by Paris de Puteo. See Cockle, *Bibliography of Military Books*, nos. 500, 865.

55. For Tartaglia's life, see Favaro, *Per la biografia di Niccolò Tartaglia;* Gabrieli, *Nicolò Tartaglia;* and Masotti, "Tartaglia (also Tartalea or Tartaia), Niccolò." Recent studies include Arend, *Die Mechanik des Niccolò Tartaglia*, Cuomo, "Shooting by the Book"; and Voss, "Between the Cannon and the Book," 358–428.

56. Tartaglia, *Nova scientia*. For a partial English translation, see Drake and Drabkin, *Mechanics in Sixteenth-Century Italy*, quotation on 63. See also Cuomo, "Shooting by the Book," 156 and 166–67; and Voss, "Between the Cannon and the Book," esp. 358–78.

57. Tartaglia, *Nova scientia*, in Drake and Drabkin, *Mechanics in Sixteenth-Century Italy*, 63–69. For Francesco Maria della Rovere in the Veneto, see esp. Concina, *La macchina territoriale*, 83–108.

58. See Arend, *Die Mechanik des Niccolò Tartaglia;* Freguglia, "Niccolò Tartaglia"; and Settle, "Tartaglia Ricci Problem." See also Voss, *Between the Cannon and the Book*, 393–428, for an insightful analysis of Tartaglia's position on the borderline between the cultures of Aristotelian physics, practical mathematics, and practical gunnery. And for Tartaglia's mathematics, see also Bortolotti, "I contributi del Tartaglia"; and Pace, *Le matematiche e il mondo*, 242–60.

59. Tartaglia, *Quesiti et inventioni diverse;* Drake and Drabkin, *Mechanics in Sixteenth-Century Italy*, 98–100.

60. Drake and Drabkin, *Mechanics in Sixteenth-Century Italy*, 100.

61. Tartaglia, *Quesiti et inventioni diverse;* Drake and Drabkin, *Mechanics in Sixteenth-Century Italy*, esp. 99, which includes a convenient summary of the treatise, of which Drake and Drabkin have only translated small parts of bks. 1 and 7, bk. 8.

62. For a thorough introduction to the dispute and a facsimile of the broadsides that were exchanged, see Ferrari and Tartaglia, *Cartelli di sfida matematica*. A narrative account of the dispute, with English translations of some of Tartaglia's account, unfortunately without source citations, is Øre, *Cardano*. Tartaglia's account of his discovery of

the solution and Cardano's long and ultimately successful attempt to get it from him are found in bk. 9 of Tartaglia, *Quesiti et inventioni diverse*, which includes the reproduction (apparently) of letters exchanged. For a discussion of the mathematics involved, see Boyer, *History of Mathematics*, 282–87. For Cardano, see Siraisi, *Clock and the Mirror*; and Grafton, *Cardano's Cosmos*.

63. See Tartaglia, *Quesiti et inventioni diverse*, fols. 107r–108v (quesito 25), for the 1530 episode.

64. Ibid., fols. 114v–115v (quesito 31); Øre, *Cardano*, 65–67.

65. Tartaglia, *Quesiti et inventioni diverse*, fols. 117r–121r (quesito 32), quotation on 119r; Øre, *Cardano*, 67–74.

66. Tartaglia, *Quesiti et inventioni diverse*, fols. 118r–122v (quesito 33), quotation on 119r. For a description of Tartaglia's visit to Cardano and his revelation of the solution, see ibid., fols. 123v–124v (quesito 34); and Øre, *Cardano*, 74–77. For Tartaglia's instruments, see Drake, "Tartaglia's Squadra and Galileo's Compasso."

67. Tartaglia, *Quesiti et inventioni diverse*, fols. 127v–128r (quesito 40), reproduces Cardano's final letter. Cardano and Ferrari's trip to Bologna in 1542 is discussed in Ferrari's second cartello; see Ferrari and Tartaglia, *Cartelli di sfida matematica*, 25–35.

68. See Ferrari and Tartaglia, *Cartelli di sfida matematica*, lxiii–lxxxix, for biographical and bibliographical information on the recipients of Farrari's cartelli; and Biagioli, "Social Status of Italian Mathematicians," for the role of social status in the dispute.

69. Ferrari and Tartaglia, *Cartelli di sfida matematica*, 25–35 (Ferrari's second cartello); Øre, *Cardano*, 93–96.

70. For the disputation in Milan, see esp. Ferrari and Tartaglia, *Cartelli di sfida matematica*, xxxiv–xl, which cites the scattered sources for information about the event; and Øre, *Cardano*, 99–107.

71. Biagioli, "Social Status of Italian Mathematicians."

72. For treatises on fortification, see Croix, "Literature on Fortification"; and Wilkinson, "Renaissance Treatises on Military Architecture." For the Venetian and Florentine contexts, see also Concina, *La macchina territoriale*; and Lamberini, *Il principe difeso*.

73. For the 1556 edition and Zanchi's biography, see Cockle, *Bibliography of Military Books*, 198 (no. 767); Pollak, *Military Architecture*, 117 (no. 73); and Promis, "Giambattista Zanchi." For the quotations, see Zanchi, *Del modo di fortificar le città*, "Al serenissimo et invittiss. Re Massimiliano d'Austria, Re di Boemia," 4–6: "molti valorosi soldati, & essercitati, & giudiciosissimi Capitani"; "invidiosi, ch' altri nell' otio, & piacere apparandola quello di honor procacciar si potesse, che con grandissime fatiche, et lunghe esperienze essi di acquistar si ingegnarono"; "diverse pruove"; "come sopra fermissime basi numero di conclusioni fondare"; "alla sola essercitatione della militia, & non alla dottrina atte[n]deano"; "più nobile, et intendenti spiriti."

74. Zanchi, *Del modo di fortificar le città*, 10: "certo modo, & universal regola di talmente edificare, et fortificare le città."

75. Ibid., "Al Molto magifico, et eccellente signore, il S. Dottor Nicola Manuali, Girolamo Ruscelli," 59–63. Ruscelli wrote, among other things, a book of secrets (under the name Alessio Piemontese) and a treatise on gunnery. See Eamon, *Science and the Secrets*

of Nature, 139–51; and for the treatise on gunnery, Cockle, *Bibliography of Military Books*, 170 (no. 663).

76 Lanteri, *Due dialoghi*, "Allo Illustre, et molto Generoso Sig. Il sig. Cavalier Marc' Antonio Moro," n.p.: "mio basso & picciol dono", "mio debile ingegno"; "quanta sia l'affettiore ch'io porto & porterò sempre à V. S. alla quale mi raccamando, & bascio le virtuose mani." See also Vivenza, "Giacomo Lanteri da Paratico." For a description of the treatise, see Lamberini, *Il principe difeso*, 130–31; Pollak, *Military Architecture*, 58–59; and for Cattaneo, see ibid , 12–13, and Olivato, "Cattaneo (Cataneo), Girolamo "

77. Lanteri, *Due dialoghi*, "Al Benigni Lettori," n.p.: "di fare alcuna operatione, dalla quale il mondo pigliasse qualche giovamento"; "una delle piú necessarie cose del mondo", "dal furore de' nimici"; "ne i quali potra ogniuno imparare il modo di disegnare le pia[n]te, cosi delle città che si vogliono fortificare"; "perche tutti que' che scrivono deono procurare di scrivere cose certe, & non false "

78. Ibid., 2: "alcuna paroletta allegra, per passar l'otio, & il caldo"; "sendo l'amicitia di noi tre, forse (come io stimo) à null'altra inferiore", "il bellissimo disegno della città"; "sommo diletto", "sentirvi à proponere cose che disputate possino rendere honore, e utile insieme "

79 Ibid., 4–6: "la perfetta cognitione di cosi bella arte può più tosto essere perfetta, & chiara con lo studio che havete detto, che con la esperienza della guerra"; "questo virtuoso desiderio è ben degno dei nobilissimi animi vostri."

80. Ibid., 44–47· "che delle lettere parimente, & delle cose della guerra sa rendere buonissimo conto", "in tranquillo & quieto stato?" "à tutte l'altre arti cosi liberali come mechanice?"; "paia cosa mechanica lo essercitarla?"

81. ibid., 48–49. "vorreste voi forse, che i nobili essercitassero quest'arte manualmente?"; "tutti i virtuosì ad haver di questa (per via di scienza non di prattica) qualche cognitione"; "non solo fosse buono humanista, ma filosofo, medico, & astrologo"; "Bastami d'havervi provato, & proverò ancho ultimamente, che l'architettura può dopo l'agricotura ottenere il primo luogo."

82. Ibid., 62–77· "sempre si habbia rispetto, si alle circostanze de i siti, come alle qualità delle materie che si hanno à adoprare nelle fabriche "

83 Ibid., 88–95.

84. Lanteri, *Duo libri . del modo di fare le fortificationi*, "Allo Illustriss. et eccellentiss. Signore, il Signor Don Alfonso da Este, Prencipe di Ferrara," n.p.: "grande utilità"; "il mio ingegno, & le mie picciole forze"; "è da biasimare chi teme troppo le alte imprese come chi nulla le teme", "la bella & utilissima materia delle fortificationi moderne"; "ò per invidia, ò per malignità, biasimare le cose altrui"; "forma, & perfettione scrivendo "

85. Ibid., "Ai Lettori Giacomo Lanteri, Salute," n p., and 1–2: "vi siano d'utile, et di giovamente parimente", "che l'intendere l'ordine delle fortificationi"; "al publico, con qualche ricordo, & con qualche regola"; "quando che . . . mi sono pervenuti alle mani da quattro ò cinque fogli di carta scritti à giusa di sommario in questa materia da un valentissimo huomo di questa professione", "siano scritti male quanto sia possibile, si che à pena se ne può intendere il sentimento; mi hanno nondimeno certificato di molti dubij."

86. Ibid., 3–6. "la quale non si puo in vero perfettamente possedere, senza la Geometria"; "tutta l'Architettura nasce da fabrica, & discorso."

87. Ibid., 15–27.

88. Ibid., 28–54, quotation on 44: "non sia arenosa, sassosa, ò sgrettolosa", "che à guisa di pasta non si maneggi bene."

89. Ibid., 55–58, 95. "sopra tutto si doverà studiare di haver de'migliori, è piu sofficienti muratori, che haver si potranno studiando sopra tutto di havergli."

90. Ibid., 87–88. "si havera sempre riguardo alle forze, & alla volontà del Principe", "spesse volte gli animi de i grandi restano offuscatisi", "non conoscono il proprio utile; ma spinti dal mal dire de' falsi corteggiani adulatori, & dall' avaritia molte volte, priveranno colui che gli haurà serviti con fede, & con lealta, non solo della gratia loro, ma anco bene spesso de i premij di molte fatiche ch' egli haurà in loro servigio fatte."

Chapter 7
Openness and Authorship II

1. Kristeller, "Modern System of the Arts"; and Rossi, *Dalle botteghe alle accademie.*

2. Goldthwaite, *Wealth and the Demand for Art,* argues that these activities stimulated the economy in that they involved large-scale building projects, with all their attendant subsidiary industries, such as brickmaking, carpentry, and stonemasonry. See also Jardine, *Worldly Goods.* Warnke, *Court Artist,* argues for the importance of the court over other settings.

3. For an introduction to the literature, see esp. Long, "Contribution of Architectural Writers"; Pagliara, "Vitruvio da testo a canone", and the classic study, Wittkower, *Architectural Principles in the Age of Humanism.*

4. See Shelby, trans. and ed., *Gothic Design Techniques,* 7–28, 61–79.

5. See ibid., 31–38, for a detailed account of these writings.

6. Ibid., 82–83.

7. Ibid.

8. Ibid., 61–79 for a discussion of Roriczer's geometry, 107–11 for the *Wimpergbüchlein* on how to construct gablets, 113–23 for *Geometria deutsch,* 28–31, 38–39, and 126–42 for Schmuttermayer, 40 for Lorenz Lechler. For Lechler, see also Seeliger-Zeiss, *Lorenz Lechler von Heidelberg und sein Umkreis.*

9. See Shelby, trans. and ed., *Gothic Design Techniques,* 25–26.

10. Ibid., 18–26. But see also Shelby, "'Secret' of the Medieval Masons," which argues that masons' "secrets" were not specific esoteric geometric techniques but the general techniques of the craft; and Rykwert, "On the Oral Transmission of Architectural Theory."

11. Rykwert, "On the Oral Transmission of Architectural Theory."

12. Hutchison, *Albrecht Dürer,* 21–26, quotation on 26. For an introduction to recent scholarship, see Eichberger and Zika, *Durer and His Culture.*

13. I have taken the count of Dürer's extant works from Durer, *Human Figure by Albrecht Dürer,* v. The classic study of Dürer, still indispensable, is Panofsky, *Life and Art of Albrecht Dürer.* See also Strieder, "Dürer. (1) Albrecht Dürer."

14. Hutchison, *Albrecht Dürer*, 14, 23–26, 57–66, 78–83.

15. See esp. Koerner, *Moment of Self-Portraiture*, 203–14.

16. Hutchison, *Albrecht Dürer*, 40–47, 78–96. For the letter, see Durer, *Schriftlicher Nachlass*, 1:58–59, no. 10, Dürer to Pirckheimer, 13 October 1506: "Hÿ pin jch ein her, doheim ein schmarotzer etc." For an English translation, see Dürer, *Writings of Albrecht Dürer*, 57–59.

17. Bonicatti, "Dürer nella storia delle idee umanistiche"; Hutchison, *Albrecht Dürer*, 48–56, Silver, "Germanic Patriotism in the Age of Durer" For Pirckheimer, see esp. Eckert and Imhoff, *Willibald Pirckheimer*, and for Celtis, Bernstein, "Celtis, Conrad."

18. Dürer, *Writings of Albrecht Dürer*, 57–58; Durer, *Schriftlicher Nachlass*, 1:58–59, no 10, Dürer to Pirckheimer, 13 October 1506.

19. Kemp, "From 'Mimesis' to 'Fantasia'"; Koerner, *Moment of Self-Portraiture*, esp. 34–51, 63–79; Moxey, *Practice of Theory*, 111–47

20 Hutchison, *Albrecht Durer*, 93–94, 98; Dürer, *Writings of Albrecht Dürer*, 55, 57; Durer, *Schriftlicher Nachlass*, 1.54–57, nos 8 and 9, Durer to Pirckheimer, 8 September and 23 September 1506.

21. See Dürer, *Writings of Albrecht Durer*, 167–69, for a useful description of the five volumes of Durer manuscripts in the British Library.

22. Ibid., 178–79, 197–98; Dürer, *Schriftlicher Nachlass*, 2:108–14.

23. See Dürer, *Painter's Manual*, 9 for evidence that the book on human proportions was finished in 1523 and for Durer's acquisition of ten books, and 25 for the manuscript copy of Alberti's *De pittura*

24. Dürer, *Writings of Albrecht Durer*, 227; Durer, *Schriftlicher Nachlass*, 1.97–100, no. 42, Durer to Pirckheimer, [1523].

25 Dürer, *Writings of Albrecht Durer*, 228; Durer, *Schriftlicher Nachlass*, 1:100

26. Durer, *Painter's Manual*, 36–37.

27. Ibid., 8 for Pirckheimer's relationship with Wilhelm von Reichenau, 16–17 for the influence of Schmuttermayer and Roriczer, 24 for Pirckheimer's translations.

28. Durer, *Etliche underricht zu befestigung der Stett*, "Durch leuchtigister grossmechtigen König . . .", n.p., and sig Air, "der doch nicht nütz gewest ist" Dürer, *Writings of Albrecht Dürer*, 262–73, translates a few passages from the treatise, including the dedication (264–65), which translation I have cited.

29. Durer, *Etliche underricht zu befestigung der Stett*, sigs D ia–E; Durer, *Writings of Albrecht Durer*, 265–73; Pollak, *Military Architecture*, 34–35 (no. 17).

30. For sixteenth-century architectural writings, see esp. Carpo, *L'architettura dell'età della stampa*, Guillaume, *Les traités d'architecture*, Hart with Hicks, *Paper Palaces*, Kruft, *History of Architectural Theory*, 7–123; Long, "Contribution of Architectural Writers"; Payne, *Architectural Treatise;* Vagnetti and Marcucci, "Per una coscienza vitruviana"; and Wiebenson, *Architectural Theory and Practice*

31. Cesariano, *Di Lucio Vitruvio Pollione de architectura*, fols. 91r–v: "naturale furibondia"; "diverse Civitate e regione"; "per inspeculare e cognoscere varii ingenii e costumi de homini: c[on]versando e Studendo asai co[n] il mio quottidiano lucro: exercendomi co[n] la pictura e Architectura· da Dio e da me stesso mi sono auxiliato." On the 1521 edi-

tion, see Vagnetti and Marcucci, "Per una coscienza vitruviana," 37–40 For Cesariano's life and activities as a painter and architect, see esp. Gatti, "L'attività milanese del Cesariano"; Gatti, "Nuovi documenti sull'ambiente familiare", Gatti, "Un contributo alla storia delle vicissitudini"; Gatti Perer and Rovetta, *Cesare Cesariano e il classicismo;* Fiore, "Cesariano [Ciserano], Cesare"; and Ludovici, "Cesariano, (Cisariano), Cesare." For a study of his sources for the commentary, see Fiore, "La traduzione vitruviana."

32 Cesariano, *De Lucio Vitruvio Pollione de architectura,* fol. 92r, "Mundi Electiva Caesaris Caesariani Configurata." For a detailed discussion of the illustration, see Krinsky, "Cesare Cesariano," 297–300.

33. Cesariano, *De Lucio Vitruvio Pollione de architectura,* fol. 91v: "ta[n]gere la co[m]mata fro[n]te de la Fortuna"; "li maximi Divi e Regi e Pri[n]cipi"; "co[n] q[ue]sti mei co[m]me[n]tarii e co[n] lo circino e regula a la Audatia. acio che dal tenebroso orbe uti Topa[m] e da le ma[n]e de la Noverca e Paup[er]tate mi sia cercato de profugare"; "Deus e natura nihil agunt frustra"; "per maxima utilitate et necessitate", "inenarrabili casi", "proru[m]pere . . lacrime."

34. Ibid., fol. 10r: "non solum la Architectura: ma ciscune altre arte"; "di opera seu fabricatione e di ratiocinatione", "ben calculate & considerate"; "saper dire e fare"; "quasi a magior opportunita"; "il parlare de quella operaria cosa con ratione."

35. Ibid , fols. 11v–12r· "per expositione trahendo il senso de la cosa como fano il periti magistri de qualche artificio che non solum con li dicti ma con li facti dimonstrano le arte per erudire li rudi operantii"; "in q[ue]sta vita nisi p[er] causa de la tractatione", "che sano operare con le tractatione se perduceno a la elegantia . p[er] essere allegati del suo sapere."

36. Ibid., fols 18r, 162v· "la excogitativa e effectrice e inventrice del operatione manuale"; "questa Machinatione [e] intellectiva cum sia causa de la formatione de li instrumenti fabrili: seu artisti opportuni ad explicare lo effecto de qualu[n]q[ue] cose che noi volemo perficere"; "non solum in larte millitare bisogna q[ue]sta ingeniosa scientia Mechanica ma in tute le liberale dimonstratione e operatione"; "Praeclari Philosophi"; "contemplatione intesa", "magne cognitione", "uno ardente desiderio de produre in opera sensibile con le proprie mane quello che con la mente havevano ratiocinatio."

37 Ibid., fols. 12r, 105v–106r: "O gra[n] bonita. de li Sapienti a lassare le piu care cose"; "ta[n]te singularissime"; "divini e non humani"; "li effecti e exempli evidenti", "in qual modo debeno usare li humani exercitii e artificii."

38. See ibid., fol. 154v, where there is a break in the commentary and a note explaining that it will be completed by Benedetto Giovo and Bono Mauro Cesariano gave his side of the dispute in a manuscript note in his own copy of the commentary, now lost. The note was transcribed and published in Pagave, *Vita di Cesare Cesariano,* 26–28, n. 1, and 23–35 See also Krinsky, introduction to *Vitruvius: De architectura,* 5–28, esp 10–11, and Vasari, *Le Vite,* 4 75 "più da bestia che da persona." For Cesariano's autograph manuscript, see Agosti, "Riflessioni su un manoscritto," 70–73; and Cesariano, *Volgarizzamento dei libri IX (capitoli 7 e 8) e X.*

39 For the study of ancient ruins, see esp. Weiss, *Renaissance Discovery*

40. "Lettera al Conte Agostino de'Landi," in Cataneo and Vignola, *Trattati,* 30–61

(quotation on 52): “è necessario per venire a qualche escellenza, non solo speculare, ma ancora porre in opera”; “onde, congiungendo i precetti de gli scrittori con gli esempi e avvertimenti che si traggon da l’opere, si sforzaranno, come meglio si può, volger gli occhi a l’una parte e a l’altra ”

41. Philander, GULIELMI/PHILANDRI CASTILIO/NII GALLI CIVIS RO. INDE.-/*cem Libros*. For Philander, see Lemerle-Pauwels, “Philander [Philandrier], Guillaume”; Lemerle, “On Guillaume Philandrier”; and Wiebenson, “Guillaume Philander’s Annotations to Vitruvius ”

42. See esp. Carpo, *L’architettura dell’età della stampa*, 49–63; Dinsmoor, “Literary Remains”; Hart and Hicks, “On Sebastiano Serlio”; Howard, “Serlio”; and Payne, *Architectural Treatise*, 111–43.

43 Serlio, *On Architecture*. trans. and ed Hart and Hicks, 252. The most convenient early edition is Serlio, *Tutte l’opere*/ D’ARCHITETTURA.

44. Serlio, *On Architecture*, 253; Howard, “Serlio,” 471 See also Hart, “Serlio and the Representation of Architecture.”

45. Howard, “Serlio.” For a list of editions and translations of Serlio’s various books, see Serlio, *On Architecture*, 470–71. For books 6, 7, and 8, see Serlio, *Architettura civile*.

46 See Barbaro, *I dieci libri dell’architettura*; and for the expanded edition of 1567, see Barbaro, *Vitruvio, I dieci libri* See also Vagnetti and Marcucci, “Per una coscienza vitruviana,” 58–62, 66–68, 70–71, 92–93, 147, 164; and Ackerman, “Daniele Barbaro and Vitruvius.” For Barbaro, see Alberigo, “Barbaro, Daniele Matteo Alvise”; and Tafuri, “Daniele Barbaro e la cultura scientifica.” And for the relationship between Barbaro and Palladio, see Forssman, “Palladio e Daniele Barbaro ”

47 Barbaro, *I dieci libri dell’architettura*, 8–9: “Bisogna . esercitio, bisogna discorso, il discorso come padre, la *fabrica* è come madre dell’Architettura”; “con la sola fabrica”, “con il solo discorso”; “sarebbe stimato cosa imperfetta, anzi monstruosa”, “E di gratia se uno havesse il sapere solamente, e usurpare si volesse il nome d”Architetto non sarebbe egli sottoposto all’offese de gli Esperti? Non potrebbe ogni manuale improverarli, e dirli che fai tu?”; “dall’altra parte se per havere un lieve essercitio, e alquanto di pratica, di si gran nome degno esser si credesse, non potrebbe uno intelligente, e litterato chiuderli la bocca, domandandoli conto, e ragione delle cose fatte?”

48. Ibid., 165: “Non solo adunque deve lo Architetto darsi con ardente desiderio alla cognitione delle lettere, ma dillettarsi di sepere come vanno le cose artificiose, investigarle, e farle affine, che la sua cognitione non resti morta, e inutile.” For the passage in Vitruvius, see *De architectura* 6.pref 4.

49. Barbaro, *I dieci libri dell’architettura*, 243· “senza farne la prova”; “perche in ogni esperienza e difficulta, dove non e stato essercitio”, “e questo molto piu facendo, e isperimentando, che leggendo.”

50. Ibid., 253–54: “Bella utile e meravigliosa pratica”; “non guardi con stupore”; “aiutato da un picciolo strumento”; “un pesa smisurato”; “con debil fune artificiosamente rivolta”; “con meraviglia”; “scienza ò arte mecanica”; “la ragione dimostratrice del modo di fare la machine”; “[i]l vulgo”, “chiamando mecanica ogni arte vile”; “si fa prima nella mente”; “poi regola le opere artificiose”; “della machinatione, e discorso.”

51. Ibid., 201: "sottilissime raagioni delle alte cose"; "de gli honori Celesti"; "al beneficio commune."

52. Ibid., 39–40: "da se stessi vanno a basso come quelli, che essendo huomini, mancar vogliono dell'ufficio della humanità"; "molte cose belle dalle genti di diversi paesi"; "gli invidiosi"; "A quelli veramente, che lodano la secretezza, direi, che quello, che appartiene alla conservatione de gli huomini, non si deve tener secreto." Elsewhere Barbaro defends his discussion of the Venetian navy and arsenal against those who would accuse him of giving aid to enemies, arguing that the strength of Venetian military strength is the result of the citizens of a free and well-defended city (163).

53. Ibid., 40: "io ho cercato imparare da ognuno, ad ogn'uno che mi ha giovato resto debitore." For an introduction to Palladio, see esp. Ackerman, *Palladio*, 19–35; Beyer, "Palladio, Andrea"; Boucher, *Andrea Palladio*; and Payne, *Architectural Treatise*, 170–213. For Trissino and Palladio, see Barbieri, "Giangiorgio Trissino e Andrea Palladio."

54. Ackerman, *Palladio*, 21–27; Boucher, *Andrea Palladio*, 9–29. For Cornaro, see Fiocco, *Alvise Cornaro*, and Puppi, *Alvise Cornaro*. For the architectural writings of Trissino and Cornaro, see Trissino, *[Dell'architettura]*; and Cornaro, *Trattato di architettura*.

55. Palladio, *I quattro libri dell'architettura*; Palladio, *Four Books of Architecture*. For Palladio's porches, see Ackerman, *Palladio*, 61–65.

56. Palladio, *Four Books of Architecture*, 6.

57. See Blunt, *Philibert de L'Orme*; Blunt, *Art and Architecture in France*, 84–94; Guillaume, "On Philibert de L'Orme", and Hoffmann, "L'Orme [Delorme], Philibert de."

58. L'Orme, *Traités d'architecture, Premier tome de l'architecture*," fol. 7v: "ta[n]t par livres, que par l'experie[n]ce qu i'en ay eu en divers lieux, et aussi par diverses oeuvres que i'ay faict faire et co[n]duites en mon temps"; fols. 47r–v: "Ie vous advise que tout ce que ie vous en propose et escris, a esté experimenté en divers lieux par mon ordonnance, advis et commandement"; fol. 11r for difficulties with servants and relatives; fol. 23v for the importance of using a model; fols. 131r–v for his study of Roman buildings.

59. Ibid., *Nouvelles inventions pour bien bastir*, fol. 37v: "si l'Architecte ou Superieur qui comma[n]de aux maistres maçons & autres ouvriers, n'en est bie[n] muny, & n'ente[n]d promptement leur theorique & pratique", "difforme & ridicule"; "esclave du maistre Maço[n], ou quelque ouvrier", "gra[n]d detriment & deshon[n]eur."

60. Ibid., "Epistre au lecteur," n.p.: "tous ceux que font profession de ladicte Architecture, comme aussitous ouvriers, et autres qui voudront faire bastiments", and ibid., *Premier tome de l'architecture*, fol. 57v: "fraternellement advertir, admonnester et prier"; "recognoistre et vouloir estudier et apprendre ce qui est requis et necessaire à leur art et estat."

61. Ibid., *Premier tome de l'architecture*, fol. 21r: "laquelle s'apprend par longue experience et pratique d'avoir mis plusieurs edifices en oeuvres"; "non tant par livres, que par long et grand usage."

62. See Piccolpasso, *I tre libri dell'arte del vasaio*, which includes a facsimile of the manuscript in the Victoria and Albert Museum, London; and Piccolpasso, *Li tre libri dell'arte del vasaio*. See also Hess, "Piccolpasso, Cipriano di Michele."

63. Piccolpasso, *I tre libri dell'arte del vasaio,* 1.xxi–xxiv, 2:xii–xiii.

64. See ibid., 1:xxiv–xxxii for a description of the manuscript and evidence within it for Piccolpasso's intention to publish, 2.4–5 for the quotation.

65. Ibid., 2.6–7.

66. Ibid., 2:42, 48

67. Ibid., 2:105.

68. For Vasari, see Kliemann, "Vasari (1) Giorgio Vasari." For Cellini, see Nova, "Cellini, Benvenuto"; and Borsellino and Camesasca, "Cellini, Benvenuto."

69. Kliemann, "Vasari (1) Giorgio Vasari."

70. Vasari, *Le Vite* A convenient translation of the technical section of Vasari's *Le Vite* is Vasari, *Vasari on Technique.* See also Rubin, *Giorgio Vasari*, and Williams, *Art, Theory, and Culture,* 29–72. For Vasari's collection, see Kliemann, "Vasari (1) Giorgio Vasari," 22

71. Nova, "Cellini, Benvenuto"; Cellini, *Autobiography*

72. For Cellini's writings, including his autobiography, see Cellini, *Opere;* and an English translation, Cellini, *Autobiography.* For a comparison of Cellini's autobiography with Vasari's *Lives,* see Rubin, *Georgio Vasari,* 24

73. See Cellini, *Treatises . . . on Goldsmithing and Sculpture,* 1–6, 11; and Cellini, "Dell'Oreficeria," 623–24, 643: "la quale non si può insegnare con lo scrivere ma e' se n'è buona parte con le parole vive e con la sperienzia detta; pure noi seguiteremo il nostro modo di ragionare." Because Ashbee's translation at times imparts a tone that I do not find in the Italian, I have sometimes retranslated, providing the Italian only in those cases.

74. Cellini, *Treatises . . . on Goldsmithing and Sculpture,* 13, 91–92, and (for jewelers) 23; Cellini, "Dell'Orefecia," 644–46, 772–75, 662.

75. Cellini, *Treatises . . . on Goldsmithing and Sculpture,* 25, 59, 61, and (for progress over the ancients) 71–73; Cellini, "Dell'Orefecia," 665–66, 725, 729: "io credo che fussi prima la practica che la teorica di tutte le scienzie, e che alla pratica se le ponesse di poi regola, a tale che la si venissi a fare con quella virtuosa ragione che si vede usare da gli uomini periti nelle belle scienzie"; "possa sapere che queste vere esperienzie io non le insegno per averle mendicate dalle altrui fatiche." Later, describing the fabrication of seals, he again refers to his own practice (746–49).

76. Cellini, *Treatises . . . on Goldsmithing and Sculpture,* 113 (clay), 114–20, 127–33 (bronze casting and furnaces), and 141 (colossal statues); Cellini, "Trattato della scultura," 793, 794–816 (bronze casting), 817–27 (furnaces), 839 (colossal statues): "e avvertiscasi che questo è un segreto mirabile, che non è stato mai usato"; "io trovai un'altra regola con questa, la quale fu fatta da me proprio, né mai intesa da altri, nata dà mia grandi studii: così l'insegno, come liberale, a quegli che aranno voglia di far bene"

77. McNab, "Palissy, Bernard"

78. Ibid. The edition of Palissy's complete works that I have used is Palissy, *Oeuvres complètes*

79. Palissy, *Admirable Discourses,* 23–24; Palissy, *Discours admirables,* in Palissy, *Oeuvres complètes,* 2:10–11.

80. See Palissy, *Admirable Discourses*, 24 and (for the cabinet labels) 233–50; Palissy, *Discours admirables*, in *Oeuvres complètes*, 2:10–11, 361–72.

81. Palissy, *Admirable Discourses*, 26–27; Palissy, *Discours admirables*, in *Oeuvres complètes*, 2:14–15.

82. Palissy, *Admirable Discourses*, 27; Palissy, *Discours admirables*, in *Oeuvres complètes*, 2:16–18.

83. Palissy, *Admirable Discourses*, 188; Palissy, *Discours admirables*, in *Oeuvres complètes*, 2:285–86.

84. Palissy, *Admirable Discourses*, 188–90, Palissy, *Discours admirables*, in *Oeuvres complètes*, 2:286–90.

85. Palissy, *Admirable Discourses*, 191–92; Palissy, *Discours admirables*, in *Oeuvres complètes*, 2:291–93.

Epilogue

1. See Galison, "Computer Simulations and the Trading Zone." Galison refers to trading between members of diverse professional disciplines, whereas I speak of trading between differently educated individuals from more loosely defined social groups.

2. Olschki, *Geschichte der neusprachlichen wissenschaftlichen Literatur*; Merton, *Science, Technology, and Society*, 136–261, Cohen, *Puritanism and the Rise of Modern Science*; Webster, *Great Instauration*; Zilsel, "Sociological Roots of Science", Zilsel, "Origins of Gilbert's Scientific Method"; Rossi, *Philosophy, Technology, and the Arts*; Crombie, *History of Science from Augustine to Galileo*; Crombie, *Styles of Scientific Thinking*. See also Rob Iliffe, "Rational Artistry"; Keller, "Mathematicians, Mechanics"; Keller, "Mathematics, Mechanics"; and Bennett, "Mechanics' Philosophy and the Mechanical Philosophy." For a discussion of the various traditions represented here, see Cohen, *Scientific Revolution*, esp. 322–27 (Olschki), 336–42 (Zilsel), and 314–21 (Merton).

3. See Galilei, *Operations of the Geometric and Military Compass*; Galilei, *Difesa contro alle calunnie ed imposture*, for Galileo's defense of his "intellectual property"; Galilei, *Sidereus Nuncius*, Gilbert, *De magnete*; Bacon, *Novum organum*; and Shapin and Schaffer, *Leviathan and the Air-Pump*. See also Dear, *Discipline and Experience*.

4. Middleton, *Experimenters*, 91; Biagioli, *Galileo, Courtier*, 358–62. See also Galluzzi, "L'Accademia del Cimento."

5. See Royal Society of London, *Philosophical Transactions* 1 (6 March 1665): 1–2, for statements concerning openness; Shapin, "House of Experiment"; and Stroup, *Company of Scientists*, esp. 199–217.

6. Principe, "Robert Boyle's Alchemical Secrecy"; Principe, *Aspiring Adept*; Dobbs, *Janus Face of Genius*.

7. Pérez-Ramos, *Francis Bacon's Idea of Science*, 38, 48, 52–53, 106, 135.

8. Iliffe, "Material Doubts."

9. Shapin, *Social History of Truth*, 355–407; Smith, *Business of Alchemy*

Bibliography

Abbreviations

ADB *Allgemeine deutsche Biographie*. Ed Otto Kaemmel 56 vols Leipzig· Duncker & Humblot, 1875–1912.

ANRW *Aufstieg und Niedergang der Römischen Welt*. Pt 1, *Von den Anfängen Roms bis zum Ausgang der Republik*, ed Hildegard Temporini, 4 vols ; pt. 2, *Principat*, ed. Hildegard Temporini and Wolfgang Haase, 37 vols. in 87 to date. Berlin· Walter de Gruyter, 1972–.

CE *Contemporaries of Erasmus: A Biographical Register of the Renaissance and Reformation* Ed. Peter G. Bietenholz and Thomas B. Deutscher 3 vols. Toronto: University of Toronto Press, 1987.

DA *The Dictionary of Art*. Ed Jane Turner. 34 vols. New York: Grove Dictionaries, 1996

DBI *Dizionario biografico degli Italiani*. Ed. Alberto M. Ghisalberti et al. 53 vols. to date. Rome· Istituto Enciclopedia Italiana, 1960–.

DMA *The Dictionary of the Middle Ages*. Ed. Joseph R. Strayer. 13 vols. New York: Charles Scribner's Sons, 1982–89.

DSB *Dictionary of Scientific Biography*. Ed Charles Coulston Gillispie. 16 vols. New York Charles Scribner's Sons, 1970–80.

KP *Der Kleine Pauly. Lexikon der Antike auf der Grundlage von Pauly's Realencyclopädie*. Ed Konrat Ziegler and Walther Sontheimer 5 vols. Stuttgart· A. Druckenmüller, 1964–75.

NDB *Neue deutsche Biographie* Ed. Historische Kommission bei der Bayerischen Akademie der Wissenschaften. 19 vols. to date. Berlin: Duncker & Humblot, 1953–.

RE *Paulys Realencyclopädie der classischen Altertumswissenschaft*. Ed. A. Pauly, G Wissowa, and D Kroll 49 vols in 58 Stuttgart· J B. Metzler, 1894–1980.

Primary Sources

Manuscripts

Florence Biblioteca Nazionale Centrale. II, I, 333 (formerly Magl. cl. XVII.33). Lorenzo Ghiberti. *I commentari.*

———. Palat. 766. Mariano Taccola. *De ingeneis.*

Göttingen Niedersachsische Staats- und Universitätsbibliothek Cod. Phil 63. Conrad Kyeser. *Bellifortis.*

Heidelberg. Universitätsbibliothek. Cod. palat. germ. 126. Treatise of Philibs Monch.

Innsbruck. Tiroler Landesmuseum Ferdinandeum. *Schwazer Bergbuch.*

London. British Museum. Codex 197 B 21 [MS Harley 3281]. Francesco di Giorgio. *Opusculum de architectura*

Madrid. Biblioteca Nacional. Leonardo da Vinci. *Madrid Codex I*

Munich. Bayerische Staatsbibliothek. Cod. icon. 222. Bartholomeus Freysleben. *Zeugbuch,* for the emperor Maximilian.

——— Cod. icon. 242. Giovanni Fontana. *Bellicorum instrumentorum liber.*

——— Codex Latinus Monacensis 197, pt. 2 Mariano Taccola. *De ingeneis*

———. Codex Latinus Monacensis 28,800 Mariano Taccola. *De machinis.*

———. Cod germ 599, fols. 66r–101v. Work on artillery by Martin Mercz

——— Cod. germ. 600 Anonymous illustrated pamphlet on gunnery

———. Cod. germ. 734, fols. 60v–71r. Work on artillery by Johann Formschneider.

Munich. Bibliothek des Deutschen Museums. MS 1949/258. Fragments of drawings on artillery.

New York. New York Public Library. Spencer MS 104. *Feuerwerkbuch* and German translation of *Bellifortis.*

Nuremberg. Germanisches Nationalmuseum. HS 719 (formerly Kr. 300). Work on artillery.

———. HS 25,801. *Kriegsmaschinen.*

Paris. Bibliothèque Nationale. Codex Parisinus Latinus 7239. Mariano Taccola. *De rebus militaribus (De machinis, 1449)*

Turin. Biblioteca Reale. Codex serie militare 383 (Cod. 14856, 14876, 14896 D[otazione] C[orona]). Copy of Francesco di Giorgio, *Opusculum de architectura.*

Vatican City Biblioteca Apostolica Vaticana. MS Urb. Lat. 281. Federico da Montefeltro's copy of Roberto Valturio, *Elenchus et index rerum militarum.*

Printed Primary Sources

Agricola, Georgius. *Ausgewählte Werke.* Ed. Hans Prescher et al. 10 vols. and suppls. Berlin: Deutscher Verlag der Wissenschaften, 1955–74

———. *Bermannus (Le mineur). Un dialogue sur les mines.* Ed. Robert Halleux and Albert Yans. Paris: Belles Lettres, 1990.

——— *Bermannus oder Über den Bergbau: Ein Dialog.* Ed. and trans Helmut Wilsdorf, Hans Prescher, and Heinz Techel Vol 2 of *Ausgewählte Werke,* ed. Hans Prescher Berlin· Deutscher Verlag der Wissenschaften, 1955.

——— *De Natura Fossilium (Textbook of Mineralogy)*. Trans. Mark Chance Bandy and Jean A. Bandy. New York. Mineralogical Society of America, 1955.

———. *De re metallica*. Trans Herbert Clark Hoover and Lou Henry Hoover. 1912. Reprint, New York: Dover, 1950.

——— *De re metallica libri XII*. Basel: H. Frobenius & N. Episcopius, 1556.

———. *Georgii Agri- / COLAE MEDICI / BERMANNUS, / SIVE DE RE ME- / tallica* Paris· Hieronymus Gormontius, 1541.

Agrippa, Henricus Cornelius ab Nettesheym. *De incertitudine et vanitate scientiarum atque artium declamatio. . .* In *Opera*, intro Richard H. Popkin, 2:1–114. Lyon, [1600?]. Reprint, Hildesheim: Georg Olms Verlag, 1970.

———. *De occulta philosophia*. Ed. Karl Anton Nowotny. Graz: Akademische Druck- u. Verlagsanstalt, 1967.

———. *De occulta philosophia libri tres* Ed. V. Perrone Compagni. Leiden. Brill, 1992.

———. *Of the Vanitie and Uncertaintie of Artes and Sciences* Trans. James Sanford Ed Catherine M. Dunn. Northridge· California State University Press, 1974

——— *Opera*. 2 vols Intro Richard H. Popkin. Lyon, [1600?]. Reprint, Hildesheim: Georg Olms Verlag, 1970.

———. *Three Books of Occult Philosophy*. Trans. J. Freake Ed Donald Tyson. St. Paul, Minn : Llewellyn, 1993.

Aineias the Tactician *How to Survive under Siege*. Trans. David Whitehead. Oxford· Clarendon, 1990.

Alberti, Leon Battista *L'architettura (De re aedificatoria)*. Ed. and trans. Giovanni Orlandi. Intro. and notes by Paolo Portoghesi. 2 vols. Milan: Il Polifilo, 1966.

———. *De pictura*. In *Opere volgari*, ed Cecil Grayson, 3.7–107 Bari: Gius. Laterza, 1973.

———. *On Painting*. Trans. John R. Spencer. Rev. ed. New Haven Yale University Press, 1966.

——— *On Painting*. Trans. Cecil Grayson. Intro. Martin Kemp. London: Penguin, 1991.

———. *"On Painting" and "On Sculpture": The Latin Texts of "De pictura" and "De statua."* Ed. and trans Cecil Grayson London. Phaidon, 1972.

———. *On the Art of Building in Ten Books* Trans. Joseph Rykwert, Neil Leach, and Robert Tavernor. Cambridge. MIT Press, 1988

Apuleius. *The Apologia and Florida of Apuleius of Madaura*. Trans. H. E Butler 1909. Reprint, Westport, Conn.· Greenwood, 1970.

——— *The Golden Ass* Trans. P G. Walsh. Oxford: Clarendon, 1994.

———. *The Isis-Book (Metamorphoses, Book XI)*. Ed and trans. J. Gwyn Griffiths. Leiden: Brill, 1975.

———. *Metamorphoses*. 2 vols. Ed. and trans. J. Arthur Hanson Loeb. Cambridge. Harvard University Press, 1980.

———. *Opuscules philosophiques (Du dieu de Socrate, Platon et sa doctrine, Du monde) et fragments*. Ed. and trans. Jean Beaujeu. Paris· Belles Lettres, 1973.

Aristotle. *Nichomachean Ethics*. Trans. W. D. Ross. Rev. J. O. Urmson. In *The Complete Works of Aristotle*, ed. Jonathan Barnes, 2:1729–1867. Rev. ed. Princeton, N J.: Princeton University Press, 1984.

Bacon, Francis *Novum organum*. Trans. and ed. Peter Urbach and John Gibson Chicago: Open Court, 1994.

Barbaro, Daniele *I dieci libri dell'architettura di M Vitruvio tradutti et commentati da Monsignor Barbaro eletto patriarca D'Aquileggi*. Venice: Francesco Marcolini, 1556.

———. *Vitruvio: I dieci libri dell'architettura tradotti e commentati da Daniele Barbaro, 1567*. With an essay by Manfredo Tafuri and an analysis by Manuela Morresi Milan: Il Polifilo, 1997

Bassignana, Pier Luigi, ed. *Le macchine di Valturio nei documenti dell'Archivio Storico AMMA*. Turin: Umberto Allemandi, 1988.

Battisti, Eugenio, and Giuseppa Saccaro Battisti. *Le macchine cifrate di Giovanni Fontana*. Turin. Arcadia, 1984

Berthelot, Marcellin Pierre Eugène, and C -E Ruelle, eds. and trans. *Collection des anciens alchimistes grecs*. 3 vols 1888. Reprint, Osnabrück. Zeller, 1967.

Betz, Hans Dieter, ed *The Greek Magical Papyri in Translation Including the Demotic Spells*. 2nd ed. Chicago: University of Chicago Press, 1992.

Biringuccio, Vannoccio *De la pirotechnia*, 1540. Ed. Adriano Carugo. Milan: Il Polifilo, 1977.

———. *The Pirotechnia of Vannoccio Biringuccio: The Classic Sixteenth-Century Treatise on Metals and Metallurgy*. Trans. and ed. Cyril Stanley Smith and Martha Teach Gnudi. 2nd ed. 1959. Reprint, New York: Dover, 1990.

Bossert, Helmuth T , and Willy F. Storck, eds *Das mittelalterliche Hausbuch nach dem Originale im Besitze des Fürsten von Waldburg-Wolfegg-Waldsee im Auftrag des deutschen Vereins für Kunstwissenschaft*. Leipzig: Seemann, 1912.

Brepohl, Erhard. *Theophilus Presbyter und die mittelalterliche Goldschmiedekunst* Vienna: Böhlau, 1987.

Bruno, Giordano. *La Cena de le Ceneri / The Ash Wednesday Supper*. Ed. and trans. Edward A. Gosselin and Lawrence S. Lerner. 1977. Reprint, Toronto: University of Toronto Press, 1995

Burnam, John M., trans and ed. *A Classical Technology Edited from Codex Lucensis, 490*. Boston. Richard G. Badger, Gorham Press, 1920.

Caley, Earle Radcliffe "The Leyden Papyrus X. An English Translation with Brief Notes." *Journal of Chemical Education* 3 (October 1926): 1149–66

———. "The Stockholm Papyrus: An English Translation with Brief Notes." *Journal of Chemical Education* 4 (August 1927) 979–1002

Cataneo, Pietro, and Giacomo Barozzi da Vignola. *Trattati con l'aggiunta degli scritti di architettura di Alvise Cornaro, Francesco Giorgi, Claudio Tolomei, Giangiorgio Trissino, Giorgio Vasari* Ed. Elena Bassi et al. Milan: Il Polifilo, 1985.

Cellini, Benvenuto. *Autobiography*. Trans. George Bull. Harmondsworth, England: Penguin, 1956

———. "Dell'Oreficeria." In *Opera*, ed Bruno Maier, 623–790. Milan: Rizzoli, 1968.

———. *Opere*. Ed. Bruno Maier. Milan. Rizzoli, 1968.

———. "Trattato della scultura." In *Opera*, ed. Bruno Maier, 791–846. Milan· Rizzoli, 1968.

———. *The Treatises of Benvenuto Cellini on Goldsmithing and Sculpture*. Trans. C. R Ashbee 1888. Reprint, New York: Dover, 1967.

Cennini, Cennino d'Andrea. *The Craftsman's Handbook, "Il Libro dell'Arte."* Trans Daniel V. Thompson Jr. 1933. Reprint, New York. Dover, 1954.

Cesariano, Cesare *Di Lucio Vitruvio Pollione de architectura libri dece traducti de latino in vulgare affigurati· Comentati et con mirando ordine insigniti* . Como· Gotardus de Ponte, 1521

———. *Volgarizzamento dei libri IX (capitoli 7 e 8) e X. di Vitruvio· De architectura· secondo il manoscritto 9–2790 Sección de Cortes della Real Accademia de la Historia Madrid.* Ed Barbara Agosti. Pisa: Scuola Normale Superiore, 1996

Cicero. *De natura deorum and Academica.* Trans. H. Rackham. Loeb. London. Heinemann, 1933

——— *On Duties.* Ed. Miriam T. Griffin and E. Margaret Atkins. Trans. E. Margaret Atkins. Cambridge. Cambridge University Press, 1991.

Columella, Lucius Junius Moderatus. *On Agriculture.* Ed. and trans. Harrison Boyd Ash, E. S. Forster, and Edward H. Heffner. 3 vols. Loeb. Cambridge. Harvard University Press, 1948–55

Copenhaver, Brian P., ed. and trans. *Hermetica: The Greek Corpus Hermeticum and the Latin Asclepius in a New English Translation with Notes and Introduction.* Cambridge: Cambridge University Press, 1992.

Cornaro, Alvise. *Trattato di architettura.* Ed. Camillo Semenzato In *Trattati con l'aggiunta degli scritti di architettura di Alvise Cornaro, Francesco Giorgi, Claudio Tolomei, Giangiorgio Trissino, Giorgio Vasari,* by Pietro Cataneo and Giacomo Barozzi da Vignola, ed Elena Bassi et al., 77–113 Milan· Il Polifino, 1985

Della Valle, Battista. *Vallo libro continente appertenentie ad capitanii, retenere & fortificare una citta co[n] bastioni. . .* Venice: [G de Gregoriis], 1524.

Diogenes Laertius. *Lives of Eminent Philosophers.* Rev. ed. 2 vols Trans. R. D Hicks. Loeb. London: Heinemann, 1938.

Drake, Stillman, and I. E. Drabkin. *Mechanics in Sixteenth-Century Italy: Selections from Tartaglia, Benedetti, Guido Ubaldo, and Galileo.* Madison: University of Wisconsin Press, 1969.

Durer, Albrecht *Etliche underricht zu befestigung der Stett / Schloss / und Flecken.* Nuremberg· [Hieronymus Andreae], 1527.

——— *The Human Figure by Albrecht Dürer· The Complete Dresden Sketchbook* Ed. and trans. Walter L. Strauss. New York· Dover, 1972.

———. *The Painter's Manual: A Manual of Measurement of Lines, Areas, and Solids by Means of Compass and Ruler Assembled by Albrecht Dürer for the Use of All Lovers of Art with Appropriate Illustrations Arranged to be Printed in the Year MDXXV.* Trans. with a commentary by Walter L. Strauss New York· Abaris, 1977

——— *Schriftlicher Nachlass.* Ed Hans Rupprich. 3 vols Berlin: Deutscher Verein für Kunstwissenschaft, 1956–69.

——— *The Writings of Albrecht Dürer.* Ed and trans. William Martin Conway Intro Alfred Werner. New York: Philosophical Library, 1958.

Ercker, Lazarus. *Beschreibung der allervornehmsten mineralischen Erze und Bergwerksarten vom Jahre 1580* Ed. Paul Reinhard Beierlein and Alfred Lange. Berlin: Akademie-Verlag, 1960

———. *Drei Schriften. Das kleine Probierbuch von 1556, Vom Rammelsberge, und dessen Bergwerk, ein kurzer Bericht von 1565; Das Münzbuch von 1563.* Ed. Paul R. Beierlein and Heinrich Winkelmann. Bochum: Vereinigung der Freunde von Kunst und Kultur im Bergbau, 1968

———. *Treatise on Ores and Assaying.* Ed. and trans Anneliese Grünhaldt Sisco and Cyril Stanley Smith Chicago: University of Chicago Press, 1951.

Fachs, Modestin. *Probier Buchlein / Darinne Grundlicher bericht vormeldet / wie man alle Metall / und derselben zugehörenden Metallischen Ertzen und getöchten ein jedes auff seine eigenschafft und Metall recht Probieren sol* Leipzig: Zacharias Berwald, 1595.

Ferrari, Lodovico, and Niccolò Tartaglia *Cartelli di sfida matematica.* Ed. Arnaldo Masotti. Brescia: Ateneo di Brescia, 1974.

Ferri, Silvio, ed *Vitruvio· Architettura (Dai libri I–VII)* Rome: Fratelli Palombi Editori, 1960.

Ficino, Marsilio. *Opera omnia* 2 vols 1576. Reprint, ed. M. Sancipriano and intro. Paul Kristeller, Turin: Bottega d'Erasmo, 1959.

———. *Théologie platonicienne de l'immortalité des âmes* Ed. and trans. Raymond Marcel 3 vols Paris: Belles Lettres, 1964–70.

——— *Three Books on Life* Ed. and trans Carol V Kaske and John R. Clark. Binghamton, N.Y.. Center for Medieval and Early Renaissance Studies, 1989.

Filarete (Antonio Averlino) *Filarete's Treatise on Architecture.* Ed. and trans. John R. Spencer 2 vols. New Haven: Yale University Press, 1965.

———. *Trattato di architettura* 2 vols Text edited by Anna Maria Finoli and Liliana Grassi. Intro. and notes by Liliana Grassi Milan. Edizioni il Polifilo, 1972

Frontinus *Kriegslisten.* Ed. and trans. Gerhard Bendz. Berlin: Akademie-Verlag, 1963

———. *The Stratagems and the Aqueducts of Rome.* Trans Charles E. Bennett. Ed. Mary B. McElwain. Loeb. London: Heinemann, 1925

Gadd, C. J., and R. Campbell Thompson. "A Middle-Babylonian Chemical Text." *Iraq* 3 (spring 1936)· 87–96

Galilei, Galileo *Difesa contro alle calunnie ed imposture di Baldessar Capra* In *Le opere di Galileo Galilei,* ed. Antonio Favaro, 2:513–601. 20 vols. 1890–1909. Reprint, Florence. G. Barbèra, 1929–39.

———. *Operations of the Geometric and Military Compass.* Trans. with an introduction by Stillman Drake Washington, D.C.. Smithsonian Institution Press, 1978.

———. *Sidereus Nuncius, or The Sidereal Messenger.* Trans. Albert van Helden. Chicago: University of Chicago Press, 1989

Garin, E., M. Brini, C Vasoli, and C. Zambelli. *Testi umanistici su l'ermetismo* Rome: Fratelli Bocca, 1955.

Gaye, Giovanni, ed *Carteggio inedito d'artisti dei secoli XIV, XV, XVI.* 3 vols. Florence: Presso Giuseppe Molini, 1839

Ghiberti, Lorenzo. *Der dritte Kommentar Lorenzo Ghibertis· Naturwissenschaften und Medizin in der Kunsttheorie der Frührenaissance.* Ed. and trans. Klaus Bergdolt. Weinheim: VCH, 1988

——— *I commentari.* Ed. Ottovio Morisani Naples: Riccardo Ricciardi Editore, 1947

———. *Lorenzo Ghibertis Denkwürdigkeiten (I commentarii)* Ed. Julius von Schlosser. 2 vols. Berlin: Julius Bard, 1912.

Gilbert, William. *De magnete.* Trans. P. Fleury Mottelay 1893. Reprint, New York Dover, 1958.

Hall, Bert S. *The Technological Illustrations of the So-Called "Anonymous of the Hussite Wars": Codex Latinus Monacensis 197, Part 1.* Wiesbaden: Ludwig Reichert, 1979

Halleux, Robert, ed and trans *Les alchimistes grecs.* Vol. 1, *Papyrus de Leyde; Papyrus de Stockholm, Fragments de recettes.* Paris. Belles Lettres, 1981

Hassenstein, Wilhelm, ed. *Das Feuerwerkbuch von 1420.* Munich. Verlag der Deutschen Technik, 1941.

Hedfors, Hjalmar. *Compositiones ad tingenda musiva, herausgegeben, übersetzt und philologisch erklärt.* Uppsala: Almquist & Wiksell, 1932

Hero of Alexandria *Heronis Alexandrini opera quae supersunt omnia* Ed. Wilhelm Schmidt et al. 6 vols Leipzig: Teubner, 1899–1914.

———. *The Pneumatics of Hero of Alexandria.* Trans. and ed. Bennet Woodcroft. 1851. Reprint, intro Marie Boas Hall, London: MacDonald, 1971.

Holt, Elizabeth Gilmore, ed. *Documentary History of Art.* Vol. 1, *The Middle Ages and the Renaissance* New York: Doubleday, Anchor Books, 1957.

Hooper, William D., and Harrison Boyd Ash, eds and trans *Marcus Porcius Cato on Agriculture, Marcus Terentius Varro on Agriculture.* Rev ed. Loeb Cambridge: Harvard University Press, 1935

Iamblichus. *On the Pythagorean Way of Life.* Ed and trans John Dillon and Jackson Hershbell. Atlanta. Scholars Press, 1991

Illinois Greek Club [W. A. Oldfather et al.], trans. *Aeneas Tacticus, Asclepiodotus, Onasander* Loeb London: Heinemann, 1923.

Kristeller, Paul O., ed. *Supplementum Ficinianum.* 2 vols. Florence. Leo S. Olschki, 1937.

Kyeser, Conrad aus Eichstätt. *Bellifortis.* Ed. and trans Götz Quarg. 2 vols. Düsseldorf: VDI-Verlag, 1967.

Lanteri, Giacomo. *Due dialoghi di M. Iacomo de' Lanteri da Paratico, Bresciano . . . del modo di disegnare le piante delle fortezze secondo Euclide* Venice: Vincenzo Valgrisi & Baldessar Costantini, 1557.

——— *Duo libri di M Giacomo Lanteri di Paratico da Brescia del modo di fare le fortificationi di terra intorno alla città, & alle castella per fortificarle.* Venice: Bolognino Zaltieri, 1559.

Leonardo da Vinci *Il Codice Atlantico della Biblioteca Ambrosiana di Milano.* Diplomatic and critical transcriptions by Augusto Marinoni. 12 vols Florence: Giunti Barbèra, 1975.

———. *Il Codice Atlantico di Leonardo da Vinci: Edizione in facsimile dopo il restauro dell'originale concervato nella Biblioteca Ambrosiana di Milano* 12 vols. New York: Harcourt Brace Jovanovich, Johnson Reprint, 1973.

———. *The Madrid Codices.* Ed. Ladislao Reti. 5 vols New York: McGraw-Hill, 1974.

———. *The Notebooks of Leonardo da Vinci.* Ed. and trans Jean Paul Richter. 2 vols. 1883 Reprint, New York: Dover, 1970.

L'Orme, Philibert de. *Traités d'architecture. Nouvelles inventions pour bien bastir et à petits fraiz (1561). Premier tome de l'architecture (1567).* Intro. Jean-Marie Pérouse de Monclos. Paris: Léonce Laget, 1988.

Lucian. *Hermotimus, or Concerning the Sects.* In *Lucian,* trans. K. Kilburn, 6:259–415. Loeb. London: Heinemann, 1959.

Luck, Georg, ed. and trans. *Arcana Mundi: Magic and the Occult in the Greek and Roman Worlds.* Baltimore: Johns Hopkins University Press, 1985.

Machiavelli, Niccolò. *The Art of War.* Trans. Ellis Farneworth. Rev. and ed. Neal Wood. 1965. Reprint, New York: Da Capo, 1990.

Mahé, Jean-Pierre, ed. and trans. *Hermès en Haute-Égypte.* 2 vols. Quebec: Les Presses de L'Université Laval, 1978–82.

Mancini, Girolamo, ed. "L'opera 'De corporibus regularibus' di Pietro Franceschi detto Della Francesca usurpata da Fra' Luca Pacioli." *Memorie della Rendiconti Accademia dei Lincei: Classe di Scienze Morali, Storiche e Filologiche,* 5th ser., 14 (1916): 441–580.

Manetti, Antonio di Tuccio. *The Life of Brunelleschi.* Intro., notes, and critical text ed. Howard Saalman. Trans. Catherine Enggass. University Park: Pennsylvania State University Press, 1970.

Marsden, Eric. *Greek and Roman Artillery: Technical Treatises.* Oxford: Oxford University Press, 1971.

Martini, Francesco di Giorgio. *Trattati di architettura ingegneria e arte militare.* Ed. Corrado Maltese. Transcription by Livia Maltese Degrassi. 2 vols. Milan: Edizioni il Polifilo, 1967.

Merrifield, Mary P. *Original Treatises on the Arts of Painting.* 2 vols. 1849. Reprint, New York: Dover, 1967.

Middleton, W. E. Knowles. *The Experimenters: A Study of the Accademia del Cimento.* Baltimore: Johns Hopkins University Press, 1971.

Milanesi, Gaetano, ed. *Documenti per la storia dell'arte senese.* Vol. 2, *Secoli XV e XVI.* 1864. Reprint, Netherlands: Davaco, 1969.

Monticolo, Giovanni, and E. Besta. *I capitolari delle arti veneziane sottoposte alla Giustizia e poi alla Giustizia Vecchia dalle origini al MCCCXXX.* 3 vols. Fonti per la storia d'Italia, 26–28. Rome: Istituto Storico Italiano, 1896–1914.

Newman, William R. *The "Summa Perfectionis" of Pseudo-Geber: A Critical Edition, Translation and Study.* Leiden: Brill, 1991.

Nock, A. D., and A.-J. Festugière, eds. and trans. *Corpus Hermeticum.* 3rd ed. 4 vols. Paris: Belles Lettres, 1972–73.

Oppenheim, A. Leo, et al. *Glass and Glassmaking in Ancient Mesopotamia: An Edition of the Cuneiform Texts Which Contain Instructions for Glassmakers with a Catalogue of Surviving Objects.* 1970. Reprint, Corning, N.Y.: Corning Museum of Glass Press, 1988.

Palissy, Bernard. *The Admirable Discourses of Bernard Palissy.* Trans. Aurèle La Rocque. Urbana: University of Illinois Press, 1957.

———. *Oeuvres complètes*. Ed. Marie-Madeleine Fragonard et al. 2 vols. Mont-de-Marsan: Editions Interuniversitaires, 1996.

Palladio, Andrea. *The Four Books on Architecture*. Trans. Robert Tavernor and Richard Schofield. Cambridge: MIT Press, 1997.

———. *I quattro libri dell'architettura*. Ed. Licisco Magagnato and Paola Marini. Intro. Licisco Magagnato. Milan: Il Polifilo, 1980.

Paracelsus (Theophrast von Hohenheim). *Four Treatises of Theophrastus von Hohenheim called Paracelsus*. Ed. Henry E. Sigerist. 1941. Reprint, Baltimore: Johns Hopkins University Press, 1996.

———. *Decem libri Archidoxis Theophrasti Germani philosophi, dicti Paracelsi magni de mysteriis naturae*. In *Sämtliche Werke*, pt. 1, *Medizinische, naturwissenschaftliche und philosophische Schriften*, ed. Karl Sudhoff, 3:91–200. Munich: R. Oldenbourg, 1930.

———. *Liber de longa vita*. In *Sämtliche Werke*, pt. 1, *Medizinische, naturwissenschaftliche und philosophische Schriften*. ed. Karl Sudhoff, 3:221–45. Munich: R. Oldenbourg, 1930.

———. *On the Miners' Sickness and Other Miners' Diseases*. Trans. George Rosen. In *Four Treatises of Theophrastus von Hohenheim called Paracelsus*, ed. Henry E. Sigerist, 43–126. 1941. Reprint, Baltimore: Johns Hopkins University Press, 1996.

———. *A Reply to Certain Calumniations of His Enemies (Seven Defensiones)*. Trans. C. Lilian Temkin. In *Four Treatises of Theophrastus von Hohenheim called Paracelsus*, ed. Henry E. Sigerist, 1–41. 1941. Reprint, Baltimore: Johns Hopkins University Press, 1996.

———. *Sämtliche Werke*. Pt. 1. *Medizinische, naturwissenschaftliche und philosophische Schriften*. Ed. Karl Sudhoff. 14 vols. Vols. 6–9, Munich: O. W. Barth, 1922–25; vols 1–5 and 10–14, Munich: R. Oldenbourg, 1928–33.

———. *Sämtliche Werke*. Pt. 2. *Theologische und religionsphilosophische Schriften*. Ed. Kurt Goldammer, 7 vols. to date. Wiesbaden: Franz Steiner Verlag, 1955–.

Philander, Guillaume. *GULIELMI* / PHILANDRI CASTILIO / NII GALLI CIVIS RO. INDE.- / *cem libros M. Vitruvii Pollionis de ar. / chitectura annotationes*. Rome: Ionnes Andreas Dossena, 1544.

Phillipps, Thomas. "Letter from Sir Thomas Phillipps . . . Addressed to Albert Way, Communicating a . . . MS. Treatise . . . entitled Mappae Clavicula." *Archaeologia* 32 (1847): 183–244.

Philo of Byzantium. *Pneumatica*. Ed. and trans. Frank David Prager. Wiesbaden: Ludwig Reichert, 1974.

Piccolpasso, Cipriano. *I tre libri dell'arte del vasaio / The Three Books of the Potter's Art*. Trans. and intro. Ronald Lightbown and Alan Caiger-Smith. 2 vols. London: Scholar Press, 1980.

———. *Li tre libri dell'arte del vasaio*. Ed. Giovanni Conti. Florence: All'Insegna del Giglio, 1976.

Pieper, Wilhelm. *Ulrich Rülein von Calw und sein Bergbüchlein*. Berlin: Akademie Verlag, 1955.

Piero della Francesca. *De prospectiva pingendi.* Ed. G. Nicco-Fassola. 2nd ed. Ed. Eugenio Battisti et al. Florence. Casa Editrice Le Lettere, 1984.

———. *Trattato d'abaco dal Codice Ashburnhamiano 280 (359*–291*) della Biblioteca Medicea Laurenzeana di Firenze.* Ed. Gino Arrighi. Pisa: Domus Galilaeana, 1970

Pingree, David, ed. *Picatrix. The Latin Version of the Ghāyat Al-ḥakīm.* London: Warburg Institute, 1986.

Plato. *Phaedrus* Trans. R Hackforth. In *The Collected Dialogues of Plato,* ed. Edith Hamilton and Huntington Cairns, 475–525. Princeton, N.J.: Princeton University Press, 1961

———. *Protagoras.* Trans C. C W Taylor. Rev. ed. Oxford: Clarendon, 1991.

Pliny. *Natural History (Naturalis historia).* 10 vols. Ed. and trans. Horace Rackham et al Loeb. Cambridge: Harvard University Press, 1938–62.

Pliny the Younger. *Letters and Panegyricus.* 2 vols. Trans. Betty Radice. Loeb. Cambridge: Harvard University Press, 1969.

Plotinus. *Enneads.* 7 vols. Trans. A H Armstrong. Loeb. Cambridge: Harvard University Press, 1966–88.

Plutarch. *Life of Marcellus.* In *Plutarch's Lives,* trans Bernadotte Perrin, vol. 5, *Agesilaus and Pompey, Pelopidas and Marcellus,* 37–523 Loeb. London: Heinemann, 1917.

———. *Life of Pericles.* In *Plutarch's Lives,* trans. Bernadotte Perrin, vol. 3, *Pericles and Fabius Maximus, Nicias and Crassus,* 37–523. Loeb. London: Heinemann, 1925.

Porphyry. "On the Life of Plotinus and the Order of His Books" In *Plotinus,* trans. A. H. Armstrong, 1:1–85. Loeb. Cambridge: Harvard University Press, 1966.

Probir buch / leyn zu Gotes lob / unnd der werlth nutz geordent. Magdeburg. Hanss Knappe den Jungeren, 1524.

Robinson, James M., ed. *The Nag Hammadi Library in English.* 3rd rev. ed. Leiden: Brill, 1988.

Rochas D'Aiglun, M. de "Traduction du traité des machines d'Athénée." In *Mélanges Graux,* 781–801. Paris: Ernest Thorin, 1884.

Schiaparelli, Luigi. *Il codice 490 della Biblioteca Capitolare di Lucca.* Rome: Pompeo Sansaini Editore, 1924.

Schneider, Rudolf, ed. and trans. "Griechische Poliorketiker. III: Athenaios über Maschinen." In *Abhandlungen der Königlichen Gesellschaft der Wissenschaften zu Göttingen, Philologisch-historische Klasse,* n s., 12 5 (1912).

Schreittmann, Ciriacus. *Probierbüchlin / Frembde und subtile Künst / vormals im Truck nie gesehen /* . Frankfurt am Main: Christian Egenolffs Erben, 1580.

Serlio, Sebastiano *Architettura civile: Libri sesto settimo e ottavo nei manoscritti di Monaco e Vienna.* Ed. Francesco Paolo Fiore. Preface and notes by Tancredi Carunchio and Francesco Paolo Fiore. Milan: Il Polifilo, 1994.

———. *On Architecture.* Vol. 1, *Books I–V of "Tutte L'Opere D'architettura et Prospetiva."* Trans and ed Vaughan Hart and Peter Hicks. New Haven: Yale University Press, 1996.

———. *Tutte l'opere /* D'ARCHITETTURA, / et PROSPETIVA, / DI SEBASTIANO SERLIO / BOLOGNESE, / DOVE SI METTONO IN DISEGNO TUTTE LE / *maniere di edificij, e si trattano di quelle cose, che sono più / necessarie à sapere gli architetti* Venice. Giacomo de' Franceschi, 1619.

Shelby, Lon R., ed. and trans. *Gothic Design Techniques: The Fifteenth-Century Design Booklets of Mathes Roriczer and Hanns Schmuttermayer* Carbondale: Southern Illinois University Press, 1977.

Sherwin-White, Adrian N. *The Letters of Pliny: A Historical and Social Commentary*. Oxford: Clarendon, 1966

Shumaker, Wayne. *Renaissance Curiosa* Binghamton, N.Y.. Center for Medieval and Early Renaissance Studies, 1982

Sisco, Anneliese Grünhaldt, and Cyril Stanley Smith, trans. and eds. *Bergwerk- und Probierbuchlein*. New York· American Institute of Mining and Metallurgical Engineers, 1949.

Smith, Cyril Stanley, and John G. Hawthorne, eds and trans *Mappae Clavicula: A Little Key to the World of Medieval Techniques. Transactions of the American Philosophical Society*, n.s., 64, pt. 4 (1974)

Spies, Gerd, ed *Technik der Steingewinnung und der Flussschiffahrt in Harzvorland in früher Neuzeit*. Braunschweig. Waisenhaus, 1992.

Sulpicius, Giovanni, ed. *L. Vitruvii Pollionis Ad Cesarem Augustum de architectura liber primus [-decimus]*. [Rome: Giorgio Herolt, 1486?].

Taccola, Mariano di Jacopo *De ingeneis: Liber primus leonis, liber secundus draconis, addenda; Books I and II, On Engines, and Addenda (The Notebook)*. Ed and trans Gustina Scaglia, Frank D. Prager, and Ulrich Montag. 2 vols. Wiesbaden. Ludwig Reichert, 1984

———. *De machinis: The Engineering Treatise of 1449*. Ed. and trans. Gustina Scaglia 2 vols. Wiesbaden. Ludwig Reichert, 1971.

———. *De rebus militaribus (De machinis, 1449)*. Ed Eberhard Knobloch. Baden-Baden: Verlag Valentin Koerner, 1984.

———. *Liber tertius de ingeneis ac edifitiis non usitatis*. Ed James H. Beck. Milan. Il Polifilo, 1969.

Tartaglia, Niccolò. *Nova scientia inventa da Nicolo Tartalea, B.* Venice: Stephano da Sabio, 1537.

——— *Quesiti et inventioni diverse de Nicolo Tartalea Brisciano*. Venice: Venturino Ruffinelli for N. Tartaglia, 1546

Theophilus. *On Diverse Arts*. Ed. and trans. John G. Hawthorne and Cyril S Smith. 1963. Reprint, New York: Dover, 1979.

———. *The Various Arts: De Diversis Artibus* Ed and trans C. R. Dodwell 1961. Reprint, Oxford: Clarendon, 1986

Trissino, Giangiorgio. *[Dell'architettura]*. Ed Camillo Semenzato. In *Trattati con l'aggiunta deble scritti di architettura di Alvise Cornaro, Francesco Giorgi, Claudio Tolomei, Giangiorgio Trissino, Giorgio Vassari*, by Pietro Cataneo and Giacomo Barozzi da Vignola, ed. Elena Bassi et al., 19–29. Milan: Il Polifilo, 1985.

Valturio, Roberto. *Elenchus et index rerum militarium. . . .* Verona: Johannes ex Verona oriundus Nicholai Cyrugie Medici Filius, 1472.

Vasari, Giorgio. *Vasari on Technique*. Trans. Louisa S. Maclehose. Ed. G. Baldwin Brown. 1907. Reprint, New York: Dover, 1960.

———. *Le Vite de' più eccellenti pittori scultori e architettori nelle redazioni del 1550 e 1568*. Ed. Rosanna Bettarini and Paola Barocchi. 6 vols. Florence: Sansoni, 1966–87.

Vitruvius. *On Architecture (De architectura)*. 2 vols. Ed. and trans. Frank Granger. Loeb. Cambridge. Harvard University Press, 1931.

———. *The Ten Books of Architecture*. Trans Morris Hickey Morgan. 1914. Reprint, New York: Dover, 1960.

———. *Ten Books on Architecture* Trans. Ingrid D. Rowland. Commentary and illustrations by Thomas Noble Howe. Additional commentary by Ingrid D. Rowland and Michael J Dewar. Cambridge. Cambridge University Press, 1999

Waltzing, Jean-Pierre. *Étude historique sur les corporations professionnelles chez les Romains* 4 vols. 1895–1900, Reprint, New York· Arno, 1979

Winkelmann, Heinrich, ed. *Schwazer Bergbuch*. Bochum· Gewerkschaft Eisenhütte Westfalia, 1956.

Xenophon. *Memorabilia and Oeconomicus*. Trans E. C. Marchant. Loeb. London: Heinemann, 1938.

———. *Scripta minora*. Trans. E. C Marchant. Loeb. Cambridge: Harvard University Press, 1946.

Zanchi, Giovan Battista de'. *Del modo di fortificar le città*. Venice: Giralamo Ruscelli, 1556.

Zimmermann, Samuel. *Probierbuch· Auff alle Metall Müntz / Ertz / und berckwerck / Dessgleichen auff Edel Gestain / Perlen / Corallen / und andern dingen mehr. . . .* Augsburg: Michael Manger, 1573.

Zosimos of Panopolis. *Les alchimistes grecs*. Vol 4, pt. 1, *Memoires authentiques*. Ed. and trans. Michèle Mertens. Paris. Belles Lettres, 1995.

———. *On the Letter Omega* Ed. and trans. Howard M. Jackson. Missoula, Mont. Scholars Press, 1978.

Secondary Sources

Ackerman, James S. "Daniele Barbaro and Vitruvius." In *Architectural Studies in Memory of Richard Krautheimer*, ed. Cecil L. Striker, 1–5. Mainz: Verlag Philipp von Zabern, 1996

———. "The Involvement of Artists in Renaissance Science." In *Science and the Arts in the Renaissance*, ed John W Shirley and F David Hoeniger, 94–129. Washington, D.C.: Folger Shakespeare Library, 1985.

———. *Palladio* Harmondsworth, England. Penguin, 1966.

Adams, Nicholas. "The Life and Times of Pietro dell'Abaco, a Renaissance Estimator from Siena (active 1457–1486)." *Zeitschrift für Kunstgeschichte* 48 (1985): 384–95

Agosti, Barbara. "Riflessioni su un manoscritto di Cesare Cesariano." In *Cesare Cesariano e il classicismo di primo cinquecento tra Milano e Como*, ed Maria Luisa Gatti Perrer and Alessandro Rovetta, 66–73. Milan: Università Cattolica del Sacro Cuore, 1996.

Aiken, Jane Andrews. "Leon Battista Alberti's System of Human Proportions." *Journal of the Warburg and Courtauld Institutes* 43 (1980): 68–96.

———. "Truth in Images: From the Technical Drawings of Ibn Al-Razzaz Al-Jazari, Campanus of Novara, and Giovanni de'Dondi to the Perspective Projection of Leon Battista Alberti." *Viator* 25 (1994): 325–59.

Alberigo, Guiseppe. "Barbaro, Daniele Matteo Alvisse." *DBI* 6:89–95.

Alexander, Shirley M. "Medieval Recipes Describing the Use of Metals in Manuscripts." *Marsyas* 12 (1964–65): 34–51.

Alford, William P. *To Steal a Book Is an Elegant Offense: Intellectual Property Law in Chinese Civilization.* Stanford: Stanford University Press, 1995.

Allen, Michael J. B. *Plato's Third Eye: Studies in Marsilio Ficino's Metaphysics and Its Sources.* Aldershot, Hampshire: Ashgate, Variorum, 1995.

Anderson, J. K. *Xenophon.* London: Duckworth, 1974.

Andrewes, A. "The Spartan Resurgence." In *The Cambridge Ancient History*, vol. 5, *Fifth Century B.C.*, ed. D. M. Lewis, John Boardman, J. K. Davies, and M. Ostwald, 464–98. 2nd ed. Cambridge: Cambridge University Press, 1992.

Anglo, Sydney. "Machiavelli as a Military Authority: Some Early Sources." In *Florence and Italy: Renaissance Studies in Honour of Nicolai Rubinstein*, ed. Peter Denley and Caroline Elam, 321–34. London: Westfield College, University of London, 1988.

Aquilecchia, Giovanni. "Bruno, Giordano." *DBI* 14:654–65.

Arend, Gerhard. *Die Mechanik des Niccolò Tartaglia. Im Kontext der zeitgenössischen Erkenntnis- und Wissenschaftstheorie.* Munich: Institut für Geschichte der Naturwissenschaften, 1998.

Argiolas, Tommaso. *Armi ed eserciti del Rinascimento italiano.* Rome: Newton Compton, 1991.

Ariès, Philippe, and Georges Duby, eds. *A History of Private Life.* Trans. Arthur Goldhammer. 5 vols. Cambridge: Harvard University Press, Belknap Press, 1987.

Armstrong, A. H. "Plotinus." In *The Cambridge History of Later Greek and Early Medieval Philosophy*, 195–271. Cambridge: University Press, 1967.

Armstrong, Elizabeth. *Before Copyright: The French Book Privilege System, 1498–1526.* Cambridge: Cambridge University Press, 1990.

Armstrong, Eva V., and Hiram S. Lukens. "Lazarus Ercker and His 'Probierbuch': Sir John Pettus and His 'Fleta Minor.'" *Journal of Chemical Education* 16 (December 1939): 553–62.

Astin, Alan E. *Cato the Censor.* Oxford: Clarendon, 1978.

Atiya, Aziz Suryal. *The Crusade of Nicopolis.* London: Methuen, 1934.

Auernheimer, Richard, and Frank Baron, eds. *Johannes Trithemius: Humanismus und Magie im vorreformatorischen Deutschland.* Munich: Profil Verlag, 1991.

Aune, David E. "Magic in Early Christianity." *ANRW*, pt. 2, *Principat.*, 23.2:1507–57.

Bagley, Paul J. "On the Practice of Esotericism." *Journal of the History of Ideas* 53 (January–March 1992): 231–47.

Baiardi, Giorgio Cerboni, Giorgio Chittolini, and Piero Floriani, eds. *Federico di Montefeltro.* 3 vols. Rome: Bulzoni Editore, 1986.

Baldwin, Barry. "Columella's Sources and How He Used Them" *Latomus* 22 (1963): 785–91.

——— "The Date, Identity, and Career of Vitruvius" *Latomus* 49 (April–June 1990): 425–34.

Baldwin, Martha. "The Snakestone Experiments: An Early Modern Medical Debate." *Isis* 86 (September 1995): 394–418

Barbieri, Franco. "Giangiorgio Trissino e Andrea Palladio." In *Convegno di studi su Giangiorgio Trissino*, ed Neri Pozza, 191–211. Vicenza: Accademia Olimpica, 1979.

Barnes, Jonathan. *The Presocratic Philosophers*. Rev ed. London: Routledge & Kegan Paul, 1982

Baron, Frank "Paracelsus und sein Drucker (1527–1539)." In *Neue Beiträge zur Paracelsus-Forschung*, ed. Peter Dilg and Hartmut Rudolph, 141–50. Stuttgart: Akademie der Diözese Rottenburg-Stuttgart, 1995

Barral i Altet, Xavier, ed. *Artistes, artisans, et production artistique au moyen agê*. 3 vols. Paris: Picard, 1986–90

Barthes, Roland "The Death of the Author." In *Image, Music, Text*, trans. Stephen Heath, 142–48. Glasgow. William Collins Sons, 1977.

Battisti, Eugenio *Brunelleschi: The Complete Work*. Trans. Robert Erich Wolf. Rev. Eugenio Battisti and Emily Lane. London: Thames & Hudson, 1981.

——— *Piero della Francesca*. 2 vols Milan: Istituto Editoriale Italiano, 1971.

Baum, Wilhelm. *Kaiser Sigismund: Hus, Konstanz und Türkenkriege*. Graz: Verlag Styria, 1993.

Baxendale, Susannah Foster. "Exile in Practice: The Alberti Family In and Out of Florence, 1401–1428." *Renaissance Quarterly* 44 (winter 1991). 720–56

Beagon, Mary *Roman Nature: The Thought of Pliny the Elder*. Oxford: Clarendon, 1992.

Beierlein, Paul R. *Lazarus Ercker: Bergmann, Hüttenmann und Münzmeister im 16. Jahrhundert*. Berlin. Akademie-Verlag, 1955.

Bengtson, Hermann. "Die griechische Polis bei Aeneas Tacticus." *Historia* 11 (October 1962): 458–68

Bennett, J. A. "The Mechanics' Philosophy and the Mechanical Philosophy," *History of Science* 24 (1986): 1–28

Benton, John F "Arnald of Villanova (ca. 1240–1311)." *DMA* 1.537–38.

Benzenhöffer, Udo. "Zum Brief des Johannes Oporinus über Paracelsus: Die bislang älteste bekannte Briefüberlieferung in einer 'Oratio' von Gervasius Marstaller." *Sudhoffs Archiv* 73 (1989) 55–63.

Berninger, Ernst H., ed. *Das Buch vom Bergbau: Die Miniaturen des "Schwazer Bergbuchs" nach der Handschrift im Besitz des Deutschen Museums in München*. Dortmund: Harenberg Kommunikation, 1980.

Bernstein, Eckhard. "Celtis, Conrad." In *Encyclopedia of the Renaissance*, ed. Paul F. Grendler, 1:380–82 New York: Charles Scribner's Sons, 1999.

Bertelli, Carlo. *Piero della Francesca*. Trans. Edward Farrelly New Haven: Yale University Press, 1992.

Berthelot, Marcellin *La chimie au moyen âge*. 3 vols. 1893 Reprint, Osnabruck· Zeller, 1967.

Bettalli, Marco. "Enea Tattico e l'insegnamento dell'arte militare." *Annali della Facoltà di Lettere e Filosofia dell'Università di Siena* 7 (1986)· 73–89

Betts, Richard J. "On the Chronology of Francesco di Giorgio's Treatises: New Evidence from an Unpublished Manuscript" *Journal of the Society of Architectural Historians* 36 (March 1977): 3–14.

Beyer, Andreas. "Palladio, Andrea." *DA* 23·861–72

Biagioli, Mario. *Galileo, Courtier. The Practice of Science in the Culture of Absolutism* Chicago· University of Chicago Press, 1993.

——— "The Social Status of Italian Mathematicians, 1450–1600." *History of Science* 27 (1989): 41–95.

Bianchi, Massimo L. *Signatura rerum: Segni, magia e conoscenza da Paracelso a Leibniz* Rome: Edizioni dell'Ateneo, 1987

———. "The Visible and the Invisible from Alchemy to Paracelsus." In *Alchemy and Chemistry in the Sixteenth and Seventeenth Centuries*, ed. Piyo Rattansi and Antonio Clericuzio, 17–50. Dordrecht: Kluwer, 1994

Bischoff, Bernhard "Die Überlieferung der technischen Literatur." In *Artigianato e tecnica nella società dell'alto medioevo occidentale*, 1:267–96 2 vols. Spoleto: Centro Italiano di Studi sull'Alto Medioevo, 1971.

Black, Antony *Guilds and Civil Society in European Political Thought from the Twelfth Century to the Present.* Ithaca, N.Y.. Cornell University Press, 1984.

Blair, John, and Nigel Ramsey, eds *English Medieval Industries: Craftsmen, Techniques, Products.* London: Hambledon, 1991

Blümlein, Kilian. *Naturerfahrung und Welterkenntnis: Der Beitrag des Paracelsus zur Entwicklung des neuzeitlichen naturwissenschaftlichen Denkens.* Frankfurt am Main: Peter Lang, 1992

Blunt, Anthony. *Art and Architecture in France, 1500–1700.* 2nd ed., rev. Harmondsworth, England: Penguin, 1973

——— *Philibert de L'Orme* London· Zwemmer, 1958

Boak, Arthur E. R. "The Book of the Prefect." *Journal of Economic and Business History* 1 (August 1929). 597–619.

———. "Guilds· Late Roman and Byzantine" In *Encyclopaedia of the Social Sciences*, ed. Edwin R. A. Seligman and Alvin Johnson, 206–8 New York. Macmillan, 1937.

———. *Manpower Shortage and the Fall of the Roman Empire in the West.* Ann Arbor: University of Michigan Press, 1955.

Bock, Gisela, Quentin Skinner, and Maurizio Viroli, eds. *Machiavelli and Republicanism* Cambridge. Cambridge University Press, 1990

Bok, Sissela *Secrets· On the Ethics of Concealment and Revelation* New York· Vintage, 1984.

Bonicatti, Maurizio. "Durer nella storia delle idee umanistiche fra quattrocento e cinquecento." *Journal of Medieval and Renaissance Studies* 1 (fall 1971)· 131–250.

Bono, James J. *The Word of God and the Languages of Man. Interpreting Nature in Early Modern Science and Medicine.* Vol 1, *Ficino to Descartes.* Madison. University of Wisconsin Press, 1995

Bornhardt, Wilhelm. *Geschichte des Rammelsberger Bergbaues von seiner Aufnahme bis zur Neuzeit.* Berlin: Preussische Geologische Landesanstalt, 1931

Borsellino, N., and E. Camesasca "Cellini, Benvenuto." *DBI* 23:440–51

Borsi, Franco *Leon Battista Alberti* Milan: Electa, 1975.

Bortolotti, Ettore, "I contributi del Tartaglia, del Cardano, del Ferrari, e della scuola matematica Bolognese alla teoria algebrica delle equazioni cubiche." *Studi e Memorie per la Storia dell'Università di Bologna* 9 (1926): 55–108.

Boucher, Bruce. *Andrea Palladio: The Architect in His Time.* New York· Abbeville, 1998.

Bourdieu, Pierre. *Distinction: A Social Critique of the Judgement of Taste* Trans Richard Nice. Cambridge: Harvard University Press, 1984.

———. *The Logic of Practice* Trans. Richard Nice. Stanford, Calif.· Stanford University Press, 1990

Bowman, Alan K., Edward Champlin, and Andrew Lintott, eds. *The Cambridge Ancient History.* Vol. 10, *The Augustan Empire, 43 B.C.–A.D. 69* 2nd ed. Cambridge: Cambridge University Press, 1996.

Boyce, Helen *The Mines of the Upper Harz from 1542 to 1589.* Menasha, Wis.: George Banta, 1920

Boyer, Carl B. *A History of Mathematics.* Ed Uta C Merzbach. Rev. ed. New York· John Wiley & Sons, 1991.

Boyle, James. *Shamans, Software, and Spleens. Law and the Construction of the Information Society* Cambridge: Harvard University Press, 1996.

Bradley, K. R. *Slaves and Masters in the Roman Empire: A Study in Social Control.* Oxford· Oxford University Press, 1987

Brann, Noel L. *The Abbot Trithemius (1462–1516): The Renaissance of Monastic Humanism.* Leiden: Brill, 1981.

Braunstein, Philippe. "A l'origine des privileges d'invention aux XIVe et XVe siècles." In *Les brevets: Leur utilisation en histoire des techniques et de l'économie,* 53–60 Paris· Centre National de la Recherche Scientifique, 1984

Brown, Peter. "Late Antiquity." In *A History of Private Life,* ed. Philippe Ariès and Georges Duby, vol. 1, *From Pagan Rome to Byzantium,* ed Paul Veyne, 234–311. Trans. Arthur Goldhammer. Cambridge· Harvard University Press, Belknap Press, 1987.

———. "The Rise and Function of the Holy Man in Late Antiquity." *Journal of Roman Studies* 61 (1971): 80–101

———. "Sorcery, Demons, and the Rise of Christianity from Late Antiquity into the Middle Ages." In *Witchcraft· Confessions and Accusations,* ed Mary Douglas, 17–45. London: Tavistock, 1970.

Brunello, Franco. *Arti e mestieri a Venezia nel medioevo e nel Rinascimento* Vicenza Neri Pozza Editore, 1980.

——— "Vannoccio Biringuccio e il trattato 'De la pirotechnia.'" In *Trattati scientifici nel veneto fra il XV e XVI secolo*, 29–37 Vicenza: Neri Pozza Editore, 1985.

Bugh, Glenn R. *The Horsemen of Athens.* Princeton, N.J : Princeton University Press, 1988.

Burd, L. Arthur "Le fonti letterarie di Machiavelli nell' 'Arte della guerra'" *Atti della R. Accademia dei Lincei: Classe di Scienze Morali, Storiche e Filologiche,* 5th ser , 4.1 (1897). 188–261

Burford, Alison "Crafts and Craftsmen " In *Civilization of the Ancient Mediterranean Greece and Rome,* ed Michael Grant and Rachel Kitzinger, 3.367–88 New York· Charles Scribner's Sons, 1988

———. *Craftsmen in Greek and Roman Society.* Ithaca, N.Y.. Cornell University Press, 1972

———. *The Greek Temple Builders at Epidauros: A Social and Economic Study of Building in the Asklepian Sanctuary, during the Fourth and Early Third Centuries B.C.* Toronto: University of Toronto Press, 1969.

Burke, Seán. *The Death and Return of the Author· Criticism and Subjectivity in Barthes, Foucault, and Derrida* Edinburgh. Edinburgh University Press, 1992.

Burkert, Walter. *Ancient Mystery Cults.* Cambridge. Harvard University Press, 1987

Butters, H C. *Governors and Government in Early Sixteenth-Century Florence, 1502–1519.* Oxford: Clarendon, 1985

Caferro, William. *Mercenary Companies and the Decline of Siena.* Baltimore· Johns Hopkins University Press, 1998.

Callebat, Louis "La prose du 'De Architectura' de Vitruve." *ANRW,* pt. 2, *Principat.,* 30.1.696–722

Campbell Thompson, R. *A Dictionary of Assyrian Chemistry and Geology.* Oxford· Clarendon, 1936.

Carpo, Mario *L'architettura dell'età della stampa· Oralità, scrittura, libro stampato e riproduzione meccanica dell'immagine nella storia delle teorie architettoniche* Milan: Editoriale Jaca Book, 1998

Castagnoli, Ferdinando. *Orthogonal Town Planning in Antiquity.* Trans. Victor Caliandro. Cambridge: MIT Press, 1971.

Catalano, Franco. *Francesco Sforza* Milan. Dall'Oglio Editore, 1983

Celato, Sergio "Enea Tattico· Il problema dell'autore e il valore dell'opera dal punto di vista militare." *Accademia Patavina di Scienze, Lettere ed Arti. Memorie* 80 (1967–68): 53–67.

Charleston, R. J., and L M Angus-Butterworth "Glass " In *A History of Technology,* ed Charles Singer et al., vol. 3, *From the Renaissance to the Industrial Revolution, c 1500–c 1750,* 206–44. New York Oxford University Press, 1957.

Chartier, Roger. *The Order of Books: Readers, Authors, and Libraries in Europe between the Fourteenth and the Eighteenth Centuries* Trans Lydia G Cochrane. Stanford, Calif : Stanford University Press, 1994.

Chavasse, Ruth "The First Known Author's Copyright, September 1486, in the Context of a Humanist Career." *Bulletin of the John Rylands University Library of Manchester* 69 (August 1986): 11–37

Chittolini, Giorgio, ed. *Gli Sforza, la chiesa lombarda, la corte di Roma: Strutture e pratiche beneficiarie nel ducato di Milano (1450–1535)* Naples: Liguori Editore, 1989.

Cichorius, Conrad. *Romische Studien: Historisches epigraphisches literargeschichtliches aus vier Jahrhunderten Roms* Leipzig. Teubner, 1922

Cicogna, Emanuele Antonio. *Delle inscrizioni veneziane* Vol. 6, pt 1. 1853. Reprint, Bologna: Forni Editore, 1970.

Clagett, Marshall "The Life and Works of Giovanni Fontana" *Annali dell'Istituto e Museo di Storia della Scienza di Firenze* 1 (1976): 5–28

Cockle, Maurice J. D. *A Bibliography of Military Books up to 1642* 2nd ed. 1957. Reprint, London: Holland Press, 1978

Cohen, H. Floris. *The Scientific Revolution: A Historiographical Inquiry*. Chicago: University of Chicago Press, 1994.

Cohen, I. Bernard, ed. *Puritanism and the Rise of Modern Science: The Merton Thesis*. New Brunswick, N.J.: Rutgers University Press, 1990

Cole, Bruce. "Titian and the Idea of Originality in the Renaissance." In *The Craft of Art: Originality and Industry in the Italian Renaissance and Baroque Workshop*, ed. Andrew Ladis and Carolyn Wood. 86–112. Athens: University of Georgia Press, 1995.

Cole, Thomas. *Democritus and the Sources of Greek Anthropology* Chapel Hill, N.C.: American Philological Association, 1967.

Colish, Marcia L. "Machiavelli's *Art of War*: A Reconsideration." *Renaissance Quarterly* 51 (winter 1998): 1151–68

Compagni, Vittoria Perrone. "'Dispersa Intentio': Alchemy, Magic, and Scepticism in Agrippa." *Early Science and Medicine* 5 (2000): 160–77

Conacher, D. J. *Aeschylus' "Prometheus Bound". A Literary Commentary*. Toronto. University of Toronto Press, 1980.

Concina, Ennio. *La macchina territoriale: La progettazione della difesa nel cinquecento veneto* Rome: Gius. Laterza & Figli, 1983

Copenhaver, Brian P "Hermes Trismegistus, Proclus, and the Question of a Philosophy of Magic in the Renaissance." In *Hermeticism and the Renaissance. Intellectual History and the Occult in Early Modern Europe*, ed Ingrid Merkel and Allan G Debus, 79–110. Washington, D.C.: Folger Shakespeare Library, 1988.

——— "The Historiography of Discovery in the Renaissance: The Sources and Composition of Polydore Vergil's 'De inventoribus rerum,' I–III." *Journal of the Warburg and Courtauld Institutes* 41 (1978): 192–214.

———. "Iamblichus, Synesius, and the Chaldaean Oracles in Marsilio Ficino's *De Vita Libri Tres*: Hermetic Magic or Neoplatonic Magic?" In *Supplementum Festivum: Studies in Honor of Paul Oskar Kristeller*, ed. James Hankins, John Monfasani, and Frederick Purnell Jr., 441–55 Binghamton, N.Y.: Medieval and Renaissance Text and Studies, 1987.

———. "Natural Magic, Hermetism, and Occultism in Early Modern Science." In *Reappraisals of the Scientific Revolution*, ed David C Lindberg and Robert S Westman, 261–301 Cambridge: Cambridge University Press, 1990.

——— "Renaissance Magic and Neoplatonic Philosophy· 'Ennead' 4.3.5 in Ficino's 'De vita coelitus comparanda.'" In *Marsilio Ficino e il Ritorno di Platone*, ed. Gian Carlo Garfagnini, 2:351–69 Florence: Leo S. Olschki, 1986.

———. *Symphorien Champier and the Reception of the Occultist Tradition in Renaissance France* The Hague· Mouton, 1978.

Cordt, Ernst *Die Gilden: Ursprung und Wesen*. Göppingen: Kümmerle Verlag, 1984.

Corso, Antonio. *Monumenti Periclei: Saggio critico sulla attività edilizia di Pericle*. Venice: Istituto Veneto di Scienze, Lettere ed Arti, 1986.

Coy, Michael W., ed. *Apprenticeship: From Theory to Method and Back Again*. Albany. State University of New York Press, 1989.

Crisciani, Chiara, and Michela Pereira. *L'arte del sole e della luna· Alchimia e filosofia nel medioevo*. Spoleto: Centro Italiano di Studi sull'Alto Medioevo, 1996

Croix, Horst de la. "The Literature on Fortification in Renaissance Italy" *Technology and Culture* 4 (winter 1963): 30–50.

Crombie, Alistair C. *The History of Science from Augustine to Galileo* 2nd rev and enl. ed 2 vols. 1959. Reprint, New York. Dover, 1995.

———. *Styles of Scientific Thinking in the European Tradition: The History of Argument and Explanation Especially in the Mathematical and Biomedical Sciences and Arts*. 3 vols. London. Duckworth, 1994

Crossgrove, William *Die deutsche Sachliteratur des Mittelalters*. Bern: Peter Lang, 1994.

Cultura, scienze e tecniche nella Venezia del cinquecento. Giovan Battista Benedetti e il suo tempo. Venice: Istituto Veneto di Scienze, Lettere ed Arti, 1987.

Cuomo, Serafina. "Shooting by the Book: Notes on Niccolò Tartaglia's *Nova scientia*" *History of Science* 35 (1997)· 155–88

Dahlmann, Hellfried. "Varroniana." *ANRW*, pt. 1, *Von den Anfängen Roms bis zum Ausgang der Republik*, 3.3–25.

D'Arms, John H. *Commerce and Social Standing in Ancient Rome*. Cambridge: Harvard University Press, 1981.

Darmstaedter, Ernst. *Berg-, Probir- und Kunstbüchlein*. Munich. Verlag der Munchner Drucke, 1926.

Daston, Lorraine, and Katharine Park. *Wonders and the Order of Nature*. New York: Zone Books, 1998.

Davis, Margaret Daly. *Piero della Francesca's Mathematical Treatises. The "Trattato d'abaco" and "Libellus de quinque corporibus regularibus."* Ravenna· Longo Editore, 1977.

Davis-Weyer, Caecilia "Panel and Wall Painting, Mosaics, Metalwork, and Other Decorative Arts" In *Medieval Latin: An Introduction and Bibliographical Guide*, ed. F A. C. Mantello and A. G Rigg, 468–73. Washington, D.C.: Catholic University of America Press, 1996.

Dear, Peter. *Discipline and Experience· The Mathematical Way in the Scientific Revolution.* Chicago: University of Chicago Press, 1995.

Debus, Allen G *The Chemical Philosophy: Paracelsian Science and Medicine in the Sixteenth and Seventeenth Centuries.* 2 vols. New York. Science History, 1977.

Dechert, Michael S A. "The Military Architecture of Francesco di Giorgio in Southern Italy." *Journal of the Society of Architectural Historians* 49 (June 1990). 161–80

Decot, Rolf "Hermann von Wied 14 January 1477–15 August 1552." CE 3:444–46.

Delebecque, Édouard. *Essai sur la vie de Xénophon.* Paris Librairie C. Klincksieck, 1957

Delia, Diana. "From Romance to Rhetoric: The Alexandrian Library in Classical and Islamic Traditions." *American Historical Review* 97 (December 1992): 1449–67.

Denley, Peter, and Caroline Elam, eds. *Florence and Italy· Renaissance Studies in Honour of Nicolai Rubinstein.* London: Westfield College, University of London, 1988.

Dill, Samuel. *Roman Society in the Last Century of the Western Empire.* 2nd rev ed New York: Meridian, 1958.

Dilley, Roy M. "Secrets and Skills: Apprenticeship among Tukolor Weavers." In *Apprenticeship. From Theory to Method and Back Again,* ed. Michael W. Coy, 181–98. Albany: State University of New York Press, 1989.

Dillon, John. *The Middle Platonists. A Study of Platonism, 80 B.C. to A.D. 220.* London: Duckworth, 1977.

Dinsmoor, William Bell. *The Architecture of Ancient Greece: An Account of Its Historic Development.* 3rd rev. ed. 1950. Reprint, New York: W. W. Norton, 1975.

———. "The Literary Remains of Sebastiano Serlio " *Art Bulletin* 24 (March 1942): 55–91

Dobbs, Betty Jo Teeter. *The Janus Faces of Genius: The Role of Alchemy in Newton's Thought.* Cambridge: Cambridge University Press, 1991.

Dodds, E. R. *The Greeks and the Irrational.* Berkeley and Los Angeles. University of California Press, 1951.

Drachmann, A. G "Ctesibius (Ktesibios)" *DSB* 3:491–92.

———. "Hero of Alexandria." *DSB* 6:310–14

———. *Ktesibios, Philon, and Heron. A Study in Ancient Pneumatics.* Copenhagen: Munksgaard, 1948

———. *The Mechanical Technology of Greek and Roman Antiquity. A Study of the Literary Sources.* Copenhagen: Munksgaard, 1963

———. "Philo of Byzantium " *DSB* 10.586–89

Drake, Stillman "Tartaglia's Squadra and Galileo's Compasso." *Annali dell'Istituto e Museo di Storia della Scienza di Firenze* 2 (1977): 35–54.

Duff, Patrick W *Personality in Roman Private Law.* Cambridge. Cambridge University Press, 1938.

Dunn, Kevin. *Pretexts of Authority· The Rhetoric of Authorship in the Renaissance Preface.* Stanford, Calif.: Stanford University Press, 1994.

Dunne, Joseph. *Back to the Rough Ground. "Phronesis" and "Techne" in Modern Philosophy and in Aristotle.* Notre Dame, Ind.: University of Notre Dame Press, 1993

Eamon, William "Court, Academy, and Printing House: Patronage and Scientific Careers in Late Renaissance Italy." In *Patronage and Institutions· Science, Technology, and Medicine at the European Court, 1500–1750*, ed. Bruce T. Moran, 25–50. Rochester, N.Y.: Boydell, 1991

———. "From the Secrets of Nature to Public Knowledge: The Origins of the Concept of Openness in Science." *Minerva* 23 (autumn 1985): 321–47.

———. *Science and the Secrets of Nature: Books of Secrets in Medieval and Early Modern Culture*. Princeton, N J.: Princeton University Press, 1994.

Easterling, P. E., and B. M. W. Knox, eds. *The Cambridge History of Classical Literature*. Vol. 1, *Greek Literature*. Cambridge. Cambridge University Press, 1985

Eckert, Willehad Paul, and Christoph von Imhoff *Willibald Pirckheimer: Dürers Freund im Spiegel seines Lebens, seiner Werke und seiner Umwelt* 2nd ed Cologne: Wienand Verlag, 1982

Edgerton, Samuel Y , Jr. *The Heritage of Giotto's Geometry: Art and Science on the Eve of the Scientific Revolution*. Ithaca, N Y.. Cornell University Press, 1991.

———. *The Renaissance Rediscovery of Linear Perspective* New York. Basic Books, 1975

Egg, Erich. "From the Beginning to the Battle of Marignano–1515." In *Guns: An Illustrated History of Artillery*, ed. Joseph Jobé, 9–36 Greenwich, Conn · New York Graphic Society, 1971.

———. *Das Handwerk der Uhr- und der Buchsenmacher in Tirol*. Innsbruck: Universitatsverlag Wagner, 1982

——— "Ludwig Lassl und Jorg Kolber. Verfasser und Maler des Schwazer Bergbuchs." *Der Anschnitt* 9 (March 1957): 15–19.

——— *Der Tiroler Geschützguss. 1400–1600*. Innsbruck: Universitätsverlag Wagner, 1961

Eichberger, Dagmar, and Charles Zika, eds *Dürer and His Culture*. Cambridge. Cambridge University Press, 1998.

Eis, Gerhard. *Mittelalterliche Fachliteratur*. Stuttgart: J. B. Metzler, 1962.

———. *Vor und nach Paracelsus: Untersuchungen über Hohenheims Traditionsverbundenheit und Nachrichten über seine Anhänger*. Stuttgart: Gustav Fischer Verlag, 1965.

Eisenstein, Elizabeth L. *The Printing Press as an Agent of Change: Communications and Cultural Transformations in Early-Modern Europe*. 2 vols. Cambridge Cambridge University Press, 1979.

Eliade, Mircea. *The Forge and the Crucible*. Trans Stephen Corrin 2nd ed. Chicago: University of Chicago Press, 1978.

———. *The Myth of the Eternal Return or, Cosmos and History*. Trans Willard R Trask Princeton, N.J.: Princeton University Press 1965.

Epstein, S. R "Craft Guilds, Apprenticeship, and Technological Change in Preindustrial Europe." *Journal of Economic History* 58 (September 1998): 684–713.

Epstein, Steven A *Wage Labor and Guilds in Medieval Europe* Chapel Hill: University of North Carolina Press, 1991.

Evans, Harry B. *Water Distribution in Ancient Rome. The Evidence of Frontinus*. Ann Arbor: University of Michigan Press, 1994.

Fabricius, E "Diades 2) Mechaniker." *RE* 5.1 (1903), col 305.

Fagan, Brian M *People of the Earth: An Introduction to World Prehistory*. 8th ed. New York: HarperCollins, 1995.

Faivre, Antoine, and Rolf Christian Zimmermann, eds. *Epochen der Naturmystik. Hermetische Tradition im wissenschaftlichen Fortschritt* Berlin: Erich Schmidt Verlag, 1979.

Falciai, Patrizia Benvenuti. *Ippodamo di Mileto: Architetto e filosofo: Una ricostruzione filologica della personalità.* Florence: Università degli Studi di Firenze, 1982.

Favaro, Antonio. *Per la biografia di Niccolò Tartaglia.* Rome: Ermanno Loescher, 1913 First published in *Archivio Storico Italiano* 71 (1913): 335–72.

Feather, John "Authors, Publishers, and Politicians· The History of Copyright and the Book Trade" *European Intellectual Property Review* 12 (1988)· 377–80

Febvre, Lucien, and Henri-Jean Martin. *The Coming of the Book: The Impact of Printing, 1450–1800* Trans. David Gerard. Ed. Geoffrey Nowell-Smith and David Wooton. London: Verso, 1990

Ferrari, Gian Arturo, and Mario Vegetti "Science, Technology, and Medicine in the Classical Tradition." In *Information Sources in the History of Science and Medicine,* ed. Pietro Corsi and Paul Weindling, 197–220. Butterworth's Guides to Information Services London: Butterworth, 1983.

Ferraro, Vittorio. "Il numero delle fonti, dei volumi e dei fatti della *Naturalis historia* di Plinio." *Annali della Scuola Normale Superiore di Pisa: Classe di Lettere e Filosofia,* 3rd ser., 5 (1975). 519–33.

Festugière, André J. *La révélation d'Hermès Trismégiste.* 4 vols. Paris· Librairie Lecoffre, 1944–54.

Field, Arthur. *The Origins of the Platonic Academy of Florence.* Princeton, N J.· Princeton University Press, 1988.

Field, J V. "Mathematics and the Craft of Painting: Piero della Francesca and Perspective." In *Renaissance and Revolution. Humanists, Scholars, Craftsmen, and Natural Philosophers in Early Modern Europe,* ed. J. V Field and Frank A. J. L. James, 73–95 Cambridge: Cambridge University Press, 1993.

———. "Piero della Francesca's Treatment of Edge Distortion." *Journal of the Warburg and Courtauld Institutes* 49 (1986): 66–90.

Finamore, John F. *Iamblichus and the Theory of the Vehicle of the Soul.* Chico, Calif.· Scholars Press, 1985.

Findlen, Paula. "Controlling the Experiment: Rhetoric, Court Patronage, and the Experimental Method of Francesco Redi." *History of Science* 31 (1993): 35–64.

Finley, Moses I. "Technical Innovation and Economic Progress in the Ancient World." In *Economy and Society in Ancient Greece,* ed. Brent D. Shaw and Richard P. Saller, 176–95. New York. Viking, 1981.

Finnegan, Ruth *Literacy and Orality: Studies in the Technology of Communication* Oxford· Basil Blackwell, 1988

Fiocco, Giuseppe. *Alvise Cornaro, il suo tempo e le sue opere.* [Venice]: Neri Pozza Editore, [1965].

Fiore, Francesco Paolo. "Cesariano [Ciserano], Cesare." DA 6:356–59.

———. "La traduzione vitruviana di Cesare Cesariano." In *Roma, centro ideale della cul-*

tura dell'antico nei secoli XV e XVI da Martino V al sacco di Roma, 1417–1527, ed. Silvia Danesi Squarzina, 458–66. Milan: Electa, 1989

Fiore, Francesco Paolo, and Manfredo Tafuri, eds. *Francesco di Giorgio architetto.* Milan: Electa, 1993.

Firpo, Luigi. *Il processo di Giordano Bruno.* Ed. Diego Quaglioni. Rome: Salerno Editrice, 1993

Forssman, Erik. "Palladio e Daniele Barbaro." *Bollettino del Centro Internazionale di Studi di Architettura Andrea Palladio* 8 (1966): 68–81.

Foucault, Michel. "What Is an Author?" In *Language, Counter-Memory, Practice: Selected Essays and Interviews*, ed. Donald F Bouchard, trans. Sherry Simon and Donald F. Bouchard, 113–38. Ithaca, N.Y.: Cornell University Press, 1977.

Fowden, Garth *The Egyptian Hermes. A Historical Approach to the Late Pagan Mind.* Princeton, N.J.: Princeton University Press, 1986.

Franceschelli, Remo. "Le origini e lo svolgimento del diritto industriale nei primi secoli dell'arte della stampa" *Rivista di Diritto Industriale* 1, pt 1 (1952): 151–95.

———. "Il primo privilegio in materio di stampe (Il privilegio concesso il 18 Settembre 1469 dal Senato Veneto allo stampatore Giovanni da Spira)." *Rivista di Diritto Industriale* 1, pt. 1 (1952). 370–74.

Fraser, Peter M "Aristophanes of Byzantion and Zoilus Homeromastix in Vitruvius. A Note on Vitruvius VII. Praef. 4–9." *Eranos* 68 (1970): 115–22.

———. *Ptolemaic Alexandria.* 3 vols Oxford. Clarendon, 1972.

Freguglia, Paolo, "Niccolò Tartaglia e il rinnovamento delle matematiche nel cinquecento" In *Cultura, scienze e techniche nella Venezia del cinquecento: Giovan Battista Benedetti e il suo tempo*, 203–16. Venice: Istituto Veneto di Scienze, Lettere ed Arti, 1987.

French, Roger *Ancient Natural History. Histories of Nature* London: Routledge, 1994.

Frontinus Gesellschaft e.V. [Gunther Garbrecht et al., eds] *Sextus Iulius Frontinus, Curator Aquarum: Wasserversorgung im antiken Rom.* Munich: R. Oldenbourg, 1982.

Frost, Frank J "Pericles and Dracontides." *Journal of Hellenic Studies* 84 (1964). 69–72.

Frumkin, Maximilian. "Early History of Patents for Invention." *Transactions of the Newcomen Society* 26 (1947–49). 47–56.

Fuhrmann, Manfred *Das systematische Lehrbuch: Ein Beitrag zur Geschichte der Wissenschaften in der Antike.* Göttingen: Vandenhoeck & Ruprecht, 1960

Gabba, Emilio "The *Collegia* of Numa: Problems of Method and Political Ideas." *Journal of Roman Studies* 74 (1984): 81–86.

———. "Scienza e potere nel mondo ellenistico." In *La scienza ellenistica*, ed. Gabriele Giannantoni and Mario Vegetti, 11–37. Pavia: Centro di Studio del Pensiero Antico, 1984.

———. "Tecnologia militare antica." In *Tecnologia economia e società nel mondo Romano*, 219–34. Como: Banco Popolare Commercio e Industria, 1980.

Gabrieli, Giovanni Battista. *Nicolò Tartaglia: Invenzioni, disfide e sfortune.* Siena: Università degli Studi, 1986.

Gadamer, Hans-Georg. *Truth and Method.* 2nd rev. ed. Trans. Joel Weinsheimer and Donald G. Marshall. New York: Continuum, 1993.
Gadol, Joan. *Leon Battista Alberti. Universal Man of the Early Renaissance.* Chicago: University of Chicago Press, 1969.
Galison, Peter. "Computer Simulations and the Trading Zone." In *The Disunity of Science: Boundaries, Contexts, and Power,* ed. Peter Galison and David J. Stump, 118–57. Stanford, Calif.: Stanford University Press, 1996.
Galluzzi, Paolo. "L'Accademia del Cimento: 'Gusti' del principe, filosofia e ideologia dell'esperimento." *Quaderni Storici* 16 (1981): 788–844.
———. "The Career of a Technologist." In *Leonardo da Vinci: Engineer and Architect,* ed. Paolo Galluzzi, 1–109. Montreal: Montreal Museum of Fine Arts, 1987.
———. "Le macchine senesi: Ricerca antiquaria, spirito di innovazione e cultura del territorio." In *Prima di Leonardo: Cultura delle macchine a Siena nel Rinascimento,* ed. Paolo Galluzzi, 15–44. Milan: Electa, 1991.
———. "Leonardo da Vinci: From the 'Elementi macchinali' to the Man-Machine." *History and Technology* 4 (1987): 235–65.
———, ed. *Leonardo da Vinci. Engineer and Architect.* Montreal: Montreal Museum of Fine Arts, 1987.
———. *Prima di Leonardo. Cultura delle macchine a Siena nel Rinascimento.* Milan: Electa, 1991.
Ganzenmüller, Wilhelm. "Paracelsus und die Alchemie des Mittelalters." In *Beiträge zur Geschichte der Technologie und der Alchemie,* 300–314. Weinheim: Verlag Chemie, 1956.
Garfagnini, Gian Carlo, ed. *Marsilio Ficino e il ritorno di Platone: Studi e documenti.* 2 vols. Florence: Leo S. Olschki, 1986.
Garin, Eugenio. *L'età nuova: Ricerche di storia della cultura dal XII al XVI secolo.* Naples: Casa Editrice A. Morano, 1969.
Garlan, Yvon. "Cité, armées et stratégie à l'époque hellénistique d'après l'oeuvre de Philon de Byzance." *Historia* 22 (1973): 16–33.
———. *Recherches de poliorcétique grecque.* Athens: Écoles Française d'Athènes, 1974.
Gasparetto, Astone. *Il vetro di Murano dalle origini ad oggi.* Venice: Neri Pozza Editore, 1958.
Gasparro, Giulia Sfameni. "La gnosi ermetica come iniziazione e mistero." *Studi e Materiali di Storia delle Religioni* 36 (1965): 43–61.
Gatti, Sergio. "L'attività milanese del Cesariano dal 1512–13 al 1519." *Arte Lombarda* 16 (1971): 219–30.
———. "Un contributo alla storia delle vicissitudini incontrate dal 'Vitruvio' del Cesariano subito dopo la sua stampa a Como nel 1521." *Arte Lombarda,* n.s., 97 (1991): 132–33.
———. "Nuovi documenti sull'ambiente familiare e la prima educazione di Cesare Cesariano." *Arte Lombarda,* n.s., 86/87 (1988): 187–94.
Gatti Perer, Maria Luisa, and Alessandro Rovetta, eds. *Cesare Cesariano e il classicismo*

di primo cinquecento tra Milano e Como. Milan: Università Cattolica del Sacro Cuore, 1996

Gause, Ute. *Paracelsus (1493–1541)*. Tübingen: J. C B Mohr, 1993

Geoghegan, Arthur T. *The Attitude towards Labor in Early Christianity and Ancient Culture*. Washington, D.C : Catholic University of America Press, 1945

Gerkan, Armin von. *Griechische Städteanlagen. Untersuchungen zur Entwicklung des Städtebaues im Altertum*. Berlin: Walter de Gruyter, 1924.

Gersh, Stephen. *Middle Platonism and Neoplatonism. The Latin Tradition*. 2 vols Notre Dame, Ind.: University of Notre Dame Press, 1986.

Gessler, A. "Merz (Mercz), Martin." In *Die Deutsche Literatur des Mittelalters. Verfasserlexikon*, ed. Wolfgang Stammler and Karl Langosch, vol. 3, cols. 368–70. Berlin. Walter de Gruyter, 1943.

——— "Monch, Philipp." In *Die Deutsche Literatur des Mittelalters. Verfasserlexikon*, ed. Wolfgang Stammler and Karl Langosch, vol. 3, col 427. Berlin: Walter de Gruyter, 1943

Gilbert, Felix "Bernardo Rucellai and the Orti Oricellari· A Study on the Origin of Modern Political Thought." *Journal of the Warburg and Courtauld Institutes* 12 (1949). 101–31.

———. *Machiavelli and Guicciardini: Politics and History in Sixteenth-Century Florence*. Princeton, N.J.. Princeton University Press, 1965.

———. "Machiavelli: The Renaissance of the Art of War." In *Makers of Modern Strategy from Machiavelli to the Nuclear Age*, ed. Peter Paret with the collaboration of Gordon A Craig and Felix Gilbert, 11–31. Princeton, N.J.: Princeton University Press, 1986

Gille, Bertrand *Engineers of the Renaissance*. Cambridge: MIT Press, 1966

——— *Les mécaniciens grecs La naissance de la technologie* Paris· Éditions du Seuil, 1980.

Gilmore, Myron P., ed. *Studies on Machiavelli*. Florence: Sansoni, 1972.

Gingerich, Owen, and Robert S Westman *The Wittich Connection· Conflict and Priority in Late Sixteenth Century Cosmology. Transactions of the American Philosophical Society* 78, pt 7 (1988).

Ginsburg, Jane C "A Tale of Two Copyrights· Literary Property in Revolutionary France and America " *Tulane Law Review* 64 (May 1990): 991–1031.

Giry, A "Notice sur un traité du Moyen Age intitulé 'De coloribus et artibus romanorum.'" *Bibliothèque de L'École des Hautes Études: Sciences Philologiques et Historiques*, fasc. 35 (1878). 209–27.

Gold, Barbara K., ed *Literary and Artistic Patronage in Ancient Rome*. Austin: University of Texas Press, 1982

Goldammer, Kurt "Magie bei Paracelsus mit besonderer Berücksichtigung des Begriffs einer 'naturlichen Magie.'" In *Magia naturalis und die Entstehung der modernen Naturwissenschaften*, ed. Albert Heinekamp and Dieter Mettler, 30–55 Wiesbaden: Franz Steiner Verlag, 1978.

Goldthwaite, Richard A. *The Building of Renaissance Florence: An Economic and Social History*. Baltimore. Johns Hopkins University Press, 1980.

———. "Schools and Teachers of Commercial Arithmetic in Renaissance Florence." *Journal of European Economic History* 1 (fall 1972): 418–33.

———. *Wealth and the Demand for Art in Italy, 1300–1600*. Baltimore. Johns Hopkins University Press, 1993.

Golinski, Jan V. "Chemistry in the Scientific Revolution: Problems of Language and Communication." In *Reappraisals of the Scientific Revolution*, ed. David C. Lindberg and Robert S Westman, 367–96. Cambridge. Cambridge University Press, 1990.

———. *Making Natural Knowledge· Constructivism and the History of Science*. Cambridge: Cambridge University Press, 1998.

Goody, Jack, and Ian Watt. "The Consequences of Literacy." *Comparative Studies in Society and History* 5 (April 1963): 304–445.

Goodyear, F. R. D. "Technical Writing." In *The Cambridge History of Classical Literature*, vol 2, *Latin Literature*, ed. E. J Kenney and W V. Clausen, 667–73. Cambridge: Cambridge University Press, 1982

Gordon, Richard. "Aelian's Peony. The Location of Magic in Graeco-Roman Tradition." *Comparative Criticism* 9 (1987)· 59–95.

Gosselin, Edward A. "Bruno's 'French Connection'. A Historiographical Debate." In *Hermeticism and the Renaissance· Intellectual History and the Occult in Early Modern Europe*, ed. Ingrid Merkel and Alan G. Debus, 166–81 Washington, D.C.: Folger Shakespeare Library, 1988.

———. "Fra Giordano Bruno's Catholic Passion." In *Supplementum Festivum. Studies in Honor of Paul Oskar Kristeller*, ed. James Hankins, John Monfasani, and Frederick Purnell Jr., 537–61. Binghamton, N.Y.. Medieval and Renaissance Text and Studies, 1987

Graff, Harvey J. *The Legacies of Literacy. Continuities and Contradictions in Western Culture and Society*. Bloomington: Indiana University Press, 1987.

Grafton, Anthony. *Cardano's Cosmos: The Worlds and Works of a Renaissance Astrologer*. Cambridge: Harvard University Press, 1999.

———. *Commerce with the Classics· Ancient Books and Renaissance Readers*. Ann Arbor: University of Michigan Press, 1997.

———. *Forgers and Critics: Creativity and Duplicity in Western Scholarship*. Princeton, N.J : Princeton University Press, 1990.

Granada, Miguel A "Thomas Digges, Giordano Bruno e il copernicanesimo in inghilterra." In *Giordano Bruno, 1583–1585: The English Experience / L'esperienza inglese*, ed. Michele Ciliberto and Nicholas Mann, 125–55. Florence: Leo S. Olschki, 1997.

Gratwick, A S. "Prose Literature " In *The Cambridge History of Classical Literature*, vol. 2, *Latin Literature*, ed. E J Kenney and W. V Clausen, 138–55. Cambridge Cambridge University Press, 1982.

Grayson, Cecil, and Giulio C. Argan. "Alberti, Leon Battista." *DBI* 1:702–13.

Greenberg, Sidney. *The Infinite in Giordano Bruno with a Translation of His Dialogue "Concerning the Cause, Principle, and One."* New York: King's Crown Press, Columbia University, 1950.

Greenblatt, Stephen, and Giles Gunn, eds. *Redrawing the Boundaries. The Transformation of English and American Literary Studies.* New York: Modern Language Association of America, 1992.

Greene, Kevin. "Technological Innovation and Economic Progress in the Ancient World: M. I. Finley Re-considered." *Economic History Review* 53 (February 2000): 25–59.

Greenstein, Jack M. "Alberti on Historia: A Renaissance View of the Structure of Significance in Narrative Painting." *Viator* 21 (1990): 273–99.

Grendler, Paul F. "Printing and Censorship." In *The Cambridge History of Renaissance Philosophy*, ed. Charles B. Schmitt et al., 25–53. Cambridge: Cambridge University Press, 1988.

———. *The Roman Inquisition and the Venetian Press, 1540–1605.* Princeton, N.J.: Princeton University Press, 1977.

———. *Schooling in Renaissance Italy: Literacy and Learning, 1300–1600.* Baltimore: Johns Hopkins University Press, 1989.

Grese, William C. "Magic in Hellenistic Hermeticism." In *Hermeticism and the Renaissance: Intellectual History and the Occult in Early Modern Europe*, ed. Ingrid Merkel and Alan G. Debus, 45–58. Washington, D.C.: Folger Shakespeare Library, 1988.

Grimal, Pierre. "Encyclopédies antiques." *Cahiers d'histoire mondiale* 9 (1966): 459–82.

Gros, Pierre. "Vitruve. L'architecture et sa théorie, à la lumière des études récentes." *ANRW*, pt. 2, *Principat*, 30.1:659–95.

Gross, Walter H. "Iktinos." *KP* 2 (1967), cols. 1361–62.

———. "Pheidias." *KP* 4 (1972), cols. 722–24.

———. "Polykleitos 5. P. Von Argos." *KP* 4 (1972), cols. 1000–1003.

Gruneisen, Henny. "Friedrich I. der Siegreiche." *NDB* 5:526–28.

Guillaume, Jean. "On Philibert de L'Orme: A Treatise Transcending the Rules." In *Paper Palaces: The Rise of the Renaissance Architectural Treatise*, ed. Vaughan Hart with Peter Hicks, 219–31. New Haven: Yale University Press, 1998.

———, ed. *Les traités d'architecture de la Renaissance.* Paris: Picard, 1988.

Gustafsson, Bo. "The Rise and Economic Behaviour of Medieval Craft Guilds: An Economic-Theoretical Interpretation." *Scandinavian Economic History Review* 35 (1987): 1–40.

Habermas, Jurgen. *The Structural Transformation of the Public Sphere. An Inquiry into a Category of Bourgeois Society.* Trans. Thomas Burger with Frederick Lawrence. Cambridge: MIT Press, 1989.

———. *Theory and Practice.* Trans. John Viertel. Boston: Beacon, 1973.

Hacker, Barton C. "Women and Military Institutions in Early Modern Europe: A Reconnaissance." *Signs* 6 (summer 1981): 643–71.

Hale, John R. "The Early Development of the Bastion. An Italian Chronology, c. 1450–c. 1534." In *Europe in the Later Middle Ages*, ed J. R Hale, J. R. L. Highfield, and B. Smalley, 466–94. London: Faber & Faber, 1965.

———. "Printing and Military Culture of Renaissance Venice." In *Renaissance War Studies*, 429–70 London: Hambledon, 1983

———. *Renaissance Fortification: Art or Engineering?* London: Thames & Hudson, 1977.

——— *War and Society in Renaissance Europe, 1450–1620*. Rev. ed. Montreal: McGill-Queen's University Press, 1998.

Hall, A. Rupert. "Guido's *Texaurus*, 1335." In *On Pre-modern Technology and Science: A Volume of Studies in Honor of Lynn White, Jr.*, ed. Bert S. Hall and Delno C West, 11–33. Malibu, Calif.: Undena, 1976.

Hall, Bert S. "'Der Meister sol auch kennen schreiben und lesen': Writings about Technology ca 1400–ca. 1600 A D and Their Cultural Implications." In *Early Technologies*, ed Denise Schmandt-Besserat, 47–58 Malibu, Calif.: Undena, 1979.

———. "Giovanni de' Dondi and Guido da Vigevano: Notes toward a Typology of Medieval Technical Writings." In *Machaut's World: Science and Art in the Fourteenth Century*, ed Madeleine P Cosman and Bruce Chandler, 127–42. Annals of the New York Academy of Sciences, 414. New York: New York Academy of Sciences, 1978

——— *Weapons and Warfare in Renaissance Europe: Gunpowder, Technology, and Tactics*. Baltimore: Johns Hopkins University Press, 1997.

Halleux, Robert. "Les ouvrages alchimiques de Jean de Rupescissa" *Histoire Litteraire de la France* 41 (1981): 241–84

———. *Les textes alchimiques* Typologie des sources du moyen âge occidental, 36 Turnhout, Belgium: Brepols, 1979

Halleux, Robert, and Paul Meyvaert. "Les origines de la 'Mappae clavicula.'" *Archives d'Histoire Doctrinale et Littéraire du Moyen Age* 62 (1987): 7–58.

Hamilton, J. R. *Alexander the Great*. Pittsburgh: University of Pittsburgh Press, 1974.

Hammond, N. G. L. *A History of Greece to 322 B.C* 3rd ed. Oxford: Clarendon, 1986.

Hammond, N. G. L., G. T. Griffith, and F W Walbank *A History of Macedonia*. 3 vols. Oxford. Clarendon, 1972–88.

Hankins, James "Cosimo De' Medici and the 'Platonic Academy.'" *Journal of the Warburg and Courtauld Institutes* 53 (1990): 144–62.

——— "The Myth of the Platonic Academy of Florence." *Renaissance Quarterly* 44 (autumn 1991) 429–75

——— *Plato in the Italian Renaissance*. 2 vols. Leiden: Brill, 1990

Hankins, James, John Monfasani, and Frederick Purnell Jr, eds *Supplementum Festivum Studies in Honor of Paul Oskar Kristeller*. Binghamton, N.Y: Medieval and Renaissance Text and Studies, 1987.

Hannaway, Owen. *The Chemists and the Word. The Didactic Origins of Chemistry*. Baltimore: Johns Hopkins University Press, 1975.

———. "Georgius Agricola as Humanist." *Journal of the History of Ideas* 53 (October–December 1992) 553–60.

——— "Laboratory Design and the Aim of Science: Andreas Libavius versus Tycho Brahe." *Isis* 77 (December 1986). 585–610

Harris, William V. *Ancient Literacy*. Cambridge: Harvard University Press, 1989

Hart, Vaughan. "Serlio and the Representation of Architecture." In *Paper Palaces: The Rise of the Renaissance Architectural Treatise*, ed. Vaughan Hart with Peter Hicks, 170–85. New Haven· Yale University Press, 1998.

Hart, Vaughan, and Peter Hicks "On Sebastiano Serlio: Decorum and the Art of Architectural Invention " In *Paper Palaces· The Rise of the Renaissance Architectural Treatise*, ed Vaughan Hart with Peter Hicks, 140–57 New Haven· Yale University Press, 1998.

Hart, Vaughan, with Peter Hicks, eds. *Paper Palaces· The Rise of the Renaissance Architectural Treatise*. New Haven· Yale University Press, 1998.

Hartt, Frederick *History of Italian Renaissance Art· Painting, Sculpture, Architecture*. Ed. David G. Wilkins. 4th ed. rev. New York: Abrams, 1994

Hathaway, Neil. "Compilatio· From Plagiarism to Compiling." *Viator* 20 (1989): 19–44.

Heinekamp, Albert, and Dieter Mettler, eds *Magia naturalis und die Entstehung der modernen Naturwissenschaften* Wiesbaden. Franz Steiner Verlag, 1978

Henschke, Ekkehard *Landesherrschaft und Bergbauwirtschaft· Zur Wirtschafts- und Verwaltungsgeschichte des Oberharzer Bergbaugebietes im 16. und 17 Jahrhundert*. Berlin: Duncker & Humblot, 1974

Hershbell, Jackson P. "Democritus and the Beginnings of Greek Alchemy." *Ambix* 34 (March 1987) 5–20

Hess, Catherine. "Piccolpasso, Cipriano di Michele " *DA* 24:732–33.

Hesse, Carla "Enlightenment Epistemology and the Laws of Authorship in Revolutionary France, 1777–93." *Representations* 30 (spring 1990): 109–37.

Hodge, Trevor A. "How Did Frontinus Measure the Quinaria?" *American Journal of Archaeology* 88 (1984)· 205–16.

Hoffmann, V "L'Orme [Delorme], Philibert de " *DA* 19:690–95.

Holmyard, E. J. *Alchemy*. 1957. Reprint, New York: Dover, 1990.

Hope, Charles. "The Early History of the Tempio Malatestiano." *Journal of the Warburg and Courtauld Institutes* 55 (1992)· 51–154

Horsfall, Nicholas. "Prose and Mime." In *The Cambridge History of Classical Literature*, vol. 2, *Latin Literature*, ed E J Kenney and W V. Clausen, 286–90, 842–45. Cambridge. Cambridge University Press, 1982.

Hoven, Birgit van den. *Work in Ancient and Medieval Thought: Ancient Philosophers, Medieval Monks and Theologians, and Their Concept of Work, Occupations, and Technology*. Amsterdam J. C. Gieben, 1996.

Howard, Deborah. "Serlio, Sebastiano " *DA* 28·466–72.

Hubicki, Włodzimierz. "Ercker (also Erckner or Erckel), Lazarus." *DSB* 4.393–94.

Hull, David L. "Openness and Secrecy in Science. Their Origins and Limitations." *Science, Technology, and Human Values* 10 (spring 1985): 4–13.

———. *Science as a Process: An Evolutionary Account of the Social and Conceptual Development of Science*. Chicago: University of Chicago Press, 1988.

Hultsch, F. "Charias, 11." *RE* 3.2 (1899), col. 2133.

Hutchison, Jane Campbell. *Albrecht Durer. A Biography*. Princeton, N.J.: Princeton University Press, 1990.

Hutchison, Keith "What Happened to Occult Qualities in the Scientific Revolution?" *Isis* 73 (June 1982): 233–53

Hyman, Isabelle, ed. *Brunelleschi in Perspective*. Englewood Cliffs, N.J : Prentice-Hall, 1974.

Iliffe, Rob. "'In the Warehouse' Privacy, Property, and Priority in the Early Royal Society." *History of Science* 30 (1992): 29–68.

———. "Material Doubts. Hooke, Artisan Culture, and the Exchange of Information in 1670s London." *British Journal for the History of Science* 28 (1995): 285–318.

———. "Rational Artistry" *History of Science* 36 (1998): 329–57.

Jähns, Max. *Geschichte der Kriegswissenschaften vornehmlich in Deutschland*. 3 vols. 1889–1900. Reprint, New York: Johnson Reprint, 1965

Janssen, Johannes. *History of the German People after the Close of the Middle Ages*. Vol. 15, *Commerce and Capital—Private Life of the Different Classes—Mendicancy and Poor Relief*. Trans. A. M. Christie. London. Kegan Paul, Trench, Trübner & Co , 1910.

Jardine, Lisa. *Worldly Goods: A New History of the Renaissance*. New York: Doubleday, 1996.

Jardine, Nicholas. *The Birth of History and Philosophy of Science: Kepler's "A Defence of Tycho against Ursus" with Essays on Its Provenance and Significance*. Cambridge: Cambridge University Press, 1984.

Jarzombek, Mark. "The Structural Problematic of Leon Battista Alberti's *De pictura*." *Renaissance Studies* 4 (September 1990): 273–85.

Jenkins, A Fraser "Cosimo de' Medici's Patronage of Architecture and the Theory of Magnificence." *Journal of the Warburg and Courtauld Institutes* 33 (1970): 162–70.

Johns, Adrian. *The Nature of the Book. Print and Knowledge in the Making*. Chicago: University of Chicago Press, 1998.

Johnson, Rozelle Parker. *Compositiones Variae from Codex 490, Biblioteca Capitolare, Lucca, Italy: An Introductory Study*. Urbana: University of Illinois Press, 1939.

Jones, Philip J. *The Malatesta of Rimini and the Papal State: A Political History*. Cambridge: Cambridge University Press, 1974.

Joshel, Sandra R *Work, Identity, and Legal Status at Rome: A Study of the Occupational Inscriptions*. Norman. University of Oklahoma Press, 1992.

Jung, Carl G. *Psychology and Alchemy* Trans. R. F. C. Hull. 2nd ed. rev. Princeton, N.J.. Princeton University Press, 1968.

Kaemmel, Otto "Plateanus: Petrus P." *ADB* 26.241–43.

Kagan, Donald. *Pericles of Athens and the Birth of Democracy*. New York: Free Press, 1991.

Kaplan, Benjamin. *An Unhurried View of Copyright*. New York: Columbia University Press, 1967.

Kaufmann, Thomas DaCosta. "Editor's Statement: Images of Rule: Issues of Interpretation " *Art Journal* 48 (summer 1989): 119–22.

Kee, Howard Clark. *Miracle in the Early Christian World: A Study in Sociohistorical Method.* New Haven: Yale University Press, 1983.

Keefer, Michael H. "Agrippa's Dilemma: Hermetic 'Rebirth' and the Ambivalences of 'De vanitate' and 'De occulta philosophia.'" *Renaissance Quarterly* 41 (winter 1988). 614–53

Kellenbenz, Hermann *The Rise of the European Economy: An Economic History of Continental Europe from the Fifteenth to the Eighteenth Century.* Rev. and ed. Gerhard Benecke. London· Weidenfeld & Nicolson, 1976.

Keller, Alexander G "Mathematicians, Mechanics, and Experimental Machines in Northern Italy in the Sixteenth Century" In *The Emergence of Science in Western Europe,* ed. Maurice Crossland, 15–34. London: Macmillan, 1975.

———. "Mathematics, Mechanics, and the Origins of the Culture of Mechanical Invention." *Minerva* 23 (autumn 1985). 348–61.

Kelley, Donald R *Foundations of Modern Historical Scholarship. Language, Law, and History in the French Renaissance.* New York: Columbia University Press, 1970

———, ed *History and the Disciplines· The Reclassification of Knowledge in Early Modern Europe.* Rochester, N Y.· University of Rochester Press, 1997.

Kemp, Martin. *Behind the Picture· Art and Evidence in the Italian Renaissance.* New Haven. Yale University Press, 1997

———. "'La diminutione di ciascun piano': La rappresentazione delle forme nello spazio di Francesco di Giorgio." In *Prima di Leonardo: Cultura delle macchine a Siena nel Rinascimento,* ed. Paolo Galluzzi, 105–12 Milan: Electa, 1991

———. "From 'Mimesis' to 'Fantasia': The Quattrocento Vocabulary of Creation, Inspiration, and Genius in the Visual Arts." *Viator* 8 (1977): 347–98.

———. *Leonardo da Vinci: The Marvellous Works of Nature and Man.* Cambridge. Harvard University Press, 1981.

——— *The Science of Art· Optical Themes in Western Art from Brunelleschi to Seurat.* New Haven: Yale University Press, 1990.

———. "The 'Super-Artist' as Genius: The Sixteenth-Century View" In *Genius: The History of an Idea,* ed Penelope Murray, 32–53. Oxford. Basil Blackwell, 1989.

Kempers, Bram. *Painting, Power, and Patronage: The Rise of the Professional Artist in the Italian Renaissance.* Trans Beverly Jackson. London: Penguin, 1984.

Kennedy, George A. "The Earliest Rhetorical Handbooks" *American Journal of Philology* 80 (April 1959). 169–78

Kenney, E. J "Books and Readers in the Roman World." In *The Cambridge History of Classical Literature,* vol. 2, *Latin Literature,* ed. E. J. Kenney and W. V. Clausen, 3–32. Cambridge: Cambridge University Press, 1982.

Kenney, E. J., and W V. Clausen, eds. *The Cambridge History of Classical Literature.* Vol. 2, *Latin Literature* Cambridge. Cambridge University Press, 1982.

Kent, F. W , and Patricia Simons with J. C. Eade, eds. *Patronage, Art, and Society in Renaissance Italy.* Canberra· Humanities Research Centre, 1987

Kerford, George B *The Sophistic Movement.* Cambridge: Cambridge University Press, 1981.

———, ed *The Sophists and Their Legacy*. Wiesbaden: Franz Steiner Verlag, 1981.

Kernan, Alvin B. *The Death of Literature*. New Haven: Yale University Press, 1990.

———. *Printing Technology, Letters, and Samuel Johnson* Princeton, N J.: Princeton University Press, 1987

Kirk, G S., J. E Raven, and M. Schofield *The Presocratic Philosophers· A Critical History with a Selection of Texts*. 2nd ed. Cambridge: Cambridge University Press, 1983.

Kirnbauer, Franz. *400 Jahre Schwazer Bergbuch, 1556–1956*. Vienna: Montan-Verlag, 1956.

Klemm, Friedrich "Valturio, Roberto" *DSB* 13:567–68.

Kliemann, Julian "Vasari. (1) Georgio Vasari." *DA* 32:10–25.

Knell, Heiner. *Perikleische Baukunst* Darmstadt: Wissenschaftliche Buchgesellschaft, 1979.

———. *Vitruvs Architekturtheorie* Darmstadt: Wissenschaftliche Buchgesellschaft, 1985

Knox, B M W. "Books and Readers in the Greek World from the Beginnings to Alexandria." In *The Cambridge History of Classical Literature*, vol. 1, *Greek Literature*, ed. P. E. Easterling and B. M. W. Knox, 1–16. Cambridge: Cambridge University Press, 1985.

Koch, Manfred *Geschichte und Entwicklung des bergmännischen Schrifttums*. Goslar: Hermann Hubener, 1963.

Koerner, Joseph Leo. *The Moment of Self-Portraiture in German Renaissance Art*. Chicago: University of Chicago Press, 1993.

Kolb, Carolyn. "The Francesco di Giorgio Material in the Zichy Codex." *Journal of the Society of Architectural History* 47 (June 1988)· 132–59

Kornemann, Ernst. "Collegium." *RE*, new ed., 4 (1901), cols. 380–480.

Koves-Zulauf, Th "Die Vorrede der plinianischen 'Naturgeschichte.'" *Wiener Studien*, n.s., 7 (1973). 134–84.

Kramer, Gerhard W *Berthold Schwarz· Chemie und Waffentechnik im 15. Jahrhundert*. Munich. Oldenbourg, 1995.

Kraschewski, Hans-Joachim. *Wirtschaftspolitik im deutschen Territorialstaat des 16. Jahrhunderts: Herzog Julius von Braunschweig-Wolfenbüttel (1528–1589)*. Cologne Böhlau Verlag, 1978.

Kraus, Paul. *Jābir Ibn Hayyān: Contribution à l'histoire des idées scientifiques dans l'Islam*. 2 vols. Mémoires présentés à l'Institut d'Égypte, 44–45. Cairo: L'Institut Français d'Archéologie Orientale, 1942–43.

Krautheimer, Richard. "Alberti and Vitruvius" In *Acts of the Twentieth International Congress of the History of Art*, vol 2, *The Renaissance and Mannerism*, 42–52 Princeton, N.J. Princeton University Press, 1982.

Krautheimer, Richard, with Trude Krautheimer-Hess. *Lorenzo Ghiberti*. 3rd ed. Princeton, N.J.: Princeton University Press, 1982

Kraye, Jill, ed. *The Cambridge Companion to Renaissance Humanism*. Cambridge· Cambridge University Press, 1996.

Krinsky, Carol Herselle "Cesare Cesariano and the Como Edition of 1521." Ph.D. diss , New York University, 1965

——— Introduction to *Vitruvius. De architectura: Nachdruck der kommentierten ersten italienischen Ausgabe von Cesare Cesariano (Como, 1521).* Munich: Wilhelm Fink Verlag, 1969.

Kristeller, Paul Oskar. *Marsilio Ficino and His Work after Five Hundred Years.* [Florence]: Leo S. Olschki, [1987].

———. "Marsilio Ficino e Lodovico Lazzarelli: Contributo alla diffusione delle idee ermetiche nel Rinascimento." In *Studies in Renaissance Thought and Letters,* 221–47. Rome: Edizioni di Storia e Letteratura, 1956.

———. "The Modern System of the Arts." In *Renaissance Thought II: Papers on Humanism and the Arts,* 163–227. New York: Harper & Row, Harper Torchbooks, 1965.

———. *The Philosophy of Marsilio Ficino.* Trans. Virginia Conant. 1943. Reprint, Gloucester, Mass.: Peter Smith, 1964.

Kroll, W. "Diades Nr. 2." *RE,* suppl. 6 (1935), cols. 25–27.

Kruft, Hanno-Walter. *A History of Architectural Theory from Vitruvius to the Present.* Trans. Ronald Taylor, Elsie Callander, and Antony Wood. New York: Princeton Architectural Press, 1994.

Kuflik, Arthur. "Moral Foundations of Intellectual Property Rights." In *Owning Scientific and Technical Information: Value and Ethical Issues,* ed. Vivian Weil and John W. Snapper, 219–40. New Brunswick, N.J.: Rutgers University Press, 1989.

Kuhlow, Hermann F. W. *Die Imitatio Christi und ihre kosmologische Überfremdung: Die theologischen Grundgedanken des Agrippa von Nettesheim.* Berlin: Lutherisches Verlagshaus, 1967.

Kummel, Hans Martin. *Familie, Beruf und Amt im spatbabylonischen Uruk.* Berlin: Gebr. Mann Verlag, 1979.

Lamberini, Daniela. *Il principe difeso: Vita e opere di Bernardo Puccini.* Florence: Editrice La Giuntina, 1990.

Landels, J. G. *Engineering in the Ancient World.* Berkeley and Los Angeles: University of California Press, 1978.

Lavin, Marilyn A. *Piero della Francesca a Rimini: L'affresco nel Tempio Malatestiano.* Bologna: Nuova Alfa, 1984.

———, ed. *Piero della Francesca and His Legacy.* Washington, D.C.: National Gallery of Art, 1995.

Lawrence, A. W. *Greek Aims in Fortification.* Oxford: Clarendon, 1979.

Lemerle, Frédérique. "On Guillaume Philandrier: Forms and Norm." In *Paper Palaces: The Rise of the Renaissance Architectural Treatise,* ed. Vaughan Hart with Peter Hicks, 186–97. New Haven: Yale University Press, 1998.

Lemerle-Pauwels, F. "Philander [Philandrier], Guillaume." *DA* 24:603–4.

León-Jones, Karen Silvia de. *Giordano Bruno and the Kabbalah: Prophets, Magicians, and Rabbis.* New Haven: Yale University Press, 1997.

Lewis, D. M., John Boardman, J. K. Davies, and M. Ostwald, eds. *The Cambridge Ancient History.* Vol. 5, *The Fifth Century B.C.* 2nd ed. Cambridge: Cambridge University Press, 1992.

Lindberg, David C., and Robert S. Westman, eds. *Reappraisals of the Scientific Revolution* Cambridge: Cambridge University Press, 1990.

Lindsay, Jack. *The Origins of Alchemy in Graeco-Roman Egypt.* New York: Barnes & Noble, 1970

Lippold, G. "Pheidias 2) Sohn des Carmides" *RE* 19 (1938), cols. 1919–35.

Lloyd, G. E. R *Magic, Reason, and Experience: Studies in the Origin and Development of Greek Science* Cambridge: Cambridge University Press, 1979.

Loewenstein, Joseph "The Script in the Marketplace." *Representations* 12 (fall 1985): 101–14.

Long, Pamela O. "The Contribution of Architectural Writers to a 'Scientific' Outlook in the Fifteenth and Sixteenth Centuries." *Journal of Medieval and Renaissance Studies* 15 (fall 1985)· 265–89.

———. "Humanism and Science." In *Renaissance Humanism: Its Sources, Forms, and Legacy,* ed. A. Rabil Jr., 3:486–512. Philadelphia: University of Pennsylvania Press, 1988.

——— "Invention, Authorship, 'Intellectual Property,' and the Origin of Patents: Notes toward a Conceptual History." *Technology and Culture* 32 (October 1991): 846–84.

———. "Openness and Empiricism· Values and Meaning in Early Architectural Writings and in Seventeenth Century Experimental Philosophy" In *The Architecture of Science,* ed. Peter Galison and Emily Thompson, 79–103. Cambridge: MIT Press, 1999.

———. "The Openness of Knowledge: An Ideal and Its Context in 16th-Century Writings on Mining and Metallurgy." *Technology and Culture* 32 (April 1991): 318–55.

———. "Power, Patronage, and the Authorship of *Ars.* From Mechanical Know-How to Mechanical Knowledge in the Last Scribal Age" *Isis* 88 (March 1997): 1–41.

Long, Pamela O., and Alex Roland. "Military Secrecy in Antiquity and Early Medieval Europe: A Critical Reassessment." *History and Technology* 11 (1994). 259–90.

Ludovici, S. Samek. "Cesariano (Ciseriano), Cesare" *DBI* 24:172–80.

MacIntyre, Alasdair. *After Virtue: A Study in Moral Theory* 2nd ed. Notre Dame, Ind.: University of Notre Dame Press, 1984.

——— *Three Rival Versions of Moral Enquiry: Encyclopaedia, Genealogy, and Tradition.* Notre Dame, Ind.: University of Notre Dame Press, 1990.

Mackenney, Richard. *Tradesmen and Traders: The World of the Guilds in Venice and Europe, c. 1250–c 1650* Totowa, N.J.: Barnes & Noble, 1987.

MacLeod, Christine "Accident or Design? George Ravenscroft's Patent and the Invention of Lead-Crystal Glass." *Technology and Culture* 28 (October 1987): 776–803.

———. *Inventing the Industrial Revolution· The English Patent System, 1660–1800* Cambridge: Cambridge University Press, 1988

MacMulten, Ramsay. *Enemies of the Roman Order: Treason, Unrest, and Alienation in the Empire.* Cambridge: Harvard University Press, 1966.

Mallett, Michael E *Mercenaries and Their Masters: Warfare in Renaissance Italy.* London· Bodley Head, 1974

———. "The Theory and Practice of Warfare in Machiavelli's Republic." In *Machiavelli and Republicanism*, ed. Gisela Bock, Quentin Skinner, and Maurizio Viroli, 173–80. Cambridge: Cambridge University Press, 1990.

Mályusz, Elemér *Kaiser Sigismund in Ungarn, 1387–1437*. Budapest. Akadémiai Kiadó, 1990.

Mandich, Giulio. "Primi riconoscimenti veneziani di un diritto di privativa agli inventori" *Rivista di Diritto Industriale* 7 (1958): 101–55

———. "Le privative industriali veneziane (1450–1550)." *Rivista del Diritto Commerciale e del Diritto Generale delle Obbligazioni* 34 (1936): 511–47.

———. "Venetian Origins of Inventors' Rights" Trans. Frank Prager *Journal of the Patent Office Society* 42 (June 1960)· 378–82

———. "Venetian Patents (1450–1550)" Trans. Frank Prager. *Journal of the Patent Office Society* 30 (March 1948). 166–224.

Manno, Antonio "Architettura e arti meccaniche nel fregio del palazzo ducale di Urbino." In *Federico di Montefeltro*, vol. 2, *Le arti*, ed. Giorgio Cerboni Baiardi, Giorgio Chittolini, and Piero Floriani, 89–104. Rome: Bulzoni Editore, 1986.

Manselli, R., et al. "2. A[rnald] v Villanova." In *Lexikon des Mittelalters*, ed. R. Auty et al., vol. 1, cols. 994–96. Munich: Artemis Verlag, 1977.

Marcel, Raymond. *Marsile Ficin (1433–1499)* Paris· Belles Lettres, 1958

Marcucci, Laura. "Giovanni Sulpicio e la prima edizione del *De architectura* di Vitruvio." *Studi e Documenti di Architettura*, no. 8 (September 1978): 185–95.

Marinoni, Augusto. "I manoscritti di Leonardo da Vinci e le loro edizioni." In *Leonardo· Saggi e ricerche*, ed. Comitato Nazionale per le Onoranze a Leonardo da Vinci nel Quinto Centenario della Nascità, 231–74. Rome: Istituto Poligrafico dello Stato, 1954.

———. "The Writer: Leonardo's Literary Legacy" In *The Unknown Leonardo*, ed Ladislao Reti, 56–85. New York McGraw-Hill, 1974.

Mark, Ira S. "The Gods on the East Frieze of the Parthenon." *Hesperia* 53 (July–September 1984): 289–342

Marsden, Eric W. *Greek and Roman Artillery: Historical Development*. Oxford. Clarendon, 1969.

———. "Macedonian Military Machinery and Its Designers under Philip and Alexander." In *Ancient Macedonia*, 2·211–23. Thessalonike: Institute for Balkan Studies, 1977.

Martin, René. "État présent des études sur Columelle." *ANRW*, pt 2, *Principat.*, 32.3:1959–79

———. *Recherches sur les agronomes latins et leurs conceptions économiques et sociales*. Paris: Belles Lettres, 1971.

Martin, Roland. *L'urbanisme dans la Grèce antique*. 2nd ed Paris: Édition A. & J Picard, 1974.

Martines, Lauro. *Power and Imagination· City-States in Renaissance Italy*. New York: Vintage, 1980

Masotti, Arnaldo. "Tartaglia (also Tartalea or Tartaia), Niccolò" *DSB* 13:258–62.
McCray, W. Patrick. *Glassmaking in Renaissance Venice: The Fragile Craft*. Aldershot: Ashgate, 1999
McKirahan, Richard D., Jr. *Philosophy before Socrates: An Introduction with Texts and Commentary*. Indianapolis: Hackett, 1994
McMullin, Ernan. "Openness and Secrecy in Science: Some Notes on Early History." *Science, Technology, and Human Values* 10 (spring 1985). 14–23.
McNab, Jessie. "Palissy, Bernard." *DA* 23:849–50.
Meiggs, Russell. *The Athenian Empire*. Oxford: Clarendon, 1972.
———. *Roman Ostia*. 2nd ed. Oxford: Clarendon, 1973
Mendelsohn, I. "Gilds in Babylonia and Assyria." *Journal of the American Oriental Society* 60 (March 1940): 68–72.
Mentasti, Rosa Barovier. "Tecnica del vetro nella Venezia del cinquecento." In *Cultura, scienze e tecniche nella Venezia del cinquecento: Giovan Battista Benedetti e il suo tempo*, 473–82. Venice: Istituto Veneto di Scienze, Lettere ed Arti, 1987
——— *Il vetro veneziano*. Milan: Electa, 1982
Merchant, Carolyn *The Death of Nature: Women, Ecology, and the Scientific Revolution*. New York: Harper & Row, 1980.
Merkel, Ingrid, and Allen G Debus. *Hermeticism and the Renaissance: Intellectual History and the Occult in Early Modern Europe*. Washington, D.C.: Folger Shakespeare Library, 1988
Merkur, Daniel. "The Study of Spiritual Alchemy. Mysticism, Gold-Making, and Esoteric Hermeneutics." *Ambix* 37 (March 1990): 35–45.
Merton, Robert K. "Priorities in Scientific Discovery: A Chapter in the Sociology of Science." *American Sociological Review* 22 (December 1957): 635–59.
———. *Science, Technology, and Society in Seventeenth-Century England*. 2nd ed. New York: Harper & Row, 1970.
———. *The Sociology of Science: Theoretical and Empirical Investigations*. Ed Norman W. Storer Chicago: University of Chicago Press, 1973
——— "Über die vielfältigen Wurzeln und den geschlechtslosen Charakter des englischen Wortes 'scientist.'" In *Generationsdynamik und Innovation in der Grundlagenforschung*, ed Peter Hans Hofschneider and Karl Ulrich Mayer, 259–94 Munich: Max-Planck-Gesellschaft, [1990].
Meyer, Eduard. *Geschichte des Altertums* Vol. 4, pt. 1, *Das Perserreich und die Griechen bis zum Vorabend des peloponnesischen Krieges*. 5th ed. Stuttgart J. G. Cotta'sche Buchhandlung, 1954.
Michel, Paul Henri. *The Cosmology of Giordano Bruno*. Trans R. E. W. Maddison. Paris: Hermann, 1973.
Miller, Edward, and John Hatcher *Medieval England. Towns, Commerce, and Crafts, 1086–1348*. London Longman, 1995.
Millon, Henry. "The Architectural Theory of Francesco di Giorgio." *Art Bulletin* 40 (September 1958): 257–61.

Minnis, A. J. *Medieval Theory of Authorship*. 2nd ed. rev. Philadelphia: University of Pennsylvania Press, 1988.

Miskimin, Harry A. *The Economy of Early Renaissance Europe, 1300–1460*. Cambridge: Cambridge University Press, 1975.

———. *The Economy of Later Renaissance Europe, 1460–1600*. Cambridge: Cambridge University Press, 1977.

Mondolfo, Rodolfo. *Polis, lavoro e tecnica*. Ed. Massimo Venturi Ferriolo. Milan: Feltrinelli, 1982.

Monticolo, Giovanni. *L'Ufficio della Giustizia Vecchia a Venezia dalle origini sino al 1330*. Venice: R. Deputazione Veneta di Storia Patria, 1892.

Moran, Bruce T. *The Alchemical World of the German Court: Occult Philosophy and Chemical Medicine in the Circle of Moritz of Hessen (1572–1632)*. Stuttgart: Franz Steiner Verlag, 1991.

———. "German Prince-Practitioners: Aspects of the Development of Courtly Science, Technology, and Procedures in the Renaissance." *Technology and Culture* 22 (April 1981): 253–74.

Morrison, J. S. "The Place of Protagoras in Athenian Public Life (460–415 B.C.)." *Classical Quarterly* 35 (January–April 1941): 1–16.

Moxey, Keith. *The Practice of Theory: Poststructuralism, Cultural Politics, and Art History*. Ithaca, N.Y.: Cornell University Press, 1994.

Muccillo, M. "Della Valle, Battista (Giovanni Battista)." *DBI* 37:728–29.

Müller, Reimar. "Sophistique et démocratie." In *Positions de la sophistique*, ed. Barbara Cassin, 179–93. Paris: Librairie Philosophique J. Vrin, 1986.

Muller-Jahncke, Wolf-Dieter. "Agrippa von Nettesheim: 'De occulta philosophia': Ein 'magisches System.'" In *Magia naturalis und die Entstehung der modernen Naturwissenschaften*, ed. Albert Heinekamp and Dieter Mettler, 19–29. Wiesbaden: Franz Steiner Verlag, 1978.

———. "The Attitude of Agrippe von Nettesheim (1486–1535) towards Alchemy." *Ambix* 22 (July 1975): 134–50.

———. "Johannes Trithemius und Heinrich Cornelius Agrippa von Nettesheim." In *Johannes Trithemius: Humanismus und Magie im vorreformatorischen Deutschland*, ed. Richard Auerheimer and Frank Baron, 29–37. Munich: Profil Verlag, 1991.

———. "Von Ficino zu Agrippa: Der Magia-Begriff des Renaissance-Humanismus im Überblick." In *Epochen der Naturmystik: Hermetische Tradition im wissenschaftlichen Fortschritt*, ed. Antoine Faivre and Rolf Christian Zimmermann, 24–51. Berlin: Erich Schmidt Verlag, 1979.

Nauert, Charles G., Jr. *Agrippa and the Crisis of Renaissance Thought*. Urbana: University of Illinois Press, 1965.

Nef, John U. "Mining and Metallurgy in Medieval Civilisation." In *The Cambridge Economic History of Europe*, vol. 2, *Trade and Industry in the Middle Ages*, ed. M. M. Postan and Edward Miller assisted by Cynthia Postan, 691–761 and 933–40. 2nd ed. Cambridge: Cambridge University Press, 1987.

Nesbit, Molly. "What Was an Author?" *Yale French Studies* 73 (1987): 229–57.

Newman, William R. *Gehennical Fire: The Lives of George Starkey, an American Alchemist in the Scientific Revolution.* Cambridge: Harvard University Press, 1994

———. "The Occult and the Manifest among the Alchemists." In *Tradition, Transmission, Transformation: Proceedings of Two Conferences on Pre-Modern Science Held at the University of Oklahoma,* ed. F. Jamil Ragep and Sally P Ragep with Steven Livesey, 173–98. Leiden: Brill, 1996.

———. "Technology and Alchemical Debate in the Late Middle Ages " *Isis* 80 (September 1989): 423–45

Nickel, Rainer *Xenophon.* Darmstadt: Wissenschaftliche Buchgesellschaft, 1979

Nova, Alessandro. "Cellini, Benvenuto." *DA* 6·139–50.

Noyes, C Reinold. *The Institution of Property: A Study of the Development, Substance, and Arrangement of the System of Property in Modern Anglo-American Law* New York· Longmans, Green, 1936

Ober, Josiah. *Mass and Elite in Democratic Athens· Rhetoric, Ideology, and Power of the People.* Princeton, N.J.: Princeton University Press, 1989.

O'Brien, Michael J. "Xenophanes, Aeschylus, and the Doctrine of Primeval Brutishness." *Classical Quarterly* 35 (1985). 264–77.

Obrist, Barbara. *Les débuts de l'imagerie alchimique (XIVe–XVe siècles).* Paris: Le Sycomore, 1982.

Ogilvie, R M *The Romans and Their Gods in the Age of Augustus* New York: W. W Norton, 1969.

Olivato, Loredana. "Cattaneo (Cataneo), Girolamo." *DBI* 22·471–73.

Olschki, Leonardo. *Geschichte der neusprachlichen wissenschaftlichen Literatur.* 3 vols. 1919–27. Reprint, Vaduz, Liechtenstein: Kraus, 1965.

Onians, John. *Bearers of Meaning· The Classical Orders in Antiquity, the Middle Ages, and the Renaissance.* Princeton, N.J.. Princeton University Press, 1988.

Oppel, John. "The Priority of the Architect: Alberti on Architects and Patrons." In *Patronage, Art, and Society in Renaissance Italy,* ed. F. W. Kent and Patricia Simons with J. C. Eade, 251–67. Canberra: Humanities Research Centre, 1987.

Oppenheim, A. Leo "Mesopotamia in the Early History of Alchemy " *Revue d'Assyriologie et d'Archéologie Orientale* 60 (1966): 29–45.

Øre, Oystein *Cardano: The Gambling Scholar.* Princeton: Princeton University Press, 1953

Ostwald, M. "Athens as a Cultural Centre." In *The Cambridge Ancient History,* vol. 5, *Fifth Century B C* , ed. D. M. Lewis, John Boardman, J K. Davies, and M. Ostwald, 306–69. 2nd ed. Cambridge· Cambridge University Press, 1992.

Ovitt, George, Jr *The Restoration of Perfection: Labor and Technology in Medieval Culture.* New Brunswick, N J.: Rutgers University Press, 1987.

Pace, Anna de *Le matematiche e il mondo: Ricerche su un dibattito in Italia nella seconda metà del cinquecento* Milan. FrancoAngeli, 1993

Pagave, Venanzio de. *Vita di Cesare Cesariano: Architetto Milanese.* Ed. C. Casati. Milan: Tipografia Pirola, 1878

Pagel, Walter. "Paracelsus als 'Naturmystiker.'" In *Epochen der Naturmystik: Hermetische Tradition im wissenschaftlichen Fortschritt*, ed Antoine Faivre and Rolf Christian Zimmermann, 52–104. Berlin: Erich Schmidt Verlag, 1979

———. *Paracelsus: An Introduction to Philosophical Medicine in the Era of the Renaissance.* 2nd rev. ed Basel: S Karger, 1982.

———. "Paracelsus, Theophrastus Philippus Aureolus Bombastus von Hohenheim " *DSB* 10·304–13.

Pagliara, Pier Nicola. "Vitruvio da testo a canone " In *Memoria dell'antico nell'arte italiana*, ed. Salvatore Settis, vol. 3, *Dalla tradizione all'archeologia*, 3–85. Turin: Giulio Einaudi, 1986.

Panofsky, Erwin. *The Life and Art of Albrecht Dürer.* 4th ed. Princeton, N.J.· Princeton University Press, 1955.

Parker, Geoffrey. *The Military Revolution: Military Innovation and the Rise of the West, 1500–1800.* Cambridge: Cambridge University Press, 1988

Partington, J. R. *A History of Greek Fire and Gunpowder* Cambridge: W. Heffer & Sons, 1960.

Pasquale, Salvatore di. "Leonardo, Brunelleschi, and the Machinery of the Construction Site." In *Leonardo da Vinci: Engineer and Architect*, ed. Paolo Galluzzi, 163–81 Montreal: Montreal Museum of Fine Arts, 1987.

Payne, Alina A. *The Architectural Treatise in the Italian Renaissance: Architectural Invention, Ornament, and Literary Culture.* Cambridge: Cambridge University Press, 1999.

Pedretti, Carlo *The Codex Atlanticus of Leonardo da Vinci: A Catalogue of Its Newly Restored Sheets.* 2 pts. New York: Harcourt Brace Jovanovich, Johnson Reprint, 1978.

———. *Leonardo da Vinci on Painting, A Lost Book (Libro A).* Berkeley and Los Angeles· University of California Press, 1964.

Pereira, Michela. *The Alchemical Corpus Attributed to Raymond Lull.* Warburg Institute Surveys and Texts, 18. London: Warburg Institute, 1989.

———. "*Medicina* in the Alchemical Writings Attributed to Raimond Lull (14th–17th Centuries)." In *Alchemy and Chemistry in the Sixteenth and Seventeenth Centuries*, ed. Piyo Rattansi and Antonio Clericuzio, 1–15. Dordrecht: Kluwer, 1994.

———. *L'oro dei filosofi: Saggio sulle idee di un alchimista del Trecento.* Spoleto: Centro Italiano di Studi sull'Alto Medioevo, 1992.

Pérez-Ramos, Antonio. *Francis Bacon's Idea of Science and the Maker's Knowledge Tradition* Oxford: Clarendon, 1988

Pesenti, T. "Dondi Dall'Orologio, Giovanni." *DBI* 41:96–104.

Pezzini, Grazia Bernini. *Il fregio dell'arte della guerra nel palazzo ducale di Urbino* Rome: Istituto Poligrafico e Zecca dello Stato, 1985.

Pfuhl, Ernst. *Malerei und Zeichnung der Griechen.* 2 vols. Munich: F. Bruckmann, 1923.

Phillips, Jeremy "The English Patent as a Reward for Invention: The Importation of an Idea." *Journal of Legal History* 3 (May 1982): 71–79.

Pietramellara, Carla Tomasini, and Angelo Turchini, eds. *Castel Sismondo e Sigismondo Pandolfo Malatesta.* Rimini: Ghigi Editore, 1985.

Placido, Domingo. "El pensamiento de Protágoras y la Atenas de Pericles." *Hispania Antiqua* 3 (1973): 29–68.

———. "Protágoras y Pericles." *Hispania Antiqua* 2 (1972). 7–19.

Poel, Marc van der. *Cornelius Agrippa: The Humanist Theologian and His Declamations.* Leiden: Brill, 1997.

Poggetto, Paolo dal, ed. *Piero e Urbino, Piero e le corti rinascimentali.* Venice: Marsilio Editori, 1992.

Poland, Franz. *Geschichte des griechischen Vereinswesens.* Leipzig: Teubner, 1909.

Pollak, Martha D. *Military Architecture, Cartography, and the Representation of the Early Modern European City. A Checklist of Treatises on Fortification in the Newberry Library.* Chicago. Newberry Library, 1991.

Pollitt, J. J. *Art and Experience in Classical Greece.* Cambridge: Cambridge University Press, 1972.

———. "Art: Archaic to Classical." In *The Cambridge Ancient History,* vol. 5, *Fifth Century B.C,* ed. D. M. Lewis, John Boardman, J. K. Davies, and M. Ostwald, 171–83. 2nd ed. Cambridge: Cambridge University Press, 1992.

———. "Gold Pastoral!" *Times Literary Supplement,* 14 April 1995, 6–7.

Popplow, Marcus. *Neu, nützlich und erfindungsreich: Die Idealisierung von Technik in der frühen Neuzeit.* Münster: Waxmann, 1998.

Post, Gaines, Kimon Giocarinis, and Richard Kay. "The Medieval Heritage of a Humanistic Ideal: 'Scientia donum Dei est. Unde vendi non potest.'" *Traditio* 11 (1955): 195–234.

Pounds, Norman J. G. "Mining." *DMA* 8:397–404.

Prager, Frank D. "Brunelleschi's Patent." *Journal of the Patent Office Society* 28 (February 1946): 109–35.

———. "Fontana on Fountains: Venetian Hydraulics of 1418." *Physis* 13 (1971): 341–60.

———. "A Manuscript of Taccola, Quoting Brunelleschi, on Problems of Inventors and Builders." *Proceedings of the American Philosophical Society* 112 (June 1968): 131–49.

Prager, Frank D., and Gustina Scaglia. *Brunelleschi: Studies of His Technology and Inventions.* Cambridge: MIT Press, 1970.

———. *Mariano Taccola and His Book "De Ingeneis."* Cambridge. MIT Press, 1972.

Price, Derek J. de Solla. *Science since Babylon.* Enl. ed. New Haven: Yale University Press, 1975.

Price, S. R. F. "The Place of Religion. Rome in the Early Empire." In *The Cambridge Ancient History,* vol. 10, *The Augustan Empire, 43 B.C.–A.D. 69,* ed. Alan K. Bowman, Edward Champlin, and Andrew Lintott, 812–47, 1114–20. 2nd ed. Cambridge: Cambridge University Press, 1996.

Price, Simon. "The History of the Hellenistic Period." In *The Oxford History of the Classical World,* vol. 1, *Greece and the Hellenistic World,* ed. John Boardman, Jasper Griffen, and Oswyn Murray, 309–31. Oxford: Oxford University Press, 1988.

Principe, Lawrence M. *The Aspiring Adept: Robert Boyle and His Alchemical Quest, Including Boyle's "Lost" "Dialogue on the Transmutation of Metals."* Princeton, N.J.: Princeton University Press, 1998.

———. "Robert Boyle's Alchemical Secrecy. Codes, Ciphers, and Concealments." *Ambix* 39 (July 1992): 63–74.

Promis, Carlo "Giambattista Zanchi." In *Biografie di ingegneri militari italiani dal secolo XIV alla metà del XVIII*, 396–403. Turin: Bocca, 1874

Prost, M. August. *Les sciences et les arts occultes au XVe siècle: Corneille Agrippa, sa vie et ses oeuvres* 2 vols. 1881–82. Reprint, Nieuwkoop, Netherlands: B. De Graaf, 1965.

Puppi, Lionello, ed. *Alvise Cornaro e il suo tempo*. Padua· Comune di Padova, 1980.

Quint, David. *Origin and Originality in Renaissance Literature: Versions of the Source*. New Haven: Yale University Press, 1983

Rabil, Albert, Jr., ed. *Renaissance Humanism· Foundations, Forms, and Legacy* 3 vols Philadelphia: University of Pennsylvania Press, 1988.

Rattansi, Piyo, and Antonio Clericuzio, eds. *Alchemy and Chemistry in the Sixteenth and Seventeenth Centuries*. Dordrecht· Kluwer, 1994.

Rawson, Elizabeth. *Intellectual Life in the Late Roman Republic*. Baltimore: Johns Hopkins University Press, 1985.

Rees, D. A. "Platonism and the Platonic Tradition " In *Encyclopedia of Philosophy*, ed. Paul Edwards, 6 333–41. New York· Macmillan and Free Press, 1967

Renfrew, Colin, and Paul Bahn. *Archaeology. Theories, Methods, and Practice*. 2nd ed. London: Thames & Hudson, 1996.

Renger, Johannes. "Notes on the Goldsmiths, Jewelers, and Carpenters of Neobabylonian Eanna." Review of *Guild Structure and Political Allegiance in Early Achaemenid Mesopotamia*, by David B. Weisberg. *Journal of the American Oriental Society* 91 (October–December 1971): 494–503.

Reti, Ladislao. "Elements of Machines." In *The Unknown Leonardo*, ed. Ladislao Reti, 265–87. New York: McGraw-Hill, 1974

——— "Leonardo da Vinci and the Graphic Arts: The Early Invention of Relief-Etching." *Burlington Magazine* 112 (April 1971): 189–95.

———, ed. *The Unknown Leonardo*. New York: McGraw-Hill, 1974

Reynolds, J. "The Elder Pliny and His Times." In *Science in the Early Roman Empire. Pliny the Elder, His Sources and Influence*, ed. Roger French and Frank Greenaway, 1–10. Totowa, N.J.: Barnes & Noble, 1986.

Rhodes, P. J. *The Athenian Empire*. Oxford· Clarendon, 1985.

——— "The Athenian Revolution." In *The Cambridge Ancient History*, vol. 5, *Fifth Century B.C.*, ed. D. M. Lewis, John Boardman, J. K. Davies, and M. Ostwald, 62–95. 2nd ed Cambridge: Cambridge University Press, 1992.

——— "The Delian League to 449 B C " In *The Cambridge Ancient History*, vol. 5, *Fifth Century B.C.*, ed D. M. Lewis, John Boardman, J. K Davies, and M. Ostwald, 34–61, 535–39 2nd ed Cambridge. Cambridge University Press, 1992.

Ricci, Corrado. *Il Tempio Malatestiano*. With an appendix by Pier Giorgio Pasini. 1924. Reprint, Rimini· Bruno Ghigi Editore, 1974.

Richards, John C. "A New Manuscript of Heraclius." *Speculum* 15 (July 1940): 255–71.

Riddle, John M. *Dioscorides on Pharmacy and Medicine*. Austin: University of Texas Press, 1985.

Robertis, Francesco M. de. *Il diritto associativo romano dai collegi della repubblica alle corporazioni del basso impero.* Bari: Gius. Laterza & Figli, 1938.

———. *Lavoro e lavoratori nel mondo romano.* 1963. Reprint, New York: Arno, 1979.

———. *Storia delle corporazioni e del regime associativo nel mundo romano.* 2 vols. Bari: Adriatica Editrice, 1973.

Romanini, Angiola Maria. "Averlino (Averulino), Antonio detto Filarete." *DBI* 4:662–67.

Romano, Dennis. *Patricians and Popolani: The Social Foundations of the Venetian Renaissance State* Baltimore: Johns Hopkins University Press, 1987.

Roosen-Runge, Heinz. *Farbgebung und Technik frühmittelalterlicher Buchmalerei: Studien zu den Traktaten "Mappae Clavicula" und "Heraclius."* 2 vols. [Munich]. Deutscher Kunstverlag, 1967.

Rose, Mark. "The Author as Proprietor: *Donaldson v. Becket* and the Genealogy of Modern Authorship." *Representations* 23 (summer 1988). 51–85.

———. *Authors and Owners: The Invention of Copyright.* Cambridge: Harvard University Press, 1993.

Rosen, Edward. *Three Imperial Mathematicians: Kepler Trapped between Tycho Brahe and Ursus.* New York: Abaris, 1986.

Rosenau, Pauline Marie. *Post-modernism and the Social Sciences: Insights, Inroads, and Intrusions.* Princeton: Princeton University Press, 1992.

Rosenberg, Charles M., ed. *Art and Politics in Late Medieval and Early Renaissance Italy, 1250–1500.* Notre Dame, Ind.: University of Notre Dame Press, 1990.

Rosser, Gervase. "Crafts, Guilds, and the Negotiation of Work in the Medieval Town." *Past and Present,* no 154 (February 1997): 3–31.

Rossi, Paolo. *Philosophy, Technology, and the Arts in the Early Modern Era.* Trans. Salvator Attanasio. Ed. Benjamin Nelson. New York: Harper & Row, Harper Torchbooks, 1970.

Rossi, Sergio *Dalle botteghe alle accademie· Realtà sociale e teorie aristiche a Firenze dal XIV al XVI secolo.* Milan: Feltrinelli, 1980.

Rotondi, Pasquale. *The Ducal Palace of Urbino· Its Architecture and Decoration.* London: Alec Tiranti, 1969.

Rowland, Ingrid D. "Vitruvius in Print and in Vernacular Translation: Fra Giocondo, Bramante, Raphael, and Cesare Cesariano " In *Paper Palaces: The Rise of the Renaissance Architectural Treatise,* ed. Vaughan Hart with Peter Hicks, 105–21. New Haven: Yale University Press, 1998.

Rubin, Patricia Lee *Giorgio Vasari: Art and History.* New Haven: Yale University Press, 1995.

Rubinstein, Nicolai. "The Beginnings of Niccolò Machiavelli's Career in the Florentine Chancery." *Italian Studies* 11 (1956). 72–91.

———. "Machiavelli and the World of Florentine Politics." In *Studies on Machiavelli,* ed Myron P. Gilmore, 3–28 Florence: Sansoni, 1972.

Ruggini, Lellia Cracco. "Le associazioni professionali nel mondo Romano-Bizantino." In *Artigianato e tecnica nell società dell'alto medioevo occidentale,* 1.59–193. 2 vols. Settimane di Studio del Centro Italiano di Studi sull'Alto Medioevo, 18. Spoleto: Presso la Sede del Centro, 1971.

Rumpf, Andreas. "Agatharchos, 2. Des Eudemos Sohn aus Samos." *KP* 1 (1964), col. 116.

———. "Classical and Post-Classical Greek Painting." *Journal of Hellenic Studies* 67 (1947): 10–21.

Ruska, J., and E. Wiedemann. "Beiträge zur Geschichte der Naturwissenschaften LXVII. Alchemistische Decknamen." *Sitzungsberichte der physikalisch-medizinischen Sozietät zu Erlangen* 56 (1924): 17–36.

Rykwert, Joseph. "On the Oral Transmission of Architectural Theory." In *Les Traités d'architecture de la Renaissance*, ed. Jean Guillaume, 31–48. Paris: Picard, 1988.

Saalman, Howard. *Filippo Brunelleschi: The Buildings.* University Park: Pennsylvania State University Press, 1993.

———. *Filippo Brunelleschi: The Cupola of Santa Maria del Fiore.* London: Zwemmer, 1980.

Sallmann, Klaus. "Der Traum des Historikers: Zu den 'Bella Germaniae' des Plinius und zur julisch-claudischen Geschichtsschreibung." *ANRW*, pt. 2, *Principat*, 32.1: 578–601.

Sandbach, F. H. "Xenophon." In *The Cambridge History of Classical Literature*, vol. 1, *Greek Literature*, ed. P. E. Easterling and B. M. W. Knox, 478–80, 788–89. Cambridge: Cambridge University Press, 1985.

San Nicolò, Mariano. *Ägyptisches Vereinswesen zur Zeit der Ptolemäer und Römer.* 2 vols. Munich: C. H. Beck'sche Verlagsbuchhandlung, 1913–15.

———. "Guilds: In Antiquity." In *Encyclopaedia of the Social Sciences*, ed. Edwin R. A. Seligman and Alvin Johnson, 204–6. New York: Macmillan, 1937.

Sarton, George. *Introduction to the History of Science.* Vol. 3, pt. 2, *Science and Learning in the Fourteenth Century.* Washington, D.C.: Carnegie Institution, 1948.

Saunders, David. *Authorship and Copyright.* London: Routledge, 1992.

Saunders, David, and Ian Hunter. "Lessons from the 'Literatory': How to Historicise Authorship." *Critical Inquiry* 17 (spring 1991): 479–509.

Sax, Julius. *Die Bischöfe und Reichsfürsten von Eichstädt, 745–1806.* Vol. 1. Landshut: Ph. Krüll'sche Universitätsbuchhandlung, 1884.

Sbriziolo, Lia. "Per la storia delle confraternite veneziane: Dalle Deliberazioni Miste (1310–1476) del Consiglio dei Dieci. *Scolae comunes*, artigiane e nazionali." *Atti dell'Istituto Veneto di Scienze, Lettere ed Arti: Classe di Scienze Morali, Lettere ed Arti* 126 (1967–68): 405–42.

Scaglia, Gustina. "Drawings of Machines for Architecture from the Early Quattrocento in Italy." *Journal of the Society of Architectural Historians* 25 (May 1966): 90–114.

———. *Francesco di Giorgio: Checklist and History of Manuscripts and Drawings in Autographs and Copies from ca. 1470 to 1687 and Renewed Copies (1764–1839).* Bethlehem, Pa.: Lehigh University Press; Cranbury, N.J.: Associated University Presses, 1992.

———. "Leonardo da Vinci e Francesco di Giorgio a Milano nel 1490." In *Leonardo e l'età della ragione*, ed. Enrico Bellone and Paolo Rossi, 225–53. Milan: Scientia, 1982.

Scarborough, John. "Hermetic and Related Texts in Classical Antiquity." In *Hermeticism and the Renaissance. Intellectual History and the Occult in Early Modern Europe*, ed. Ingrid Merkel and Alan G Debus, 19–44. Washington, D C.: Folger Shakespeare Library, 1988.

Scheible, Heinz. "Johann Reuchlin of Pforzheim, 1454/5–30 June 1522." *CE* 3:145–50.

Schiappa, Edward. *Protagoras and Logos: A Study in Greek Philosophy and Rhetoric.* Columbia: University of South Carolina Press, 1991.

Schick, Kathy D., and Nicholas Toth. *Making Silent Stones Speak: Human Evolution and the Dawn of Technology* New York. Simon & Schuster, 1993.

Schlam, Carl C *The Metamorphoses of Apuleius: On Making an Ass of Oneself.* Chapel Hill: University of North Carolina, 1992.

Schlatter, Richard. *Private Property The History of an Idea.* 1951. Reprint, New York: Russell & Russell, 1973

Schmandt-Besserat, Denise, ed. *Early Technologies* Malibu, Calif.. Undena, 1979.

Schmidt, Heinrich "Heinrich der Jüngere, Herzog von Braunschweig-Lüneburg-Wolfenbuttel." *NDB* 8:351–52.

Schmidtchen, Volker *Bombarden, Befestigungen, Buchsenmeister: Von den ersten Mauerbrechern des Spätmittelalters zur Belagerungsartillerie der Renaissance. Eine Studie zur Entwicklung der Militartechnik* Düsseldorf: Droste Verlag, 1977

Schmitt, Charles B., Quentin Skinner, Eckhard Kessler, and Jill Kraye, eds *The Cambridge History of Renaissance Philosophy.* Cambridge. Cambridge University Press, 1988

Schneider, Karin, ed. *Die deutschen Handschriften der Bayerischen Staatsbibliothek, München* Vol. 5, pt. 4. New ed. Wiesbaden. Otto Harrassowitz, 1984

Schuermans, H. "Verres 'façon de Venise' fabriqués aux Pays-Bas." *Bulletin des Commissions Royales d'Art et d'Archéologie* 27 (1888): 197 301

Schurmann, Astrid. *Griechische Mechanik und Antike Gesellschaft: Studien zur staatlichen Förderung einer technischen Wissenschaft.* Stuttgart: Franz Steiner Verlag, 1991.

Sealey, Raphael. *A History of the Greek City States, ca. 700–338 B.C.* Berkeley and Los Angeles: University of California Press, 1976.

Seeliger-Zeiss, Anneliese. *Lorenz Lechler von Heidelberg und sein Umkreis. Studien zur Geschichte der spätgotischen Zierarchitektur und Skulptur in der Kurpfalz und in Schwaben.* Heidelberg. C. Winter, 1967.

Serbat, Guy. "Pline l'Ancien: État présent des études sur sa vie, son oeuvre et son influence." *ANRW*, pt 2, *Principat.*, 32.4:2069–2200

Settle, Thomas B. "Brunelleschi's Horizontal Arches and Related Devices." *Annali dell'Istituto e Museo di Storia della Scienza di Firenze* 3 (1978): 65–80.

———. "The Tartaglia Ricci Problem: Towards a Study of the Technical Professional in the Sixteenth Century." In *Cultura, scienze e tecniche nella Venezia del cinquecento Giovan Battista Benedetti e il suo tempo*, 217–26. Venice: Istituto Veneto di Scienze, Lettere ed Arti, 1987.

Shackelford, Jole "Tycho Brahe, Laboratory Design, and the Aim of Science: Reading Plans in Context." *Isis* 84 (June 1993): 211–30.

Shapin, Steven. "The House of Experiment in Seventeenth-Century England." *Isis* 79 (September 1988): 373–404.

———. *A Social History of Truth: Civility and Science in Seventeenth-Century England.* Chicago. University of Chicago Press, 1994.

Shapin, Steven, and Simon Schaffer. *Leviathan and the Air-Pump: Hobbes, Boyle, and the Experimental Life.* Princeton. Princeton University Press, 1985.

Shaw, Gregory *Theurgy and the Soul: The Neoplatonism of Iamblichus.* University Park: Pennsylvania State University Press, 1995.

Shelby, Lon R. "The 'Secret' of the Medieval Masons " In *On Pre-Modern Technology and Science: Studies in Honor of Lynn White, Jr*, ed. Bert S Hall and Delno C. West, 201–19 Malibu, Calif.: Undena, 1976

Shumaker, Wayne. *The Occult Sciences in the Renaissance. A Study in Intellectual Patterns.* Berkeley and Los Angeles: University of California Press, 1972.

Shumate, Nancy. *Crisis and Conversion in Apuleius' Metamorphoses.* Ann Arbor. University of Michigan Press, 1996.

Silberstein, Marcel. *Erfindungsschutz und merkantilistische Gewerbeprivilegien.* Zurich: Polygraphischer Verlag, 1961.

Silvano, Giovanni. "Florentine Republicanism in the Early Sixteenth Century." In *Machiavelli and Republicanism,* ed. Gisela Bock, Quentin Skinner, and Maurizio Viroli, 41–70. Cambridge: Cambridge University Press, 1990.

Silver, Larry. "Germanic Patriotism in the Age of Dürer." In *Dürer and His Culture,* ed. Dagmar Eichberger and Charles Zika, 38–68 Cambridge: Cambridge University Press, 1998.

Simmel, Georg. "The Sociology of Secrecy and of Secret Societies." *American Journal of Sociology* 11 (January 1906): 441–98

Siraisi, Nancy. *The Clock and the Mirror: Girolamo Cardano and Renaissance Medicine.* Princeton, N.J.. Princeton University Press, 1997.

Sirks, Boudewijn. *Food for Rome: The Legal Structure of the Transportation and Processing of Supplies for the Imperial Distributions in Rome and Constantinople.* Amsterdam: J. C. Gieben, 1991.

Skinner, Quentin "Language and Social Change." In *Meaning and Context: Quentin Skinner and His Critics,* ed James Tully, 119–32 Princeton, N.J : Princeton University Press, 1988.

Skydsgaard, Jens Erik *Varro the Scholar: Studies in the First Book of Varro's De re rustica.* Analecta Romana Instituti Danici IV Supplementum. Copenhagen: Munksgaard, 1968

Smith, Christine. *Architecture in the Culture of Early Humanism. Ethics, Aesthetics, and Eloquence, 1400–1470* Oxford: Oxford University Press, 1992.

Smith, Jonathan Z. *Map Is Not Territory. Studies in the History of Religions.* Chicago: University of Chicago Press, 1993.

Smith, Pamela H. *The Business of Alchemy: Science and Culture in the Holy Roman Empire.* Princeton: Princeton University Press, 1994.

Spencer, John R. "Filarete's Description of a Fifteenth Century Italian Iron Smelter at Ferriere." *Technology and Culture* 4 (spring 1963): 201–6.

Speyer, Wolfgang. *Die literarische Fälschung im heidnischen und christlichen Altertum. Ein Versuch ihrer Deutung.* Munich. Beck, 1971.

Spiegel, Gabrielle M. *The Past as Text: The Theory and Practice of Medieval Historiography.* Baltimore: Johns Hopkins University Press, 1997.

Spies, Gerd. "Werkzeuge, Geräte und Maschinen in braunschweigischen Steinbrüchen." In *Museum und Kulturgeschichte: Festschrift für Wilhelm Hansen,* ed. Martha Bringemeier et al., 233–44. Münster: Aschendorff, 1978.

Spitz, Lewis W. "The *Theologia Platonica* in the Religious Thought of the German Humanists." In *Middle Ages–Reformation–Volkskunde: Festschrift for John G. Kunstmann,* 118–33. Chapel Hill: University of North Carolina Press, 1959.

Stäcker, Thomas. *Die Stellung der Theurgie in der Lehre Jamblichs.* Frankfurt am Main: Peter Lang, 1995.

Stammler, Wolfgang, and Karl Langosch, eds. *Die Deutsche Literatur des Mittelalters. Verfasserlexikon.* 5 vols. Berlin: Walter de Gruyter, 1933–55.

Starn, Randolph, and Loren Partridge. *Arts of Power: Three Halls of State in Italy, 1300–1600.* Berkeley and Los Angeles: University of California Press, 1992.

Stemplinger, Eduard. *Das Plagiat in der griechischen Literatur.* Leipzig: Teubner, 1912.

Stephens, J. N. *The Fall of the Florentine Republic, 1512–1530.* Oxford: Clarendon, 1983.

Stewart, Andrew. "The Canon of Polykleitos: A Question of Evidence." *Journal of Hellenic Studies* 98 (1978): 122–31.

Stillman, John Maxson. *The Story of Alchemy and Early Chemistry.* 1924. Reprint, New York: Dover, 1960.

Stimmel, Eberhard. "Die Familie Schütz: Ein Beitrag zur Familiengeschichte des Georgius Agricola." *Abhandlungen des Staatlichen Museums für Mineralogie und Geologie zu Dresden* 11 (1966): 377–417.

Stöckle, Albert. "Berufsvereine." *RE,* new ed., suppl. 4 (1924), cols. 155–211.

Strieder, Peter. "Dürer. (1) Albrecht Dürer." *DA* 9:427–45.

Stroup, Alice. *A Company of Scientists: Botany, Patronage, and Community at the Seventeenth-Century Parisian Royal Academy of Sciences.* Berkeley and Los Angeles: University of California Press, 1990.

Sudhoff, Karl. *Paracelsus: Ein deutsches Lebensbild aus den Tagen der Renaissance.* Leipzig. Bibliographisches Institut A.G., 1936.

Suhling, Lothar. *Aufschliessen, Gewinnen und Fördern: Geschichte des Bergbaus.* Reinbek bei Hamburg: Rowohlt Taschenbuch Verlag, 1983.

———. "Georgius Agricola und der Bergbau. Zur Rolle der Antike im monanistischen Werk des Humanisten." In *Die Antike-Rezeption in den Wissenschaften während der Renaissance,* ed. August Buck and Klaus Heitmann, 149–65. Weinheim: Acta Humaniora, 1983.

Svennung, J. *Compositiones Lucenses: Studien zum Inhalt, zur Textkritik und Sprache.* Uppsala Universitets Årsskrift, 5. Uppsala: A.-B Lundequistska Bokhandeln, 1941.
Swanson, Heather. *Medieval Artisans: An Urban Class in Late Medieval England.* Oxford: Basil Blackwell, 1989.
Syme, Ronald. "Pliny the Procurator." *Harvard Studies in Classical Philology* 73 (1969): 201–36.
Tabanelli, Mario. *Sigismondo Pandolfo Malatesta: Signore del medioevo e del Rinascimento.* Faenza: Stabilimento Grafico Fratelli Lega, 1977.
Tabarroni, Giorgio. "Vitruvio nella storia della scienza e della tecnica." *Atti della Accademia delle Scienze dell'Istituto di Bologna. Classe di Scienze Morali, Memorie* 66 (1971–72): 1–37.
Tafuri, Manfredo "'Cives esse non licere': The Rome of Nicholas V and Leon Battista Alberti: Elements toward a Historical Revision." *Harvard Architecture Review* 6: (1987). 60–75.
———. "Daniele Barbaro e la cultura scientifica veneziana del'500." In *Cultura, scienze e tecnica nella Venezia del cinquecento: Giovan Battista Benedetti e il suo tempo,* 55–81. Venice. Istituto Veneto di Scienze, Lettere ed Arti, 1987.
Tallett, Frank. *War and Society in Early-Modern Europe, 1495–1715.* London: Routledge, 1992.
Tambiah, Stanley Jeyaraja. *Magic, Science, Religion, and the Scope of Rationality.* Cambridge: Cambridge University Press, 1990.
Tavernor, Robert. *On Alberti and the Art of Building.* New Haven: Yale University Press, 1998.
Thomas, Rosalind. *Oral Tradition and Written Record in Classical Athens.* Cambridge: Cambridge University Press, 1989.
Thompson, Daniel V. "Theophilus Presbyter: Words and Meaning in Technical Translation." *Speculum* 42 (April 1967). 313–39.
Thomson, David *Renaissance Architecture: Critics, Patrons, Luxury.* Manchester: Manchester University Press, 1993.
Thorbecke, Aug. "Ruprecht III." *ADB* 29:716–26.
Thorndike, Lynn. *A History of Magic and Experimental Science* 8 vols. New York: Columbia University Press, 1923–58.
Thraede, Klaus. "Das Lob des Erfinders: Bemerkungen zur Analyse der Heuremata-Kataloge." *Rheinisches Museum fur Philologie* 105 (1962): 158–86.
Thrupp, Sylvia L "The Gilds." In *The Cambridge Economic History of Europe,* vol 3, *Economic Organization and Policies in the Middle Ages,* ed. M. M. Postan, E. E. Rich, and Edward Miller, 230–80, 624–34. Cambridge: Cambridge University Press, 1963.
Tod, Marcus N. *Sidelights on Greek History.* Oxford: Basil Blackwell, 1932
Toledano, Ralph. *Francesco di Giorgio Martini: Pittore e scultore.* Trans. Massimo Parizzi Milan: Electa, 1987.
Tommasoli, Walter. *La vita di Federico da Montefeltro (1422–1482)* Urbino: Argalìa, 1978.
Toomer, G. J. "Vitruvius, Pollio." *DSB* 15, suppl 1: 514–21.

Treggiari, Susan *Roman Freedmen during the Late Republic.* Oxford· Clarendon, 1969.
Trinkaus, Charles. "Giovanni Pico della Mirandola (24 February 1463–17 November 1494)." *CE* 3.81–84.
———. *In Our Image and Likeness. Humanity and Divinity in Italian Humanist Thought.* 2 vols. 1970 Reprint, Notre Dame, Ind.: University of Notre Dame Press, 1995.
Tsuji, Shigeru. "Brunelleschi and the Camera Obscura: The Discovery of Pictorial Perspective." *Art History* 13 (September 1990): 276–92
Tucci, U. "Biringucci (Bernigucio), Vannoccio." *DBI* 10:625–31.
Tully, James, ed. *Meaning and Context: Quentin Skinner and His Critics.* Princeton, N.J.: Princeton University Press, 1988
Turcan, Robert. *Les cultes orientaux dans le monde romain.* Paris· Belles Lettres, 1989.
Turner, E. G. *Athenian Books in the Fifth and Fourth Centuries B.C* London: H. K. Lewis, 1952.
Vagnetti, Luigi, and Laura Marcucci "Per una coscienza vitruviana: Regesto cronologico e critico delle edizioni, delle traduzioni e delle ricerche più importanti sul trattato latino *De architectura libri* X di Marco Vitruvio Pollione." *Studi e Documenti di Architettura,* no. 8 (September 1978)· 11–184.
Van Engen, John "Theophilus Presbyter and Rupert of Deutz· The Manual Arts and Benedictine Theology in the Early Twelfth Century." *Viator* 11 (1980): 147–63.
Vasoli, Cesare. "La *prisca theologia* e il neoplatonismo religioso." In *Il neoplatonismo nel Rinascimento,* ed. Pietro Prini, 83–101. Rome: Istituto della Enciclopedia Italiana, 1993.
Verrier, Frédérique "Machiavelli e Fabrizio Colonna nell'*Arte del Guerra.* Il polemologo sdoppiato." In *Niccolò Machiavelli· Politico storico letterato,* ed Jean-Jacques Marchand, 175–87 Rome· Salerno Editrice, 1996
Verzone, Paolo. "Cesare Cesariano." *Arte Lombarda* 16 (1971)· 203–10.
Veyne, Paul. "The Roman Empire." In *A History of Private Life,* ed Philippe Ariès and Georges Duby, trans Arthur Goldhammer, vol. 1, *From Pagan Rome to Byzantium,* ed Paul Veyne, 6–233. Cambridge: Harvard University Press, Belknap Press, 1987.
Vickers, Michael, and David Gill. *Artful Crafts· Ancient Greek Silverware and Pottery.* Oxford: Clarendon, 1994
Vismara, Luigi. "Il pensiero militare di Niccolò Machiavelli." *Rivista Militare* 25 (November 1969): 1439–50
Vivenza, Gloria "Giacomo Lanteri da Paratico e il problema delle fortificazioni nel secolo XVI" *Economia e Storia* 4 (1975): 503–38.
Voss, Mary J. "Between the Cannon and the Book: Mathematicians and Military Culture in Sixteenth-Century Italy." Ph D diss., Johns Hopkins University, 1994.
Vryonis, Speros, Jr. "Byzantine *Demokratia* and the Guilds in the Eleventh Century" *Dumbarton Oaks Papers* 17 (1963) 287–314
Walker, D. P *Spiritual and Demonic Magic from Ficino to Campanella.* 1958. Reprint, Notre Dame, Ind.: University of Notre Dame Press, 1975
Wallace-Hadrill, Andrew, ed. *Patronage in Ancient Society.* London: Routledge, 1989.

———. "Pliny the Elder and Man's Unnatural History." *Greece and Rome* 37 (April 1990): 80–96.

Walsh, P. G. "Apuleius" In *The Cambridge History of Classical Literature*, vol. 2, *Latin Literature*, ed. E. J. Kenney and W V. Clausen, 667–73. Cambridge. Cambridge University Press, 1982.

Warnke, Martin. *The Court Artist: On the Ancestry of the Modern Artist.* Trans. David McLintock. Cambridge Cambridge University Press, 1993

Webster, Charles. *From Paracelsus to Newton: Magic and the Making of Modern Science* Cambridge: Cambridge University Press, 1982.

———. *The Great Instauration· Science, Medicine, and Reform, 1626–1660.* London: Duckworth, 1975.

———. "Paracelsus on Natural and Popular Magic." In *Atti del Convegno Internazionale su Paracelso, Roma, 17–18 dicembre 1993*, ed. Istituto Paracelso. Rome. Edizioni Paracelso, 1994.

Weeks, Andrew *Paracelsus· Speculative Theory and the Crisis of the Early Reformation.* New York· State University of New York Press, 1997.

Wefers, Sabine. *Das politische System Kaiser Sigmunds* Stuttgart. Franz Steiner Verlag, 1989.

Wegener, Hans. *Beschreibendes Verzeichnis der deutschen Bilder-Handschriften des späten Mittelalters in der Heidelberger Universitäts-Bibliothek.* Leipzig: Verlagsbuchhandlung J. J. Weber, 1927

Weisberg, David B. *Guild Structure and Political Allegiance in Early Achaemenid Mesopotamia.* New Haven. Yale University Press, 1967

Weiss, Roberto *The Renaissance Discovery of Classical Antiquity.* 2nd ed. Oxford: Basil Blackwell, 1988.

Wendel, Carl, and Willi Gober "Das griechisch-römische Altertum" In *Handbuch der Bibliothekswissenschaft*, vol. 3 1, *Geschichte der Bibliotheken*, ed Alloys Römer et al., 51–145 2nd ed. Wiesbaden· Otto Harrassowitz, 1955.

Wenke, Robert J. *Patterns in Prehistory. Humankind's First Three Million Years.* 3rd ed. New York: Oxford University Press, 1990,

Westermann, Ekkehard. *Das Eislebener Garkupfer und seine Bedeutung für den europäischen Kupfermarkt:* 1460–1560. Cologne. Böhlau Verlag, 1971.

Westermann, W. L. "Apprentice Contracts and the Apprentice System in Roman Egypt." *Classical Philology* 9 (1914)· 295–315.

Westfall, Carroll William. *In This Most Perfect Paradise: Alberti, Nicholas V. and the Invention of Conscious Urban Planning in Rome, 1447–1455.* University Park Pennsylvania State University Press, 1974.

Wheeler, Everett L. "*Hoplomachia* and Greek Dances in Arms." *Greek, Roman, and Byzantine Studies* 23 (autumn 1982). 223–33

———. "The *Hoplomachoi* and Vegetius' Spartan Drillmasters." *Chiron* 13 (1983). 1–20

White, John. *The Birth and Rebirth of Pictorial Space.* 3rd ed. Cambridge· Harvard University Press, Belknap Press, 1987

———. *Perspective in Ancient Drawing and Painting*. London· Society for the Promotion of Hellenic Studies, 1956.

White, Kenneth D. "Roman Agricultural Writers I. Varro and His Predecessors." *ANRW*, pt. 1, *Von den Anfängen Roms bis zum Ausgang der Republik*, 4·440–97.

White, Lynn, Jr. "Kyeser's 'Bellifortis': The First Technological Treatise of the Fifteenth Century." *Technology and Culture* 10 (July 1969): 436–41.

———. "Medieval Engineering and the Sociology of Knowledge" In *Medieval Religion and Technology: Collected Essays*, 317–38. Berkeley and Los Angeles: University of California Press, 1978.

———. "Theophilus Redivivus" In *Medieval Religion and Technology: Collected Essays*, 93–103. Berkeley and Los Angeles. University of California Press, 1978.

Whitney, Elspeth. *Paradise Restored: The Mechanical Arts from Antiquity through the Thirteenth Century. Transactions of the American Philosophical Society*, n.s , 80, pt. 1 (1990).

Wiebenson, Dora, ed. *Architectural Theory and Practice from Alberti to Ledoux*. N.p.: Architectural Publications, n.d.; distributed by University of Chicago Press, 1982.

——— "Guillaume Philander's Annotations to Vitruvius." In *Les traités d'architecture de la Renaissance*, ed. Jean Guillaume, 67–74. Paris. Picard, 1988.

Wilkinson, Catherine. "Renaissance Treatises on Military Architecture and the Science of Mechanics." In *Les traités d'architecture de la Renaissance*, ed. Jean Guillaume, 466–74. Paris: Picard, 1988.

Williams, Raymond. *Keywords: A Vocabulary of Culture and Society*. Rev. ed. New York: Oxford University Press, 1983.

Williams, Robert. *Art, Theory, and Culture in Sixteenth-Century Italy. From Techne to Metatechne*. Cambridge· Cambridge University Press, 1997.

Wilsdorf, Helmut. *Georg Agricola und sein Zeit*. Vol. 1 of *Ausgewählte Werke*, by Georgius Agricola, ed. Hans Prescher. Berlin: Deutscher Verlag der Wissenschaften, 1956

Wilsdorf, Helmut, and Werner Quellmalz. *Bergwerke und Hüttenanlagen der Agricola-Zeit*. Suppl. 1 of *Ausgewählte Werke*, by Georgius Agricola, ed. Hans Prescher. Berlin. Deutscher Verlag der Wissenschaften, 1971

Winkler, John J *Auctor and Actor: A Narratological Reading of Apuleius's "Golden Ass."* Berkeley and Los Angeles: University of California Press, 1985.

Wissell, Rudolf. *Des alten Handwerks Recht und Gewohnheit*. Ed. Ernst Schraepler. 2nd ed. rev. 6 vols. Berlin: Colloquium Verlag, 1971.

Wittkower, Rudolf. *Architectural Principles in the Age of Humanism*. 4th ed. London: Academy Editions, 1973.

Woodmansee, Martha. "The Genius and the Copyright· Economic and Legal Conditions of the Emergence of the 'Author.'" *Eighteenth-Century Studies* 17 (summer 1984): 425–48.

Woods-Marsden, Joanna. "Images of Castles in the Renaissance. Symbols of 'Signoria'/ Symbols of Tyranny" *Art Journal* 48 (summer 1989): 130–37

Wright, D. R. Edward. "Alberti's *De pictura:* Its Literary Structure and Purpose." *Journal of the Warburg and Courtauld Institutes* 47 (1984): 52–71.

Wycherley, R. E. *How the Greeks Built Cities.* 2nd ed. London: Macmillan, 1962.

———. "Rebuilding in Athens and Attica." In *The Cambridge Ancient History,* vol 5, *Fifth Century B.C.,* ed. D. M Lewis, John Boardman, J. K. Davies, and M. Ostwald, 206–22. 2nd ed. Cambridge: Cambridge University Press, 1992.

Yates, Frances A. *Giordano Bruno and the Hermetic Tradition.* Chicago: University of Chicago Press, 1964.

Zaccagnini, Carlo. "Patterns of Mobility among Ancient Near Eastern Craftsmen." *Journal of Near Eastern Studies* 42 (October 1983): 245–64.

———. "Le tecniche e le scienze." In *L'Alba della civiltà. Società, economia e pensiero nel vicino oriente antico,* vol. 2, *L'Economia,* ed. Mario Liverani et al., 293–341. Turin· Unione Tipografico-Editrice Torinese, 1976.

Zambelli, Paola. "Cornelius Agrippa, ein kritischer Magus." In *Die okkulten Wissenschaften in der Renaissance,* ed August Buck, 65–89. Wiesbaden: Otto Harrassowitz, 1992.

———. "Magic and Radical Reformation in Agrippa of Nettesheim." *Journal of the Warburg and Courtauld Institutes* 39 (1976): 69–103.

———. "Platone, Ficino e la magia." In *Studia Humanitatis: Ernesto Grassi zum 70. Geburtstag,* ed. Eginhard Hora and Eckhard Kessler, 121–42. Munich: Wilhelm Fink Verlag, 1973.

———. "Scholastic and Humanist Views of Hermeticism and Witchcraft." In *Hermeticism and the Renaissance: Intellectual History and the Occult in Early Modern Europe,* ed. Ingrid Merkel and Alan G. Debus, 125–53. Washington, D.C.: Folger Shakespeare Library, 1988.

———. "Umanesimo magico-astrologico e raggruppamenti segreti nei platonici della preriforma." In *Umanesimo e esoterismo,* ed. Enrico Castelli, 141–74. Padua. Casa Editrice Dott. Antonio Milani, 1960

Zecchin, Luigi. "Giorgio Ballarin all'insegna del San Marco." In *Vetro e vetrai di Murano: Studi sulla storia del vetro,* 2:159–62 Venice: Arsenale Editrice, 1987

———. *Vetro e vetrai di Murano: Studi sulla storia del vetro.* 3 vols. Venice: Arsenale Editrice, 1987.

Zervas, Diane Finiello. "Filippo Brunelleschi's Political Career." *Burlington Magazine* 121 (October 1979). 630–39

Ziebarth, Erich. *Das griechische Vereinswesen.* 1896. Reprint, Wiesbaden: Dr. Martin Sandig, 1969.

Ziegler, Konrat. "Plagiat." *KP* 4 (1972), cols. 879–81.

———. "Plagiat." *RE* 20.2 (1950), cols. 1956–97.

———. "Polyidos, 6" *RE* 21.2 (1952), cols. 1658–59.

Zika, Charles. "Agrippa of Nettesheim and His Appeal to the Cologne Council in 1533: The Politics of Knowledge in Early Sixteenth-Century Germany." In *Humanismus in Köln / Humanism in Cologne,* ed. James V. Mehl, 119–74. Cologne: Böhlau Verlag, 1991.

———. "Reuchlin's *De Verbo Mirifico* and the Magic Debate of the Late Fifteenth Century." *Journal of the Warburg and Courtauld Institutes* 39 (1976): 104–38.

Zilsel, Edgar. "The Origins of Gilbert's Scientific Method." In *Roots of Scientific Thought: A Cultural Perspective*, ed. Philip P. Wiener and Aaron Noland, 219–50. New York: Basic Books, 1957.

———. "The Sociological Roots of Science." *American Journal of Sociology* 47 (January 1942): 544–62.

Zwijnenberg, Robert. *The Writings and Drawings of Leonardo da Vinci: Order and Chaos in Early Modern Thought.* Trans. Caroline A. van Eck. Cambridge: Cambridge University Press, 1999.

Index

Page numbers for illustrations are in boldface

Lightning Source UK Ltd.
Milton Keynes UK
UKOW06f2212170615

253681UK00001B/12/P